The Glaciation of High Asia

Matthias Kuhle

The Glaciation of High Asia

From the Last Ice Age to the Present

Springer

Matthias Kuhle (deceased)
Göttingen
Germany

Additional material to this book can be downloaded from http://extras.springer.com.

The term 'ibid.' refers to the same source cited in the previous reference in the text.

ISBN 978-3-030-08481-3 ISBN 978-3-319-77566-1 (eBook)
https://doi.org/10.1007/978-3-319-77566-1

Softcover re-print of the Hardcover 1st edition 2018

Printed on acid-free paper

Cover figure: From 5500 m asl, 33°50′38.81″N 89°5′14.25″E, facing NNE in the western piedmont of the Purog Kangri Massif. The photo shows the second outlet glacier tongue south of the Purog Kangri-, ice-capped west side. The ice of the glacier tongue descends with a decameter-thick, steep margin. In the foreground the stony-gravelly-sandy-clayey ground moraine surface is visible, which is characteristic for the majority of surfaces in Tibet, i.e. for nearly 2 million km^2. In the current location, the surface was only freed from glacier ice very recently, i.e. ca. 65–180 years ago. Analog photo 28/09/2014 by M. Kuhle

This Springer imprint is published by the registered company Springer International Publishing AG part of Springer Nature
The registered company address is: Gewerbestrasse 11, 6330 Cham, Switzerland

Preface

When this book is opened, this will probably coincide with an interest in climate and the climate of our prehistory. Glaciation and a cold climate belong together by association. The immense lowland glaciations, originating from the north and covering vast areas of North America and Eurasia which today belong to the central zones of the human ecumene, have become the paradigmatic symbol of the Ice Age climate. Descending from the polar latitudes, the inland ice covers developed and progressed southward more or less in parallel to the circles of latitude, thereby eventually reaching considerable dimensions.

In High Asia, by contrast, at subtropical latitudes, the existence of glaciers primarily depends on the absolute altitudes (i.e., meters above sea level) of the land surface. Here, the topography, i.e., the distribution of the areas in relation to the absolute altitude, was an essential determining factor for the development of glaciers during the Ice Age. The advance and retreat of an inland ice sheet here do not proceed from north to south, parallel to the circles of latitude, but from high to low altitudes and thus parallel to the isohypses, whose progression is not reliant on planetary regularity but on the geological phenomenon of the genesis of the Himalayan-Tibetan orogen.

The author's interest in the current and former glacier cover up to its maximum extent lies in the striking climatic contrast between hot monsoon-tropical and subtropical lowlands of South to Central Asia and the hypsometric change up to and above of the snow line (equilibrium line altitude, or ELA). In the latitudes where the highest average annual temperatures on earth exist, High Asia reaches altitudes of such an extent that the largest extra-Arctic glaciers of over 50 km in length and in great density are found there.

Glaciers are dynamic equilibrium systems that react very sensitively to changes in mass balance and which, therefore, represent important indicators of climate change. Ideally, in the case of stationary glaciers, the accumulation of water in the form of snow and ice in the firn-ice region above the ELA equals precisely the loss of water through melting in the ablation zone, i.e., the glacier tongue, below the ELA. Any changes in this balance can be expressed through shifts of this equilibrium line, which may be observed as either advances or retreats of the glacier tongue. These fluctuations of the tongue's terminal positions can thus serve as indicators of both synchronic and diachronic climate change. A decrease in temperature by 0.5–0.6 °C results in a lowering of the ELA by 100 m. The glacier tongue then flows down to a lower altitude by 150–200 m.

An important factor to be considered in the context of High Asia is that we are dealing here with a climatic process that—specifically in High Asia—runs in parallel with the isohypses. This means that in a flat plateau landscape such as Tibet, an ELA depression of 100 m may result in a dramatic ice expansion while, at the same time, it may in steep terrain such as the Himalayas only result in minor changes of the glaciers' terminal positions. Put differently: the magnitude of climatic change will not be measured through absolute changes in glacier area due to glacier expansion or contraction, but through the shifting (downward or upward) of the absolute position of the ELA.

Nevertheless, the absolute size of the ice surface is still important in the context of the feedback loops existing between glacier surfaces and the climate. In this context, the thermal loss due to melting energies must be mentioned. These are significantly more important due to the subtropically increased proportion of sublimation in High Asia compared with high geographical latitudes. Moreover, there is a strong cooling effect through the albedo from larger glacier surfaces extending across high altitudes. Finally, we must also consider the increase of glacier thicknesses and their feedback influence on the enlargement of the surfaces of the nourishing areas.

Accordingly, the glaciations of High Asia are of particular interest due to the interaction of climate and relief. On the basis of this interaction, one can predict the maximum glacial snow line depression (movement to lower altitude) and the enlargement of glacier surfaces caused by it, as well as the very large cooling effect resulting from it. The supraregional, global importance of this effect must not be underestimated. It is the albedo of enlarged snow, névé, and ice surfaces in subtropical regions, i.e., the reflection of the maximum global radiation, which in addition would be strengthened by the great absolute altitude and which would thus—given an inland glaciation—cause an overall cooling of the earth's atmosphere.

The phenomenon of terrestrial glaciation provides an ideal body of physical evidence whose calibration is known through extant glaciers and which never changes throughout earth's history. It, therefore, provides the methodologically necessary, stable frame of reference that is needed for the observation of the global climate change from the past to the present.

Göttingen, Germany Matthias Kuhle

Acknowledgements

I thank my wife Sabine Kuhle for her cooperation related to this work and Juliane Jörgens and Anneliese Kuhle for translating the book manuscript. Furthermore, I am grateful to Google Earth Pro for allowing me generous access to satellite images of my research area.

Contents

About the Author

Matthias Kuhle (*20 April 1948 in Berlin; † 25 April 2015 near Yaruphant, Nepal) was an internationally renowned German geographer at the Georg-August-Universität Göttingen, where he was Professor of Geography and High Mountain Geomorphology from 1983 until his death in 2015. His main research area was the geomorphological reconstruction of a High Glacial inland ice sheet in High Asia, including Central Tibet. He studied geography, German philology, and philosophy at the Freie Universität Berlin, and completed his Ph.D. on former glaciations in southern Iran. Subsequently, he extended his research focus to include the extreme high mountain regions of High Asia, most notably the Himalayas and Karakoram. Between 1973 and 2015, he conducted more than 44 scientific expeditions in these regions and other high mountain areas around the world. Another of his research interests, shared with his wife Sabine Kuhle, was the philosophy of science and the cultural evolution of Western logic. On April 25, 2015, he unexpectedly died during an earthquake in Nepal.

1 Method

1.1 Preliminary Methodological Remarks

If we speak of the glacial (former) to current glaciation, in that precise order, then this is a matter of chronology. The progress of scientific discovery, however, took place in a reverse manner. Since the first observations of the early research travelers, one always started from the present, i.e., from the current glaciers. Only later the view into the past emerged. One started from the current glacier margins, and then, based on former moraines left behind next to those current glacier margins, reconstructed the position of the next-older ice margin. This must have been relatively simple because the state of preservation of accumulations that were built up only recently is nearly complete. With an increasing distance from the current ice margins, in contrast, the chronological distance is enlarged. Due to the concomitant duration of a non-glacigenic, different morphodynamics, this led to the successive destruction of the increasingly older relics of past glacier stages. The past ice margins that lie farthest from the current glaciers are therefore the oldest. These are the glacial indicators of the Last Glacial Maximum.

In order to be able to assess the entire potential and the significance of High Asia for global climate change, one must first reconstruct the extent of the Last Glacial Maximum (LGM). Only once we know the entire amplitude between the maximal former glaciation and the extant glaciers are we in a position to assess properly whether the latter are the relics of a former complete inland glaciation that today have retreated to refuge centers among the highest mountain massifs, or whether today's glaciers already lie very close to the demarcations of the former maximal glaciation. Some experts are of the opinion that the maximal expansion of the glaciers in High Asia occurred during the Interglacials because, so they argue, only then could the necessary monsoonal precipitation have been provided for their accumulation. According to this view, during the global ice age cycles, during which the summer monsoon was interrupted, the Tibetan Plateau was turned into no more than dry tundra; however, no considerable glacier advances took place. This position is still held by some experts in the field today. It is mainly based on a dating technique, the so-called TCN method, that has been developed over the past two decades and which measures the proportion

M. Kuhle, *The Glaciation of High Asia*,
https://doi.org/10.1007/978-3-319-77566-1_1

of cosmogenic nuclides in rock surfaces. The results of this age determination for former glacial stages as they have been assembled so far are, however, contradictory and do not lead to any clear picture with regard to climate change in High Asia. They neither provide regionally nor chronologically coherent patterns and hitherto also do not yield any connection to the global Ice Age and Interglacials as they were reconstructed on the basis of the Nordic glaciations (Chevalier et al. 2011; Kirchner et al. 2011; Heyman 2014; Owen and Dortch 2014).

Notwithstanding the diverging opinions regarding the question of whether the TCN method may *at all* be methodologically applicable in High Asia, since so far a reliable calibration for this region has not been possible (Owen and Dortch 2014: 35), there nevertheless exists a general consensus regarding at least one basic fact. Namely, all experts acknowledge, at least in principle, that even before the application of special dating techniques can take place, the total extent of the maximal glaciation of a region must be clarified first. The only possible method for such a reconstruction comes from the classical field of glacial geology. The following investigation therefore rests on this method.

1.2 Methods in the Sense of Working Techniques

The methods, i.e., working techniques as they are relied upon in the present study, were developed over nearly 200 years since the 19th century. However, they are not the subject of this book. They have been described in numerous papers, discussed and summarized in hand- as well as textbooks, and have time and again been complemented by numerous authors (e.g., von Klebelsberg 1948; Woldstedt 1961; Machatchek 1955, 1964; Louis 1968; Flint 1971; Kuhle 1991; Bennett and Glasser 2009; Ehlers 2011). Moreover, in extensive publications on the former glaciation of High Asia since 1974, the present author has detailed the methods underlying his findings numerous times (e.g., Kuhle 1990a, 1999, 2005, 2007, 2013, 2014; Kuhle and Kuhle 2010).

As an introduction to the discussion, the reader shall merely be presented with a short review of the fundamental methodological condition.

The question of whether or not an area was glaciated in the past can be answered by analyses of the positions of geomorphological features in the field. The characteristics of glaciations include both erosive and accumulative field forms as well as the sediments from depositions. Evidence for field forms and depositions is accumulated from separate form elements, i.e., isolated indicators, pointing either in favor of or against a genesis by former glaciers. Their spatial, three-dimensional combination either corresponds to or contradicts a genesis due to a past glaciation.

Decisive in these cases is whether or not:

1. the individual elements of forms and depositions are elements of glacigenic forms and depositions;
2. there exists a physical-functional connection to a former glacier.

An example demonstrating such a physical-functional relationship is Fig. 1.1: namely, the arrangement of the positions of separate glacigenic key forms such as glacial striations on the valley flank, roches moutonnées within a tongue-basin form, large erratic boulders at high positions on the valley flank, lateral and terminal moraine-like forms shaped like ramparts consisting of diamictically mixed loose rock; further, fluvial accumulations lying connected to the endmoraine-like ramparts that must be diagnosed as ice-marginal ramps (IMRs; German "Bortensander and Sanderschürzen"). In this specific example, the arrangement of the individual indicators describes a glacier tongue basin, the deepest former ice margin of which can be reconstructed unambiguously and preciesely by terminal moraines and connected IMRs (Fig. 1.1).

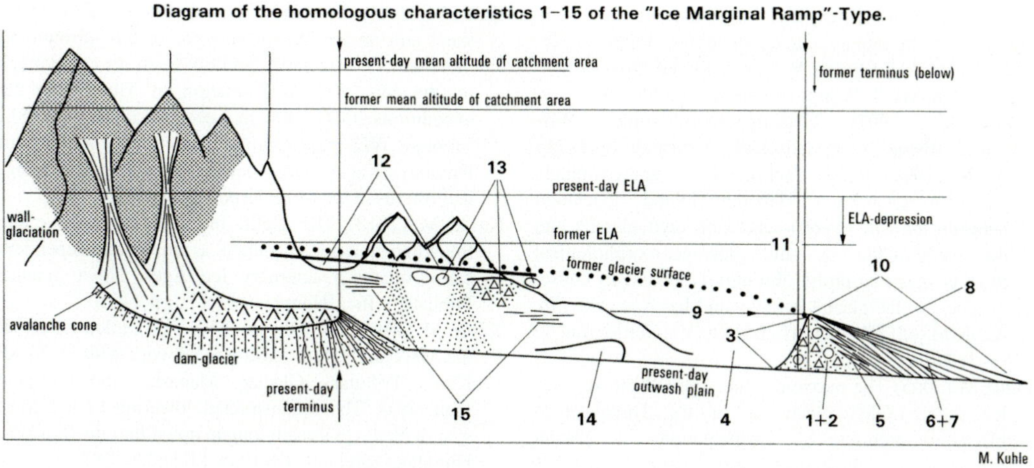

Fig. 1.1 Scheme of homologous characteristics of glacial indicators (after Kuhle 1989, 1990a, b). Field points 1–15 are homologous features of the glacial landscape, the spatial arrangement of which provides circumstantial evidence of the high probability of their past formation due to a glacier advance. (1) There appear dissimilar rocks with very different grain sizes, and the material is edged to rounded. (2) This material (1) is non-stratified, i.e., thoroughly mixed. (3) The material is formed like a rampart with an extremely steep slope upvalley. (4) A tongue-basin-shaped excavation area encompasses the area upvalley of the rampart. (5) The material on the outer slope is stratified and sorted according to grain size; it is interlocked in a serrated pattern with non-sorted material of the rampart and thus confirms a syngenetic development. (6) Due to fluvial processes, the downvalley material of the rampart has been sorted and arranged in layers by water according to grain size and is arranged in layers. (7) The material (6) is dominated by rounded to very well rounded components with the orientation of their longitudinal axes at right angles to the direction of transport. (8) The stratigraphic sequences decrease from the core of the rampart up to the surface. (9) The outer slope of the rampart has been developed by fluvial gravel layers; it is mainly a recent gravel floor and cannot be explained by current topography. (10) Small valleys, inset into the surface of the cone-shaped mantle, begin without widening into a catchment area as an indentation on the wall culmination and do not correspond with recent climate conditions. (11) The past snowline (ELA, GWL) with its lowering against the recent, i.e., the current snowline. This lowering is in a strict proportion to the height of the mountain catchment area and the altitude of that rampart (1–10); i.e., with very high mountains this rampart is situated well below and far away in the mountain foreland. At lower catchment areas, however, it is situated higher, i.e., at the same time closer to the mountains or even—at least upvalley—closer to the catchment area. (12) Polish lines and transfluence passes. (13) On transfluence passes and rock shoulders there are chaotically mixed deposits and erratic boulders, which due to their large size and great distances of transport can only be explained by glacier transport. (14) In the tongue basin there are roches moutonnées and polished rock faces. (15) On these roches moutonnées or on the valley flanks one can find glacier scratches, the arrangement of which follows the valley incline

References

Bennett MR, Glasser NF (2009) Glacial geology—ice sheets and landforms, 2nd edn. Wiley-Blackwell, Chichester, UK

Chevalier M-L, Hilley G, Tapponier P, Van Der Woerd J, Liu-Zeng J, Finkel RC, Ryerson J, Li H, Liu X (2011) Constraints on the late Quaternary glaciations in Tibet from cosmogenic exposure ages of moraine surfaces. Quatern Sci Rev 30:528–554

Ehlers J (2011) Das eiszeitalter. Spektrum Akademischer Verlag, Heidelberg

Flint RF (1971) Glacial and Quaternary geology. Wiley and Inc., New York, pp 1–892

Heyman J (2014) Paleoglaciation of the Tibetan Plateau and surrounding mountains based on exposure ages and ELA depression estimates. Quatern Sci Rev 91:30–41

Kirchner N, Greve R, Stroeven AP, Heyman J (2011) Paleoglaciological reconstructions for the Tibetan Plateau during the last glacial cycle: evaluating numerical ice sheet simulations driven by GCM-ensembles. Quatern Sci Rev 30:248–267

Kuhle M (1989) Ice-marginal ramps: an indicator of semiarid piedmont glaciations. GeoJournal 18:223–238

Kuhle M (1990a) The probability of proof in geomorphology—an example of the application of information theory to a new kind of glacigenic morphological type, the ice-marginal ramp (Bortensander). Geo J 21 (3):195–222

Kuhle M (1990b) Ice marginal ramps and alluvial fans in semi-arid mountains: convergence and difference. In: Rachocki AH, Church M (eds) Alluvial fans—a field approach, vol 3. Wiley, Chichester, pp 55–68
Kuhle M (1991) Glazialgeomorphologie. Wissenschaftliche Buchgesellschaft, Darmstadt, pp 1–213
Kuhle M (1999) Reconstruction of an approximately complete Quaternary Tibetan inland glaciation between the Mt. Everest- and Cho Oyu massifs and the Aksai Chin. A new glaciogeomorphological SE-NW diagonal profile through Tibet and its consequences for the glacial isostasy and Ice Age cycle. In: Kuhle M (ed) Tibet and High Asia (V). GeoJournal 47 (1–2):3–276
Kuhle M (2005) The maximum Ice Age (Würmian, Last Ice Age, LGM) glaciation of the Himalaya—a glaciogeomorphological investigation of glacier trim-lines, ice thicknesses and lowest former ice margin positions in the Mt. Everest-Makalu-Cho Oyu massifs (Khumbu- and Khumbakarna Himal) including informations on late-glacial-, neoglacial-, and historical glacier stages, their snow-line depressions and ages. In: Kuhle M (ed) Tibet and High Asia (VII): Glaciogeomorphology and former glaciation in the Himalaya and Karakorum. GeoJournal 62(3–4):193–650
Kuhle M (2007) Critical approach to the methods of glacier reconstruction in High Asia (Qinghai-Xizang (Tibet) Plateau, West Sichuan Plateau, Himalaya, Karakorum, Pamir, Kuenlun, Tienshan) and discussion of the probability of a Qinghai-Xizang (Tibetan) inland ice. J Mt Sci 4(2):91–123
Kuhle M (2010) New indicators of a former Tibetan ice sheet and an ice stream network in the surrounding mountain systems: new field observations and datings on the SE-, S- and W-margin of Tibet from the expeditions 2004–2009. In: Leech ML, Klemperer SL, Mooney WD (eds) Online proceedings for the 24th Himalaya-Karakoram-Tibet Workshop. U.S. Geological Survey, Open-File Report 2010-1099
Kuhle M (2013) The Uplift of High Asia above the Snowline and Its glaciation as an albedo-dependent cause of the Quaternary Ice Ages. Nova Science Publishers Inc., Hauppauge, NY, pp 1–240
Kuhle M (2014) The Glacial (MIS 3-2) outlet glacier of the Marsyandi Nadi-icestream-network with its Ngadi Khola Tributary Glacier (Manaslu- and Lamjung Himalaya). The reconstructed lowering of the Marsyandi Nadi ice stream tongue down into the Southern Himalaya foreland. J Mt Sci 11(1):236–287
Louis H (1968) Allgemeine geomorphologie. Obstsche Reihe, vol 3, Walter de Gruyter & Co., Berlin, pp 1–522
Machatschek F (1955) Das Relief der Erde, vol 1, Borntraeger, Berlin, pp 1–531
Machatschek F (1964) Geomorphologie. 8. Auflage ed. bearb. von Graul H, Rathjens G, Stuttgart, pp 1–209
Owen LA, Dortch JM (2014) Nature and timing of Quaternary glaciations in the Himalayan-Tibetan orogen. Quatern Sci Rev 88:14–54
von Klebelsberg R (1948/49) Handbuch der Gletscherkunde und Glazialgeologie, vols 1 & 2, Springer, Wien, pp 1–1028
Woldstedt P (1961) Das Eiszeitalter, 3rd edn, vol 1, pp 1–374

2 Research Areas

The focus of this section is summarized in Fig. 2.1, which depicts a sample of the more exemplary research areas studied by the author since 1973. This sample is presented here in chronological sequence (see above) from the High Glacial maximum glaciation and its proofs via the Late Glacial and Holocene recessional stages to glacier fluctuations from the historical and present periods. In the course of our discussion, we will refer to various findings in the respective localities that over the past four decades have been published in detail.

The recessional stages reconstructed in this context are minor glacier advances. However, none of these reach the ice margins of the maximum glaciation of the respective area.

It is not possible to reconstruct the entire glacial history from the Ice Age (LGM = Last Glacial Maximum; LGP = Last Glacial Period; MIS 3-2 = Marine Isotope Stage 3-2) via the Late Glacial and the Neoglacial (Holocene) glacier stages to the current glacier margins in all of these research areas. This is tied to the nature of the increasing elevation of the snowline (ELA) that has accompanied the retreat of the glaciers to the present. Only in high mountain areas do we still find historical and current glaciers. In slightly lower areas, glaciers did still exist during the Neoglacial (Holocene). Still lower areas of High Asia possess larger surfaces. They show hills and flatly undulating, vastly extended high plateau regions. During the Late Glacial, terrains such as these were still covered by glacier ice. During the Ice Age, in addition to those high plateau regions, even low valley bottoms of the marginal areas of the plateau were filled with glacier ice. During the LGM the ELA was so low that in some places marginal outlet glaciers of the high plateaus even flowed down as far as into the forelands (see Kuhle 2011a; glacier maps downloadable at: http://booksite.elsevier.com/9780444534477/).

From this past to current interaction between relief and ELA—that is, the climate—we must conclude that the current glaciers, which are mountain glaciers, mostly are located far from the large-scale LGM glaciers that once dominated Central Tibet. This means that proof of the maximum glaciation can be discerned best in those remote areas where currently no glaciers exist anymore. Accordingly, these research areas are located the greatest distance from the higher mountains and from the still-glaciated research areas. It may seem conceivable in some cases to trace back the plateau glaciation to nourishing areas of mountains located on the plateau itself. This possibility, however, becomes less and less conceivable as the horizontal distance between those mountains and the lower high plateau regions of Central Tibet increases.

The characters and symbols in the figures in this chapter can also be viewed in the online version of the figures at http://extras.springer.com/.

M. Kuhle, *The Glaciation of High Asia*,
https://doi.org/10.1007/978-3-319-77566-1_2

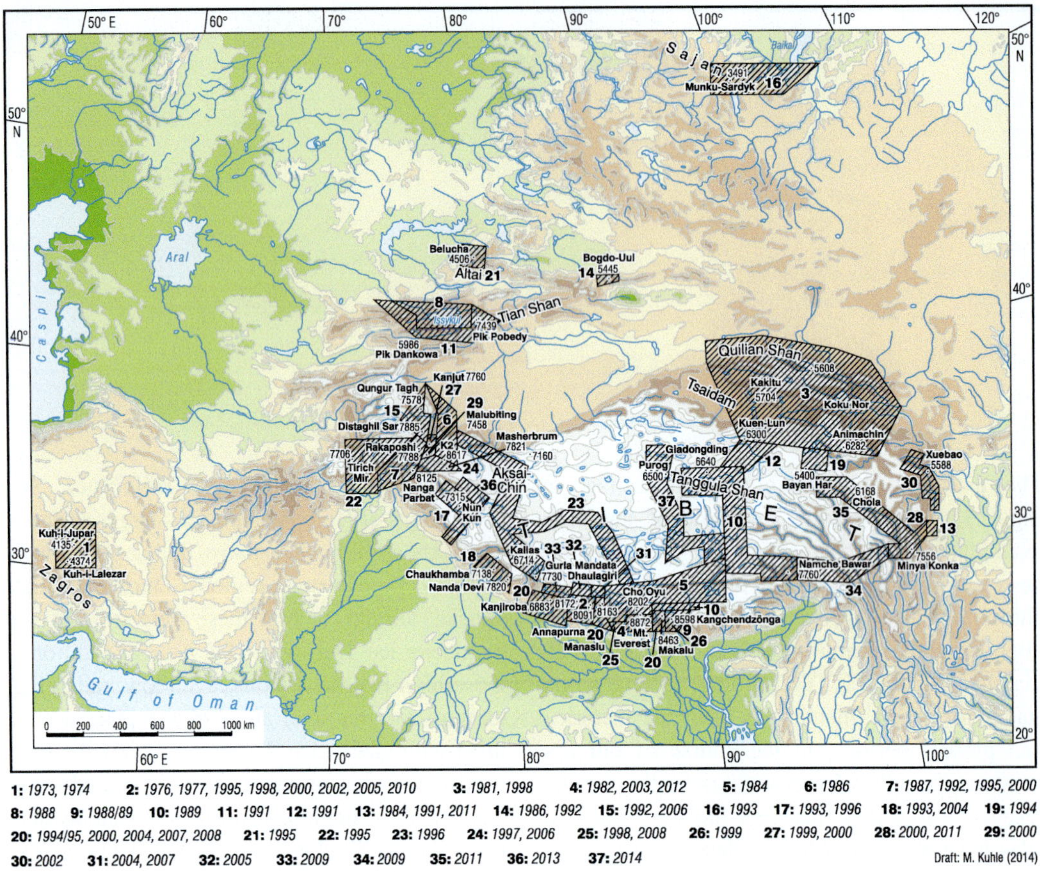

Fig. 2.1 Author's exploration regions in High Asia and surrounding mountain areas

To conclude, the logic of this coupling between the height of the snowline (ELA) and the relief as a condition of glacier development for the reconstruction of the chronological glacier history from the LGM to the present does not allow a *random* choice of the study area. Rather, the research areas themselves must complete one another with regard to the central aspect explained above. Accordingly, the refutations or proofs of a large-scale and thus Glacial to Late Glacial glaciation on the plateau have to be carried out far away from the current glaciers. However, the glaciation that at an interglacially elevated ELA still existed in the more recent past must be investigated in the steep relief of the highest mountain areas.

The mountains sitting upon the plateau must have nourished plateau glaciation by providing an additional influx of ice. Only along the plateau margins, where the breakthrough valleys leading down from the Tibetan, Tian Shan, and Pamir plateaus cut through the fringing mountains, did the ice inflow from the mountains not increase the overall ice sheet. On systematic grounds, we will begin with precisely such a research area.

Fig. 2.2 Satellite image, Google Earth Pro: center 34° 1′ 0.0″N 78° 26′ 22.9″E; http://extras.springer.com: Character A; Nos. 1–10

2.1 The Glaciation of the Pangong Co Area in WSW Tibet

The focus of this section is on Fig. 2.35. On an expedition in 1996, the eastern margin of the 170 km long valley occupied by the Pangong Co (Tso; lake) was investigated at the widening of the lake (Nako Tso). The results show a High to Late Glacial glaciation providing evidence of an ELA depression (ΔELA) of at least 700 m (Kuhle 1999: 6, 7, 43–45, 47, 260–262, Fig. 2: No. 143, Fig. 32; 2013a: 157–158).

First we will proceed in this book with additional new field and laboratory results from the 30 km long, southwesternmost section of the Pangong Co Valley. The lake has its water level at 4246 m asl (Fig. 2.2, see also http://extras.springer.com: Character A: 4246 m asl). At the locality of Fig. 2.2 (Character A: No. 1, see also http://extras.springer.com) at 33° 49′ 20.57″N 78° 34′ 15.31″E, at 4410 m asl, and in its surroundings there are erratic granite boulders (Fig. 2.6; granite/granodiorite) on outcropping mylonite (Fig. 2.3a, b: Nos. 7a, b). The granite boulders in part are extremely large, i.e., they

Fig. 2.3 **a**, **b** Mineralogic rock determination of sample 25.08.13/2: No. 7a is fracture of a hut-sized erratic boulder of granite/granodiorite and No. 7b is phyllite/mylonite bedrock (analysis of rocks and documentation: A. Vollbrecht). Locality: orographically left in the Pangong Co (Tso) valley 4338 (4430 aneroid measurement) m asl 33° 49′ 28.14″N 78° 34′ 8.18″E and 33° 49′ 20.57″N 78° 34′ 15.31″E, 4410 m asl, ca. 70 m above of Pangong Tso. Sampling by M. Kuhle

Fig. 2.3 (continued)

reach dimensions up to the size of a hut (Fig. 2.4). Their positions are up to 200 m above the lake level on rounded rock ridges and provide proof of an ice infilling of the Pangong Co lake basin. An ice infilling such as this could not have been formed by local mountain glaciers from the southwest, even if their ELA had undergone a depression of ca. 600 m. At a depression of this magnitude (ΔELA), the local glacier tongue termini would have just reached the lake level. This can be demonstrated from the following calculation:

$$\mathrm{Sdepr} = (\mathrm{tim} - \mathrm{tif})/2, \quad (2.1)$$

where tim = the current terminus of the glacier tongue; tif = the former terminus of the glacier tongue; and Sdepr = ELA depression (equilibrium line depression).

The current glacier tongue is located at tim = 5450 m asl; the past glacier tongue is situated **at the level of the lake surface** at tif = 4246 m asl:

$$\mathrm{Sdepr} = (5450 - 4246)/2 \quad (2.2)$$

$$\mathrm{Sdepr} = 1204/2 \quad (2.3)$$

$$\mathrm{Sdepr} = 602 \quad (2.4)$$

$$\text{ELA-depr} = 602\ \mathrm{m} \quad (2.5)$$

At a maximum lake depth of ca. 30 m, an ELA depression of ca. 615–620 m would be necessary in order for the local glaciers to reach

Fig. 2.4 From 4430 m asl (aneroid measurement), 33° 49′ 28.14″N 78° 34′ 8.18″E (Fig. 2.2: No. 1), facing south; (● and ◎) marks light, variably colored, erratic granite-granodiorite boulders on dark mylonite bedrock (♦) (see Fig. 2.3). The up to hut-sized granite/granodiorite boulders, transported long distances, are distributed across the entire slope up to the ridge crest (light spots in background). The erratic granite/granodiorite boulders are heavily hydrothermally altered and weathered (◎), so that their crystals have fallen from their surfaces and cover the ground surrounding the boulders up to centimeters in thickness. Thus they exhibit extremely high surface erosion. Its duration in time—i.e., the erosion rate—is not known. A TCN-age dating of the boulders for this reason alone is not possible. Analog (film) photo 25/08/2013 by M. Kuhle

glacier stadium		gravel field (Sander)	approximated age (YBP)	ELA-depression (m)
-I	= Riß (pre-last High Glacial maximum)	No. 6	150 000 - 120 000	c. 1400
0	= Würm (last High Glacial maximum)	No. 5	60 000 - 18 000	c. 1300
I - IV	= Late Glacial	No. 4 - No. 1	17 000 - 13 000 or 10 000	c. 1100 - 700
I	= Ghasa-stadium	No. 4	17 000 - 15 000	c. 1100
II	= Taglung-stadium	No. 3	15 000 - 14 250	c. 1000
III	= Dhampu-stadium	No. 2	14 250 - 13 500	c. 800 - 900
IV	= Sirkung-stadium	No. 1	13 500 - 13 000 (older than 12 870)	c. 700
V - 'VII	= Neo-Glacial	No. -0 - No. -2	5 500 - 1 700 (older than 1 610)	c. 300 - 80
V	= Nauri-stadium	No. -0	5 500 - 4 000 (4 165)	c. 150 - 300
VI	= older Dhaulagiri-stadium	No. -1	4 000 - 2 000 (2 050)	c. 100 - 200
'VII	= middle Dhaulagiri-stadium	No. -2	2 000 - 1 700 (older than 1 610)	c. 80 - 150
VII - XI	= historical glacier stages	No. -3 - No. -6	1 700 - 0 (= 1950)	c. 80 - 20
VII	= younger Dhaulagiri-stadium	No. -3	1 700 - 400 (440 resp. older than 355)	c. 60 - 80
VIII	= stadium VIII	No. -4	400 - 300 (320)	c. 50
IX	= stadium IX	No. -5	300 - 180 (older than 155)	c. 40
X	= stadium X	No. -6	180 - 30 (before 1950)	c. 30 - 40
XI	= stadium XI	No. -7	30 - 0 (= 1950)	c. 20
XII	= stadium XII	No. -8	+0 - +30 (1950 - 1980)	c. 10 - 20
XIII	= recent glacier stage	No. -9	+30 - +50 (1980 - 2000)	c. ? negative
XIV	= present glacier stage	No. -10	+50 - +60 (2000 - 2010)	c. ? negative

Draft: M. Kuhle

Fig. 2.5 Table: Glacier stages and ELA (Equilibrium Line Altitude = altitude of glacier snowline) of the mountains in High Asia, i.e., in and surrounding Tibet (Himalaya, Karakorum, Ladakh and Zanskar ranges, eastern Zagros, Hindukush, eastern Pamir, King Ata Tagh, Tian Shan [with Tersky Alatau, Dankowa, Kungai Alatau, Kirgise Shan, Altai and Bogdo Uul, Quilian Shan, Kakitu, Nan Shan], Kuenlun [with Animachin, Xuebao, Gangdise Shan, Nganclong Kangri and Shakham (Shakangsham) Kangri], Purog Kangri, Zaqungngomar, Tanggula Shan and Gladongding Massif, Bayan Har, Chola Shan, Ladake Shan, Nyainquentanglha, Namche Bawar, Shaluli Shan, Minya Konka) from the pre-Last High Glacial (pre-LGM) to the present-day glacier margins and corresponding sandrs (glaciofluvial gravel fields and gravel field terraces) with their approximate age (after Kuhle 1974–2014c)

the bottom of the lake basin. Further, in order to fill the entire lake basin with their glacier ice, a local ΔELA of 700–800 m would be needed. In High Asia an ΔELA such as this has to be classified as belonging to the last Late Glacial stage, the Sirkung stage (IV) (cf. Fig. 2.5 Table 1; Fig. 2.6). At an ELA depression (ΔELA) such as this there still existed inland glaciation during the Late Glacial Stage IV in Central Tibet, located east to southeast of the area under discussion. With its large, dome-shaped glacier surface, it lay above the

Fig. 2.6 From ca. 4435 m asl, 33° 49′ 20.71″N 78° 34′ 15.15″E (Fig. 2.2: No. 1), facing west; (●) marks light, erratic granite/granodiorite boulders on mylonite bedrock (♦) (see Fig. 2.3). The up to hut-sized granite/granodiorite boulders, transported over long distances, are distributed across the entire slope and mountain ridge (🡇). (⏶) marks a 4480 m high, glacigenically rounded mountain ridge. (- - - -) shows the past minimum glacier level. Analog photo 25/08/2013 by M. Kuhle

ELA. Therefore, at the time, the nourishing areas of the inland ice still by far exceeded those of the local hanging glaciers at the Pangong Co. The primary filling of the lake basin by an inland ice glacier tongue from the east is thus very likely (Fig. 2.7).

From here further northwest up to the end of the Pangong Co, further erratic granite boulders

Fig. 2.7 370° panorama from ca. 4445 m asl, 33° 49′ 19.53″N 78° 34′ 9.97″E (Fig. 2.2 on the left of No. 1). (↓) mark light, erratic granite/granodiorite boulders on mylonite bedrock (see Fig. 2.3). The up to hut-sized granite/granodiorite boulders, transported over long distances, are distributed across the entire slope and mountain ridge. (◓) mark roches moutonnées and glacigenically rounded mountain cupolas towering up to 4480 m asl, i.e., 240 m above the current lake level of Pangong Co. (■) are meter- to decameter-thick ground moraine covers and pedestals. (- - - -) marks the past glacier level in the vicinity of Pangong Co up to a height of 6100–6200 m asl. Analog photo 25/08/2013 by M. Kuhle

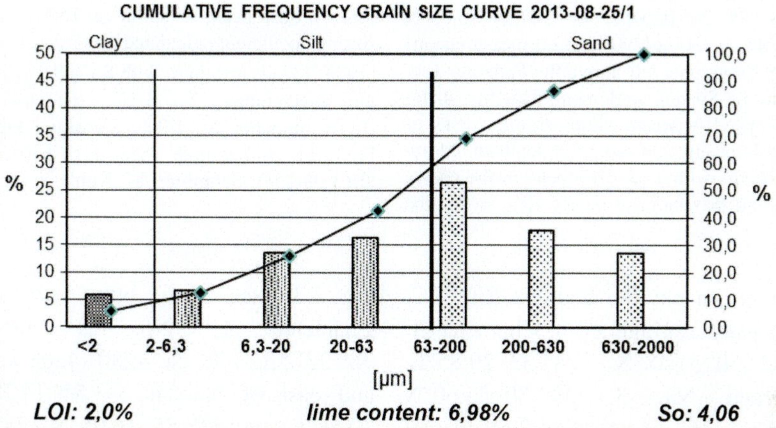

Fig. 2.8 (Grain size diagram 2013-08-25/1): Granulometry of the fine material matrix of ground moraine at 33° 50′ 28.30″N 78° 33′ 53.71″E, ca. 4280 (4268 aneroid measurement) m asl, on the orographic left valley flank of Pangong Tso upvalley from the settlement of Man (Maan), 10 m above the lake. The ground moraine is associated with meter-sized, erratic granite-boulders on clay-schist bedrock. Exposure and sampling depth ca. 0.5 m. The maximum occurs with a ca. 26% peak in the fine sand. The grain-size cumulative curve has a monomodal course, and clay content is 6%. The sorting coefficient calculated according to von Engelhardt (1973: 133) is So = 4.06 (So = $\sqrt{Q_3/Q_1}$); the lime content is 6.98%; the loss on ignition (LOI) amounts to 2.0%. Locality: Fig. 3: No. 2. See also Fig. 11a/b: No. 11; 2013-08-25/1. Sampling: M. Kuhle

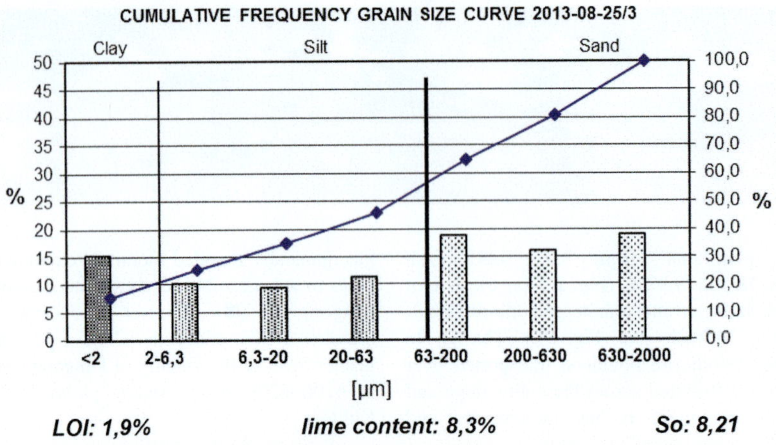

Fig. 2.9 (Grain size diagram 2013-08-25/3): Granulometry of the fine material matrix of ground moraine at 33° 55′ 14.79″N 78° 26′ 10.58″E or 33° 55′ 16.05″N 78° 26′ 7.78″E, 4334–4500 (4290 aneroid measurement) m asl, ca. 150 m orographic left above the Pangong Tso, downvalley from the Spatnik settlement. This fine material matrix is situated between white, erratic gneiss or granite boulders. Exposure and sampling depth ca. 2.5 m. The maximum occurs with a ca. 19% peak in the coarse sand, with a second maximum of nearly 19% in the fine sand. The grain-size cumulative curve has a trimodal course (Dreimanis and Vagners 1971). The clay content forms a third peak, which at ca. 15% is rather high. The sorting coefficient calculated according to von Engelhardt (1973: 133) is also high with So = 8.21 (So = $\sqrt{Q_3/Q_1}$); the lime-content is 8.3%; the loss on ignition (LOI) amounts to 1.9%. Locality: Fig. 2.2: No. 4; Fig. 2.11 foreground. See Fig. 2.10a, b: No. 12; 2013-08-25/3. Sampling: M. Kuhle

have been found on phyllite bedrock (Fig. 2.2, see also http://extras.springer.com: Character A: No. 2: 33° 50′ 51.48″N 78° 33′ 29.85″E, 4296 m asl and No. 3: 33° 50′ 53.60″N 78° 32′ 29.48″E, 4358 m asl) (Fig. 2.6). In part they are embedded within a till sheet (Figs. 2.11 and 2.15; Fig. 2.2 around No. 4). The fine material matrix of this till has been mapped in the field and the sediment analytically examined and confirmed in the laboratory. The granulometric findings are shown by Figs. 2.8 and 2.9 (samples 25.08.13/1 and /3), and the additional morphoscopic SiO_2 grain analyses of the medium sand fraction can be taken from Fig. 2.10 (Nos. 11; 12: samples 25.08.13/1 and /3; localities seen in Fig. 2.2, see also http://extras.springer.com: Character A: No. 2: 33° 50′ 28.30″N 78° 33′ 53.71″E, ca. 4280 (4268 aneroid) m asl and east of No. 4: 33° 55′ 14.79″N 78° 26′ 10.58″E or 33° 55′ 16.05″N 78° 26′ 7.78″E, 4334–4500 (4290 aneroid) m asl).

On the orographic right side of the valley, i.e., on the eastern lateral slopes of Pangong Co, there is evidence of glacigenic flank polishing, partly covered by till. Several of the polishing forms resulted in truncated spurs. They are evident in the form of glacigenically shaped triangular slopes caused by flank polishing (Fig. 2.2, see also http://extras.springer.com: Character A: No. 5: 33° 57′ 32.75″N 78° 27′ 52.10″E). On

(a)

Sample No. / *Probennummer*	Date / *Datum*	counted quartz grains of the medium sand grading fraction / *ausgezählte Quarzkörner der Mittelsandfraktion*	Freshly weathered/ glacially crushed / *frisch verwittert/ glazigen gebrochen* [%]	lustrous (fluvially polished) / *fluvial poliert* [%]	dull (aeolian) / *äolisch mattiert* [%]	Remarks / *Anmerkungen*
1	1993-07-21/1	224	46,0	49,1	4,9	High portion of quartz grains / viel Quarz enthalten
2	1993-07-21/2	128	58,9	35,1	6,0	
3	1993-07-22/1	224	55,3	39,8	4,9	
4	1993-07-22/2	224	56,2	36,7	7,1	
5	1993-08-15/1	218	78,4	12,9	8,7	
6	1993-08-15/2	218	91,7	5,6	2,7	
7	1993-08-15/3	14	71,4	28,6	0,0	Small portion of quartz grains / wenig Quarz enthalten
8 Lab.No. 2895	2013-08-19/2	41	97,6	0	2,4	*Other grains partly rounded (not the quartz)/ andere Materialien teilweise kantengerundet*
9 Lab.No. 2896	2013-08-20/1	112	100,0	0	0	*All quartz and other grains are glacially crushed / alle Bestandteile scharfkantig gebrochen*
10 Lab.No. 2897	2013-08-20/3	187	100,0	0	0	*High portion of quartz grains / viel Quarz enthalten*
11 Lab.No. 2898	2013-08-25/1	62	95,2	4,8	0	*Small portion of quartz grains / wenig Quarz enthalten*
12 Lab.No. 2899	2013-08-25/3	99	100,0	0	0	*All quartz and other grains are glacially crushed and sharp edged / alle Bestandteile scharfkantig gebrochen*
13 Lab.No. 2900	2013-08-27/1	87	100,0	0	0	*All quartz and other grains are glacially crushed / alle Bestandteile scharfkantig gebrochen*
14 Lab.No. 2901	2013-08-28/2	147	100,0	0	0	*Many grains are crusted with clay, some grains partly rounded (not the quartz)/ Viele Tonanhaftungen an den Körnern dieser Probe, andere Materialien teilweise kantengerundet.*
15 Lab.No. 2902	2013-08-29/1	59	100,0	0	0	*Quartz grains partly very fine shattered, / Quarzkörner sehr feingliedrig zersplittert*
16 Lab.No. 2903	2013-08-29/3	129	100,0	0	0	*All quartz and other grains are glacially crushed and sharp edged / alle Bestandteile scharfkantig gebrochen*
17 Lab.No. 2904	2013-08-30/1	213	100,0	0	0	*High portion of quartz grains / Viele Quarzkörner enthalten*
18 Lab.No. 2905	2013-08-31/1	347	100,0	0	0	*High portion of quartz grains. All quartz and other grains are glacially crushed and sharp edged / Viele Quarzkörner enthalten, alle Bestandteile scharfkantig gebrochen*
19 Lab.No. 2906	2013-09-01/1	295	100,0	0	0	*High portion of quartz grains. All quartz and other grains are glacially crushed and sharp edged / Viele Quarzkörner enthalten, alle Bestandteile scharfkantig gebrochen*
20 Lab.No. 2907	2013-09-01/2	97	100,0	0	0	*All quartz and other grains are glacially crushed and sharp edged / Alle Bestandteile scharfkantig gebrochen*

Sampling: M. Kuhle

Fig. 2.10 a, b Morphometric quartz grain* analysis of 20 representative samples: Indus-Valley, Pangong-Tso, Shyok Valley, Siachen Glacier valley (Nubra Valley) 1993-07-21/1 to 1993-08-15/3 and 2013-08-19/2 to 2013-09-01/2 *Quartz grains are the counted quartz grains of the medium sand fraction. Sampling: M. Kuhle

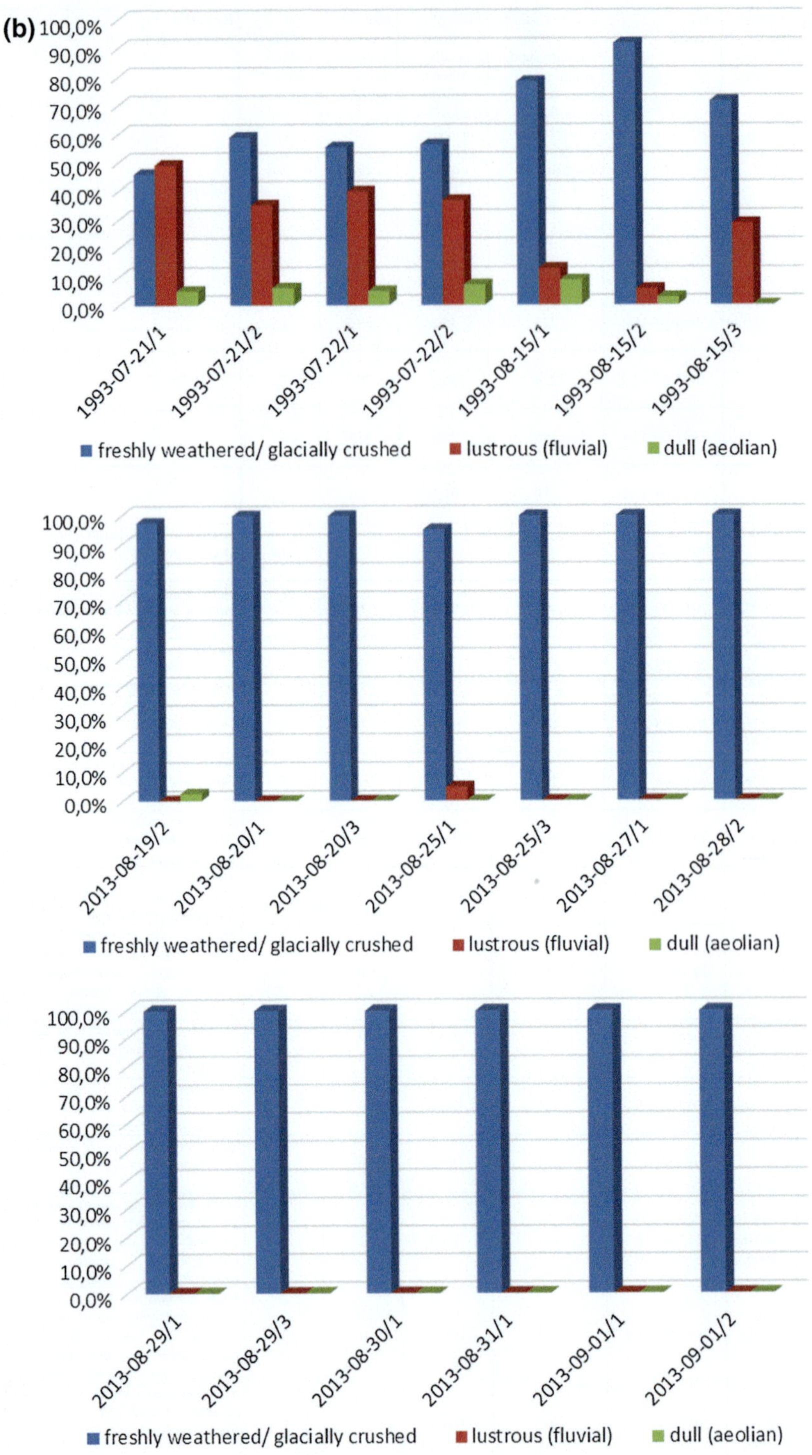

Fig. 2.10 (continued)

Fig. 2.11 From 33° 55′ 14.79″N 78° 26′ 10.58″E or 33° 55′ 16.05″N 78° 26′ 7.78″E, 4334–4500 (4290 aneroid measurement) m asl, ca. 150 m on the orographic left above Pangong Tso (Fig. 2.2: No. 4) facing west. (■) marks ground moraine (see Figs. 2.9 and 2.10). (◁) denotes a debris-flow cone. (◘) indicates glacio-fluvial gravel floors (sandrs) and debris flows. (☗) marks a classic roche moutonnée. (⊠) marks a break and slide in the ground moraine cover. (- - - -) marks the High Glacial glacier level at about 6100–5900 m asl. Analog photo 26/08/2013 by M. Kuhle

the orographic right slopes remnants of till covers are preserved (Fig. 2.12). Their thickness reaches in part a magnitude of over 10 m. On the surface and close to it, the till covers have been dislocated downslope periglacially and fluvially (Fig. 2.13). In other places the till sheets are cut by fluvial rills (Fig. 2.14). North of the northwest end of the lake, the layers of sedimentary rocks

Fig. 2.12 From ca. 4445 m asl, 33° 49′ 19.53″N 78° 34′ 9.97″E (Fig. 2.2 on the left of No. 1), facing east. (▲ left) marks a glacigenically rounded mountain ridge on the orographic right bank of Pangong Co (Fig. 2.2 on the right of A). (▲ on the right) denotes a glacial knob or roche moutonnée, undercut by the lake level (see Fig. 2.152). (■) mark ground moraine covers that have been dislocated solifluidally and fluvially downvalley. (- - - -) is the High Glacial glacier level at about ca. 6100 m asl. Analog photo 25/08/2013 by M. Kuhle

inclined steeply to vertically have been polished from approximately east to west. Up to ca. 4900 m asl they develop a classic glacigenic abrasion of the layer edges ("Schichtkopfstreifenschliff").

On the orographic left valley flank (above the western lakeshore) in the area of the outcropping granite rocks, glacigenic flank polishing can be observed (Fig. 2.15) up to an altitude of ca. 5750 m. This locally highest level of the polish line falls away to the Pangong Valley, starting below the lake, i.e., downvalley, as far as to the end of the lake, i.e., down to ca. 5240 m asl (Fig. 2.16). The same level of the polish line continues in the Pangong Valley further downward with a rounded, large-scale remnant of a valley shoulder in the bedrock at ca. 5180 m asl (Fig. 2.16). However, in the cross-profile of the end of the lake there is a still higher glacigenically rounded flat area. It is located at about 5900–6050 m asl (Fig. 2.2, see also http://extras.springer.com: Character A: No. 6: 33° 55′ 35.21″N 78° 21′ 48.01″E, 6040 m asl). This is the highest located evidence for the

Fig. 2.13 From ca. 4445 m asl, 33° 49′ 19.53″ N78° 34′ 9.97″E (Fig. 2.2 left of No. 1), facing NNE looking toward the orographic right shore of Pangong Co (Fig. 2.2 on the right of No. 2). (▲) marks glacigenic flank-polishing that reaches approximately up to the summit of the mountain ridge at 4640 m asl. (■) shows a ground moraine cover, overlying this flank-polishing. (- - - -) indicates the High Glacial glacier surface that towered above the ca. 5400 m high mountains in this section of the picture. Analog photo 25/08/2013 by M. Kuhle

minimum height of the High Glacial (Last High Glacial Maximum, LGM, i.e., Last Glacial Period, LGP) glacier level at the Pangong Co (Fig. 2.15).

On the two valley slopes of the Pangong Valley there are till covers lying downvalley from the Pangong Co (Figs. 2.11, 2.15, and 2.16). Accordingly, a great northwest Tibetan

Fig. 2.14 From ca. 4445 m asl, 33° 49′ 19.53″N, 78° 34′ 9.97″E (Fig. 2.2 left of No. 1), facing NNW toward a glacigenically abraded mountain ridge (▲) (Fig. 2.2: No. 3) with light, erratic ground moraine overlay (■ black) on dark bedrock. This cover reaches higher on the side facing the Pangong Co. A partly transparent veil of ground moraine extends upward to the summit of the mountain ridge at ca. 4630 m asl (above ▲). Since the deglaciation, it has been modified solifluidally. (■ white) marks a decameter-thick ground moraine. Analog photo 25/08/2013 by M. Kuhle

Fig. 2.15 Picture taken from ca. 33° 57′ 27.84″N 78° 25′ 38.35″E from ca. 4270 m asl looking into the orographic left flank of the valley of the Pangong Co (lake) (Fig. 2.2 above No. 4) facing SW. The up to ca. 5090 m slope is covered by ground moraine (■) as far as to its summit (☗), which has been rounded/abraded by inland ice (- - - - - marks the minimum height of the ice level). This attains a decameter-thickness (■ black). It consists of polymictic, partly rounded, erratic boulders (■ white on the lower left and right) and lies on dark phyllite bedrock. (⊠) marks breaks and slides in the ground moraine cover. Analog photo 26/08/2013 by M. Kuhle

Fig. 2.16 Ca. 180° panoramic photograph at 33° 57′ 42.02″N 78° 23′ 36.75″E from ca. 4320 m asl (Fig. 2.2 on the right above No. 6) from the end of Pangong Co (lake) (in the background of the right quarter of the section) looking down the Pangong Valley to the NW (left half). The valley shows trough-shaped cross-profiles. (■) mark up to decameter-thick covers of ground moraine as far as above 5000 m asl. (□) signifies the gravel bottom of the valley. (◁) is a debris-flow fan from an orographic left side-valley. The High Glacial glacier surface (- - - -) extends up to at least 5900 m asl. Analog photo 24/08/2013 by M. Kuhle

outlet glacier flowed down this valley, which after a distance of ca. 50 (48.54) km downvalley from the Pangong Co joined the Shyok Valley at 3600 m asl (Fig. 2.17). In so doing, it overflowed in the junction area of three bar mountains (“Riegels”) with altitudes of 4540, 4520,

Fig. 2.17 Ca. 180° panoramic photograph from 4210 m asl at ca. 34° 7′ 2.25″N 78° 5′ 20.73″E across the lower Pangong Valley from upvalley facing SSE (center of the picture) up to its junction with the Shyok Valley in the NE (left quarter of the picture). The valley is trough-shaped (∪). (■) mark up to decameter-thick ground moraine covers up to 4900 m asl; they contain meter-sized, edged, and rounded-edged boulders (⭘). (◓) indicate glacigenic abrasion forms up to 5000 m asl. The two (◓) on the left are riegel-mountains, reshaped into roches moutonnées. (⇗) is the glacigenically abraded transfluence pass that at 4300 m asl leads to the Shyok Valley. (⇓) is the upper beginning of a gorge-stretch that due to subglacial meltwater in the Late Glacial has been incised further as a subglacial gorge-form. (▷) marks a debris-flow cone from dislocated material of ground moraine. (- - - -) is the High Glacial glacier level that must have been situated clearly above the abrasion forms at 5500 m asl. Analog photo 26/08/2013 by M. Kuhle

Fig. 2.18 Ca. 180° panoramic photograph from 4200 m asl at ca. 34° 7′ 8.35″N 78° 4′ 13.94″E (Fig. 2.54: Character D on the right of No. 3) above the junction of the lowest orographic left side valley in the Pangong Valley: from facing west up to the side valley as far as facing east to the two roches moutonnées (the two ◓ on the right) in the lower Pangong Valley. This valley landscape was filled up to an ice level (- - - -) of over 5500 m asl with glaciers of the Tibetan inland ice. This is valid, too, for the side valley (left half of the picture) that currently does not contain a glacier anymore and nowhere reaches a height of 6000 m. (■) mark up to decameter-thick ground moraine covers that contain meter-sized, polymictic, rounded, and edged boulders of phyllite and granite (⭘). (◓) are abraded rock-faces. (⇓) is the thalweg of the side valley, that has freshly cut the glacigenic valley form. Analog photo 26/08/2013 by M. Kuhle

and 4065 m respectively, and obviously polished and rounded them. The lowest "riegel" is completely covered with moraine, and the surfaces of the two higher ones show remnants of till (Figs. 2.17 and 2.18). In this junction area and in the valley chamber of Darbuk, the outlet glacier of the Pangong Valley has left behind till sheets on both valley flanks up to a height of 4900 m (Fig. 2.17). Isolated moraine remnants have even been preserved up to ca. 5100 m asl. In addition,

Fig. 2.19 Ca. 115° panoramic photograph from 4140 m asl at ca. 34° 1′ 2.40″N 78° 16′ 57.20″E across the junction of an orographic right hanging side valley facing NE and the Pangong main valley facing SW upvalley. The broad Pangong Valley shows characteristics of a trough and a gravel-floor (□). (☗) marks a glacigenically polished rock threshold in steeply lying layers of metamorphites. (⇓) is a small, subglacially developed meltwater gorge across that rock threshold. (■) mark up to decameter-thick ground moraine covers containing meter-sized, polymictic edged and rounded boulders. (▷) marks a debris-flow cone from dislocated ground moraine-material. (- - - -) indicates the High Glacial glacier level, which must have run above the ground moraines and glacigenic abrasion forms. Analog photo 24/08/2013 by M. Kuhle

one can speculate that it is also till that lies as an accumulation in a saddle at between 5100 and 5200 m asl east above the settlement of Darbuk, i.e., northwest above Muglib. From here, the valley chamber of Muglib, an ice transfluence must have occurred from the Pangong Valley into the Shyok Valley adjacent to the north. All these accumulations that stratigraphically have to be classified as being from the Late Glacial prove already a minimum height of the ice level of 1050–1350 m above the current gravel floor and contour line in the thalweg. Accordingly, the highest LGM polish lines must have run clearly higher. The highest remnants of flank polishing in the valley section under discussion are situated at 5250–5500 m asl (Figs. 2.17 and 2.19). The highest rock smoothing and rounding by glacigenic flank polishing (see above) mentioned from the area of the end of the lake proves an LGM glacier level at 6050 m asl (Fig. 2.2, see also http://extras.springer.com: Character A: No. 6: 33° 55′ 35.21″N 78° 21′ 48.01″E, 6040 m asl; Fig. 2.19). Here the gravel soil of the valley bottom lies at 4270 m, i.e., 1770 m lower, so that there must have existed an ice thickness of nearly 2000 m. The gravel bottom of the Shyok Valley lies 670 m lower in the junction area of the Pangong Valley, so that the contemporary ice level must be assumed to have lain there at ca. 5380 (6050 − 670) m. That is a minimum value. A decrease of the thickness of the glacier ice as far down as to the Shyok Valley does not have to be assumed, because the surrounding mountain groups are still strongly glaciated even today, and thus during glacial times they probably added to an extremely strong ice supply from all sides. Indeed, in the Shyok Valley we can observe rounded truncated spurs up to ca. 5200–5400 m asl. At least on one glacigenically triangle-shaped slope up to 5400 m it is even possible to diagnose with some probability an overlay of till (Fig. 2.2, see also

Fig. 2.20 Shyok Valley facing upwards to the east. Ground moraine cover (■ above on the left) between 4700 and ca. 5200 m asl at 34° 17′ 43.95″N 78° 5′ 32.54″E (Fig. 2.23 E of No. 1) on glacigenic flank abrasion (▲ below) on the orographic right valley flank of the Shyok Valley. Further flank-polishes (▲ on the right) and ground moraine covers (■ above on the right) are situated on the orographic left side of the valley up to 4350 m asl. (■ below) is a ground moraine pedestal that is exposed up to a height of nearly 100 m (34° 18′ 32.07″ N 77° 55′ 33.04″E; for locality, see Fig. 2.23 on http://extras.springer.com: Character B: No. 1; see also Fig. 2.25 (■ on the very lower left) and Fig. 2.24). (□) marks a glaciofluvial gravel ledge on ground moraine on the valley floor. (- - - -) marks a Late Glacial ice level. Analog photo 28/08/2013 by M. Kuhle

http://extras.springer.com: Character A: No. 7: 34° 10′ 51.77″N 78° 12′ 5.11″E) (cf. Figs. 2.20 and 2.21).

In the section of the Pangong Valley under discussion here, glacigenic flank polishings are observed in many places. One example is the

Fig. 2.21 The 6140 m summit is situated at 34° 16′ 36.03″N 78° 10′ 28.82″E looking east up the Shyok Valley. The eastern glacier is about to advance, reaching down to ca. 5450 m asl and evidences a current orographic ELA at ca. 5725 m on a west-facing slope. (■) provide proof of ground moraine covers on the valley flanks up to ca. 5230 m asl. (▲) marks a past glacigenic flank abrasion that further above, at a decreasing slope incline, is covered by ground moraine. (∩) signifies a typical glacigenic triangular slope that currently is being destroyed by wall-rills. (⊠) marks a section of the slope that since deglaciation has been roughened by weathering. Analog photo 28/08/2013 by M. Kuhle

dark 1.33 km long and ca. 335 m high rock threshold polished like a roche moutonnée on the orographic right side upvalley of Muglib (34° 1′ 48.96″N 78° 17′ 34.08″E, 4468 m asl; Fig. 2.19). A second example is the 2.3 km extended glacigenic abrasion of the layer edges

Fig. 2.22 Facing NE along the orographic right flank of the Pangong Valley looking upward, from ca. 4040 m asl, at 34° 2′ 31.82″N 78° 13′ 19.33″E (Fig. 2.2 on the left below No. 8). (◉) mark remnants of subglacially developed potholes on the rockwall. The valley flank shows remnants of glacigenic flank-polishings (⏶). (⊠) mark sites of weathering that have developed since deglaciation, roughening the igneous bedrock. (▷) are debris cones created below, which cover ground moraine; in the foreground this ground moraine (■) is visible and uncovered; in and on its fine material matrix there are meter-sized polymictic boulders of granite and phyllite. They are rounded and round-edged, but also show breaking-edges (●). (- - - -) indicates the minimum level of the High Glacial glacier cover above the preserved glacigenic abrasion-roundings at ca. 5400 m asl. Analog photo 26/08/2013 by M. Kuhle

("Schichtkopfstreifenschliff") reaching above up to 400 m on the orographic left valley slope downvalley of Muglib (Fig. 2.2, see also http://extras.springer.com: Character A: No. 8: 34° 3′ 17.20″N 78° 14′ 13.10″E) (see also Fig. 2.22).

2.2 The High Glacial Outlet Glacier of the Shyok Valley in Connection with the Pangong Co Outlet Glacier on the Western Margin of Tibet

The primary focus of this section is Fig. 2.35. The Pangong Valley outlet glacier (Fig. 2.2, see also http://extras.springer.com: Character A) came down from the western region of the Tibetan inland ice sheet, flowing through the valley which is today occupied by the Pangong Co and joined together with the Shyok Valley glacier. The ice stream expanded through the Shyok Valley in continuation from the Pangong Valley, and thus represents one and the same outlet glacier.

Along the Shyok Valley (Fig. 2.23, see also http://extras.springer.com: Character B) this High Glacial ice stream has left behind numerous glacio-geomorphical indicators (Figs. 2.20 and 2.21) clearly proving its extension. It is a huge trough valley with a till cover up to decameter in thickness. A representative sample of till (Fig. 2.24 sample 27.8.13/1) has been taken 30 km downvalley from the junction of the Pangong Valley (Fig. 2.23: Character B: No. 1: 34° 18′ 29.87″N 77° 55′ 55.91″E, 3374 m asl). These till sheets—interspersed with large boulders of up to many meters in length, which are polymictic and partly erratic, and have rounded, round-edged, and at least partly edged shapes—show a clayey matrix (for a granulometric analysis, see Fig. 2.24) and consist of glacially crushed SiO_2 grains of the medium sand fraction (Fig. 2.10: No. 13). On the valley flanks of the valley cross-profile from which samples have been taken, till sheets like these are reaching up to 4630 m on the orographic right valley flank and on the orographic left one up to 4300 m asl, i.e., ca. 950–1280 m above the thalweg at

Fig. 2.23 Satellite image, Google Earth Pro: center 34° 30′ 14.7″N 78° 1′ 9.1″E; http://extras.springer.com: Character B; Nos. 1–10

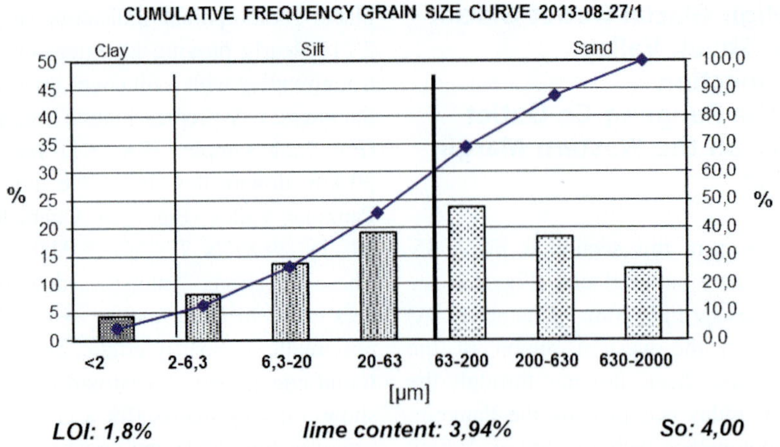

Fig. 2.24 (Grain size diagram 2013-08-27/1): Granulometry of the fine material matrix of ground moraine containing material from a rockfall at 34° 18′ 29.87″N 77° 55′ 55.91″E, 3374 (3255 aneroid measurement) m asl in the Shyok Valley ca. 10 (9.82) km upvalley from the junction of the Agham (Agym) valley, on the orographic left valley flank at a depth of 50 m, taken from an exposure. This fine material matrix is situated between polymictic, meter-sized, in part erratic boulders transported partly downvalley. The maximum occurs at the ca. 24% peak in the fine sand. The clay content amounts to ca. 4.5%. The sorting coefficient calculated according to von Engelhardt (1973, p. 133) is So = 4.0 (So = $\sqrt{Q_3/Q_1}$); the lime content is 3.94%; the loss on ignition (LOI) amounts to 1.8. Locality: see Fig. 2.23 on http://extras.springer.com: Character B, No. 1; Fig. 2.25 (■ on the very lower left); Fig. 2.20. Morphoscopy see Fig. 2.10a, b: No. 13; 2013-08-27/1. Sampling: M. Kuhle

Fig. 2.25 Ca. 200° panoramic photograph from 3380 (3260 aneroid measurement) m asl, at 34° 18′ 29.87″N 77° 55′ 55.91″E, in the Shyok Valley, ca. 10 (9.82) km upvalley from the junction of the Agham (Agym) Valley, from the orographic left valley flank (Fig. 2.23: No. 1) from upvalley facing east via facing north into the right valley flank, up to facing west seen downvalley. (■ and ↑) mark up to several decameter-thick ground moraine covers and pedestal ground moraines up to a height of ca. 4700 m asl (Fig. 2.23 beside No. 3). (■ on the lower far left) marks the spot where the sample of Fig. 2.24 has been taken. (- - - -) points to the High Glacial glacier level at over 5400 m asl (Fig. 2.29). (▻) marks a ca. 20 m thick debris-flow fan, which consists of rearranged ground moraine material. (⊠) is a current area of weathering above the outer bank of the Shyok River. (□) marks the here 1 km wide valley bottom with mega-flood sediments and up to house-sized boulders. Analog photo 27/08/2013 by M. Kuhle

3350 m asl (Fig. 2.25). Throughout the downvalley section spanning 10 km down to the junction of the Agham Valley, on the orographic right side, remnants of till covers have been registered as far down as to the valley shoulder at at least 4870 m asl (Figs. 2.26 and 2.27, Fig. 2.23, see also http://extras.springer.com: Character B: No. 2). According to Google Earth, remnants of till can even be recorded still ca. 100 m higher up on the connected flat face of a valley shoulder. Up to ca. 4200 m asl, the thickness of the till sheets is still so immense and

Fig. 2.26 Ca. 215° panoramic photograph from ca. 3330 m asl, at 34° 18′ 52.03″N 77° 53′ 5.87″E, in the Shyok Valley, ca. 5.4 km upvalley from the junction of the Agham Valley, from the orographic left valley flank (Fig. 2.23 on the left below No. 3) from upvalley facing ESE via facing north into the right valley flank, up to facing WSW downvalley into the orographic left valley flank. (■) mark up to several decameter-thick ground moraine covers and pedestal ground moraines up to a height of ca. 4800 m asl (Fig. 2.23 on the right above No. 1). (☗) indicates glacigenic abrasion-roundings. (- - - -) indicates the High Glacial glacier level exceeding 5600 m asl (see Fig. 2.29). (⊠) mark current areas of slippage on outer banks, undercut subglacially and subaerially by the Shyok River. (Δ and ⊳) mark debris- and debris-flow cones, partly from redeposited ground moraine material. (□) points to the here 0.6 km wide valley floor covered by glacio-fluvial gravels. Analog photo 27/08/2013 by M. Kuhle

Fig. 2.27 Ca. 225° panoramic photograph from ca. 3275 m asl, at 34° 20′ 53.44″N 77° 47′ 35.64″E, in the Shyok Valley, ca. 3.9 km downvalley from the junction of the Agham Valley, from the valley bottom (Fig. 2.23: No. 4) from downvalley facing WNW, via facing NE into the orographic right valley flank, further via facing ESE upvalley up to facing south into the orographic left valley flank. (■) mark up to several decameter-thick ground moraine covers and pedestal ground moraines as far as a height of ca. 4600 m asl (Fig. 2.23 on the right above No. 1); up to this height these ground moraines mainly lie on glacigenic abrasion-roundings (see Fig. 2.29). (- - - -) marks the High Glacial glacier level exceeding 5600 m asl. (⊠) shows an area of weathering on an outer bank that is undercut by the Shyok River. (□ black) marks the here ca. 2.4 km wide valley bottom consisting of glaciofluvial gravels and mega-flood sediments. (□ white) are up to ca. 380 m thick limnic and glaciofluvial terraces, in part with glaciofluvially reshaped ground moraine cores; they are frayed by erosion-rills. Analog photo 27/08/2013 by M. Kuhle

their superficial state of preservation so fresh and continuous, that it has to be classified as being from a Late Glacial stage (Fig. 2.23, see also http://extras.springer.com: Character B: No. 3), perhaps even from the youngest Late Glacial Stage IV (Fig. 2.5).

Since the deglaciation in the Late Glacial, the till covers were not only deformed and removed by downslope denudation and rill development, but also cut through by the thalwegs of the hanging sidevalleys as far down as to the bedrock (Figs. 2.25, 2.26, and 2.27). The 1–2 km-wide gravel floor proves that the Shyok River also undercuts the till on its alternating outer slopes and lets them crumble away. At the same time it transports away the matrix and smaller components of this moraine cover (Figs. 2.25, 2.26, and 2.27).

In addition, intensive lateral erosion in the till material above the valley bottom has taken place due to numerous megafloods. They were caused by periodical outbursts of glacier lakes in the upper Shyok Valley, such as for instance outbursts of the Chong Kundum (Kumdan) Glacier in 1926, 1929, and 1932 (Gunn 1930; Mason et al. 1930; Visser 1935a, b, 1938). Up to the junction of the Nubra Valley down into the Shyok Valley—the confluence area of these two valleys is situated 36–40 km away from the junction of the Agham Valley further downward at 34° 32′ 23.68″N 77° 35′ 2982″E (see Fig. 2.23, see also http://extras.springer.com: Character B: 3120 m asl)—Visser (1935a: 38) has already observed megaflood indicators of the water surge. Much further upvalley and close to the position of that ice-dammed lake, the traces of the megafloods reached a height of 40–50 m above the level of the current gravel bottom of the Shyok Valley. For the ca. 1 km wide valley cross-profile of the Shyok Valley, only somewhat upvalley of the junction of the Nubra Valley (Fig. 2.23, see also http://extras.springer.com: Character B; Fig. 2.50), dislocated moraine material has been reported on the valley flanks, scoured out to a depth/size of up to 20 m or even more (ibid.: 208). The present author has reconstructed the position of the locality with the help of Visser's route (Kuhle 2012b: 199–200). Furthermore, Visser already reported on the destruction of farmland and settlements "many hundred kilometers" down the Shyok and Indus valleys (Visser 1935a: 38–39).

In the valley chamber connected downvalley to the inflow of the Agham Valley, there are numerous up to hut-sized (the larger boulders are 6.5–8.4 m long; Fig. 2.23, see also http://extras.springer.com: Character B: No. 4), round-edged moraine boulders from the last bigger megaflood (1932) (Fig. 2.27), distributed in the shape of fans. Up to ca. 10–30 m above the current Shyok River level, the till material at the footslope of the valley flanks has been undercut freshly and steeply like a wall over a distance of several kilometers. Along extensive stretches, this has occurred so intensively that the bedrock at the foot of the slope has been exposed (Figs. 2.25, 2.26, and 2.27). Here, too, the megafloods from the outbreaks of the ice-dammed lakes must have been a cause.

During the High Glacial, the Shyok Valley and its side valleys where filled by glaciers which in many places overflowed the intermediate valley ridges. This will be introduced and proved using the side valleys of the Agham Valley (Fig. 2.27) as an example: the till cover on the orographic left side of the valley leading down from the Wari pass (Wari La, also Waring La; ca. 5270, 5300, or 5317 m asl) to the north, and on the intermediate valley ridge up to ca. 5400 m asl (Fig. 2.28). The till sheet covers the mountain ridge at a thickness of several meters from ca. 5100 m asl (34° 11′ 55.84″N 77° 50′ 6.38″E) onwards, sloping down to the north. In the upper region, this covering consists only of patches and then further down of an increasingly continuous cover nearly uninterrupted by bedrock (Fig. 2.23, see also http://extras.springer.com: Character B: No. 5).

On the orographic right flank (Figs. 2.28 and 2.29) of this valley as well as in the western, parallel valleys, the extent of the till covers reaches the same altitude. Above the ELA these thick ground moraine covers could not be deposited. This was caused by the discharge of the ice at a positive mass balance above the ELA. Therefore the till accumulation begins only below the ELA. The current ELA runs in N-exposition at ca. 5650 m asl. This means that

Fig. 2.28 Ca. 210° panoramic photo from ca. 4460 (4350 aneroid measurement) m asl at 34° 11′ 18.94″N 77° 51′ 14.19″E (Fig. 2.23 on the right of No. 5; Fig. 2.54: Character D below No. 5) in the side valley of the North Wari pass and the left side valley of the Agham Valley; from facing west into the orographic left valley flank via facing north downvalley into the Shyok main valley, up to the ESE into the left valley flank, i.e., its small side valleys. (■) are meter- to decameter-thick ground moraine covers, which on the orographic left side reach up to ca. 5400 m asl into the intermediate valley ridge. The side valley has a trough-form. (☗) mark glacigenic abrasion-roundings, that also in the orographic right Shyok valley flank in many places are covered with ground moraine and reach a height of 5400 m asl (see Fig. 2.29). (- - - -) marks the High Glacial glacier level exceeding 5600 m asl. The ground moraine in the side valley contains edged, round-edged, and rounded granite-boulders with a length of up to several meters (**O**). Analog photo 27/08/2013 by M. Kuhle

the ELA which genetically belongs to these till covers must be classified as being from an expositional ELA depression (ΔELA) of ca. 550 m. This corresponds with a climatic ΔELA of ca. 600–700 m that mediates all expositional ELAs arithmetically (average calculation). It is the most insignificant Late Glacial ΔELA in large parts of High Asia (Fig. 2.5). Thus the large-scale till covers at and below 5100 m asl (see above) have to be classified as being from Stage IV (Sirkung stage, Fig. 2.5). Accordingly, the pertinent glacier surfaces were situated upvalley at an altitude exceeding 5100 m. Due to the semiarid climate at this altitude, the glaciers were probably cold based and thus the ice was frozen to the underlying ground surface. These conditions might have led to an increasing thickness of the glacier. In addition, the glacier thickness was also enlarged due to ice temperatures of below ca. −6 °C, from the level of the ELA upwards (Kuhle 1985b, 2012a: 174, 176, 177, 191; 2014a). The reason is the increase in the viscosity (slow movement) of the ice. These factors make a glaciation probable, which covered the local relief forms extensively, even until the late Late Glacial (Stage IV). Today, however, there are only snowfields and some cirque glaciers in this mountain group.

The ice discharge of these side glaciers flowing down to the north, such as the Wari La-N-valley glacier, must have filled the valley of Agham. The Agham Valley glacier joined the Shyok outlet glacier. The Agham Valley glacier was dammed up here, at the abutment of the main valley glacier, up to its ice level (Fig. 2.23, see also http://extras.springer.com: Character B: No. 3-2). This damming brought about a ca. 6.5 km wide transfluence of the Agham Glacier and its most northern tributary, the Digar side glacier into the Shyok Valley. The transfluence is confirmed by several meter- to decameter-thick till sheets on the crest ridge north of the Digar

Fig. 2.29 From the Wari La (pass, Waring La; Fig. 2.54: Character D below No. 5) from ca. 5270 (5300 or 5317) m asl at 34° 6′ 23.24″N 77° 50′ 1.66″E facing NNE into the side valley of the Wari pass and its elongation in the Agham Valley downwards to the Shyok Valley and beyond, looking toward the northwest Karakorum with the 6394 m-peak. (◓) mark glacigenic abrasion-forms, among them a glacial transfluence pass (◓ on the left) with ground moraine overlays (■) up to somewhat above 5300 m asl. (∩) indicate glacigenic triangular slopes developed by past flank-polishing. (- - - -) points to a geomorphologically immediately duplicable past glacier level at about 5700 m asl. It is ascribed to the Late Glacial. The levels of the synchronous valley glaciers of the hanging trough valleys (∪) have been adjusted according to that of the main valley glacier. Analog photo 23/08/2013 by M. Kuhle

settlement up to ca. 4910 m asl (Fig. 2.23 east and west of No. 6: 34° 17′ 43.59″N 77° 46′ 45.68″E; Fig. 2.27 on the far right). Here, we find moraines from the Late Glacial side valley glacier inflow, as the Shyok outlet- and main glacier must have made this inflow possible by an already lowered ice level (see below). Possibly, this level of the Shyok Glacier was located at ca. 4200 m asl—a level that would be attributed to Stage IV (see above).

2.2.1 Theory on the Accumulation of Till: A Digression in Order to Understand the Example of the Shyok Valley Outlet Glacier

Up to this point, the reconstruction of the Shyok outlet glacier introduced here has referred to the upper margin of the occurrences of till sheets (see above). However, these have come into being only in the course of the melting process, i.e., during the Late Glacial. This occurred at a time when the ELA must have already been in the process of rapid uplift. At a maximum depression of the ELA of about 1200–1300 m (Fig. 2.5) evidenced in many places for the LGM in semiarid and subtropical High Asia (Norin 1932; Wiche 1958; Kuhle among others 1974, 1982a, 1985c, 1987a, 1988d, 1994a, 1997, 1998a, 2001a, 2012c; Schroeder-Lanz 1986), the accumulation of till sheets can only be expected to occur at an altitude of below ca. 3500 m asl. Upvalley—above the ELA—glacial erosion took place, i.e., ground polishing of the bedrock of the valley bottom and flank polishing in the rock of the valley slopes. But this process of erosion, which due to the high mechanical pressures of the superimposed ice load kept the valley bottoms and lower valley flanks free from till, occurred not only above the ELA; it also happened further downvalley far below the ELA. The cause of this is that the increasing ice load was superimposed all the way down to the valley bottom. Accordingly, a till cover could have persisted from the High Glacial period only much further downvalley. Due to the syngenetic, subglacial removal of the till, caused by glacial ground polishing, the deposition of moraines must have occurred primarily on the valley flanks. This applies to locations where valley flanks at 3500 m continue to extend both higher and at the same time far below the ELA toward the valley floor.

In the cross-profile near Agham, however, the valley bottom lies only ca. 200 m (at ca. 3300 m asl) below the ELA (at 3500 m asl). Accordingly, we have to expect High Glacial moraines only very much further downvalley (Fig. 2.23, see also http://extras.springer.com: Character B: Nos. 1–4 and 6). These will appear as high glacial till covers on the upper flanks of the Shyok Valley.

Valley cross-profiles with a valley bottom located so low that only their upper flanks tower above the position of the ELA at 3500 m are situated ca. 232 km farther down the Shyok Valley, 5 km before its junction area with the Indus Valley (Kuhle 2010, 2012a: 180, Photo 1). There the current valley bottom with loose rock material lies at ca. 2300 m asl. Its surface consists of river gravels of the current Shyok River. Below, in the underlying beds, there very likely exists Late Glacial ground moraine as a continuation of the same ground moraine covers that can be observed on the lower slopes. At 35° 14′ 23.82″N 75° 58′ 45.37″E, for instance, on the orographic right valley flank at up to ca. 3760 m asl, moraine-like accumulations have been observed (Kuhle 2013a: 177). For the reasons stated above, they have to be classified as very likely originating from the High Glacial (LGM = MIS (marine isotope stage) 3-2 = Stage 0; Fig. 2.5).

Given these theoretical considerations, the highest till covers must have been deposited increasingly higher on the valley slopes with the passage of time.

Accordingly, they have to be classified as belonging to the oldest Late Glacial (Stage I). Further up the valley, they must be attributed to at least the youngest, the late Late Glacial Stage IV.

However, a further aspect has to be considered. The ELA, moving upwards from Stage 0 to Stage IV, from ca. 1300 m ΔELA to only 700 m ΔELA, is followed by a decrease in glacier thickness. This has consequences with regard to a decrease in height up to which the till sheets are deposited on the valley flanks, that is, below the ELA. Thus, for every glacier stage there exists a valley cross-profile for which, following the valley downwards, the highest level of till cover sinks beneath that of the next-older stage. This applies even though the level height of the deposition of till upvalley correlates with a higher situated ELA, and thus with younger glacier stages.

The younger stages situated upvalley, with their ELA lying at higher altitudes, consequently produce till sheets at a greater absolute altitude. The older glaciers that reach further down, however, belong to stages with a lower ELA, so that they produce till covers only further down, at a minor altitude level. The latter glaciers, however, are thicker. On the valley flanks their ice towers relatively further up, so that their moraine covers reach their highest level further downvalley than those of the glaciers of the following recessional stages.

Where the intersection of the older and younger moraine levels is located can only clearly be recognized in the case of former valley or cirque glaciers that are only a few kilometers long and thus steep.

For the very long, only slightly inclined outlet glaciers of Tibet, which flowed through the Karakoram Range, such as the 300 km long Shyok Glacier under discussion, this intersection cannot be determined definitely.

Therefore the classification of the till sheet levels from the High to the Late Glacial (Stages 0–IV; Fig. 2.5) on the flanks of the Shyok Valley can only be made vaguely and with reservation. The best guideline for this is provided by the state of conservation of the till covers.

All considerations point to the fact that also the high occurrences of till covers in the valley chamber of Agham attest to a glacier level belonging to the Late Glacial. Even the highest-positioned occurrences of till at 4870 m (Figs. 2.26 and 2.27, Fig. 2.23, see also http://extras.springer.com: Character B: No. 2) and up to ca. 5000 m asl must have lain clearly below the glacier level of the Shyok outlet glacier during the High Glacial (LGP = Last Glacial Period; LGM = Stage 0). The perfectly preserved till sheet level at 4200 m asl (Fig. 2.23, see also http://extras.springer.com: Character B: No. 3) for this reason has been classified as belonging to the youngest Late Glacial stage (Sirkung Stage IV) (see Sect. 2.2).

If the uppermost till sheet level reaches ca. 4900 m and this inevitably must be below the ELA (see above), then an ΔELA of 800–900 m at maximum has to be estimated. This means that this highest till sheet level must be classified as belonging to Stage III, i.e., to the medium Late Glacial (Dhampu stage, Fig. 2.5).

During the older two Late Glacial stages and during the LGM, the level of the Shyok outlet glacier was situated higher. Thus it was lying far above the accompanying ELAs. Accordingly, the ground moraine covers belonging to these older stages *must* be lacking.

Therefore we have to look for different indicators of older to maximum ice levels. Polish and abrasion demarcation lines make it possible to reconstruct the highest ice levels.

2.2.2 Continuation of the Reconstruction of the High Glacial Shyok Valley Outlet Glacier in Connection with the Pangong Co Outlet Glacier on the Western Margin of Tibet

In the following, some polish and abrasion demarcation lines are examined in order to register thicknesses of levels of ice stream networks and glaciers.

Rounded summit cupolas on mountain ridges are appropriate indicators to record the minimum height of glacigenic abrasion. In the area of the intermediate valley ridge between the Shyok Valley and the valley of Agham, a higher glacigenically rounded rock flattening at ca. 5550 m asl is preserved (Fig. 2.23, see also http://extras.springer.com: Character B: No. 7: 34° 16′ 25.51″N 77° 54′ 57.35″E; Figs. 2.27, 2.28, and 2.29). A rounded mountain ridge is still about 250 m higher (34° 15′ 23.66″N 77° 56′ 52.84″E). In this area and at this altitude, the edge of the Shyok outlet glacier passed over into the local mountain glaciation so that no exact demarcation can be reconstructed. Accordingly, this intermediate valley ridge was overridden up to ca. 5800–5900 m asl by the Shyok Glacier, locally enforced by local glaciation. A correspondingly high maximum glacier level on the orographic left side, in the form of a well-rounded 5900 m high mountain cupola (34° 20′ 55.29″N 77° 42′ 33.44″E) and neighboring rounded transfluence passes, are preserved. Further downvalley at the inflow of the Nubra Valley (Fig. 2.23: Character B), only small-scale abrasion roundings can be observed in the not very resistant phyllite outcroppings at a maximum altitude of ca. 5500 m (34° 29′ 13.92″ N 77° 33′ 9.23″E).

On the orographic right valley side, glacigenic roundings are preserved on rock cupolas and mountain ridges flanking the side-valley exits at an altitude of 5570–6050 m asl; for example, north to east above the settlements of Rongdu (at 34° 27′ 33.51″N 77° 48′ 58.29″E, Fig. 2.23: No. 8 and to the north), Tsati (at 34° 32′ 12.81″N 77° 47′ 10.88″E) and Therit (at 34° 33′ 48.58″N 77° 44′ 24.58″E and 34° 37′ 12.38″N 77° 40′ 46.63″E). These example locations are situated NNE to east of Character B in Fig. 2.23. Here, too, in the area of the orographic right Shyok Valley flank, the margin of the High Glacial Shyok outlet glacier passed over into the glaciation of the local mountain group.

During the Ice Age this local glaciation flowed towards the Shyok outlet- and main valley glacier. This is the mountain group of the 6730 m high Arganglas Kangri, even today with up to 14 km long valley glaciers.

From these abrasion levels, ice thicknesses of ca. 2000–2800 m were calculated for the Shyok outlet glacier. In this calculation, the drop of the level and loss in height of the High Glacial glacier surface from the abrasion levels in the side-valley exits (e.g., Fig. 2.23: No. 8) towards the center of the Shyok Valley glacier are taken into account. At the same time, this process has been nearly compensated for by the postglacial heightening of the valley bottom by deposits of gravels (see Fig. 2.23, see also http://extras.springer.com: Character B: No. 4, Fig. 2.27). Therefore, this heightening of several hundred meters also had to be factored into the equation.

The ice thickness of ca. 2000–2800 m is confirmed by the strong glacigenic flank polishing on both valley sides. Trough-shaped cross-profiles are preserved on a large scale on the two valley flanks. An especially clearly preserved example is situated above the Rongdu settlement (Fig. 2.23, somewhat left below No. 8). The concave trough-shaped cross-profile is located on a truncated mountain spur, polished by the thick valley glacier that still lay here during at least the late Late Glacial (Fig. 2.30). The triangle-shaped slope that thus has been developed is situated between two side-valley exits. Between the settlements of Tsati and Rongdu there are four of these triangle-shaped slopes covered with Late Glacial ground moraine (Fig. 2.30, right half). All four together and elongated upwardly have formed a united High Glacial glacigenic triangular slope. This culminates at about the level of No. 8 in Fig. 2.23 at 5900 m. The large, glacigenically formed, triangle-shaped slope, situated between two larger side valleys, has been strongly reshaped since the deglaciation. This occurred synchronously with the shifting of the ice level lower. As a result of this exposure, the upper slope areas were already strongly cut into and roughened by weathering, rock fall, and water discharge through slope rills (Fig. 2.30).

The same kind of concave, glacigenically formed, triangle-shaped slopes as seen in the valley cross- profile flank the Shyok Valley 70 km further downvalley as far as to the gorge section near Yagulung settlement. They continue over an even greater distance, that is, up to Khaplu and the inflow into the Indus Valley (Kuhle 2010, 2012a, 2013a). Because the Shyok Valley as far as to the aforementioned gorge section is flat and at most several meters wide, an impressive trough valley form is preserved here.

The rest of the side valleys of the Shyok main valley, like the N-Khardung La valley, were synchronously filled with side-valley glaciers (Fig. 2.23, see also http://extras.springer.com: Character B: south of B). Just as in the case of the described Agham Valley (see above), their

Fig. 2.30 Ca. 200° panoramic photograph from ca. 3360 m asl, 34° 28′ 9.82″N 77° 43′ 2.99″E (Fig. 2.23 west [left] of No. 8), from the orographic left valley flank of the Shyok Valley facing SE upvalley (on the right) via facing NE into the orographic right valley flank, up to facing NW along the left valley flank seen downvalley. (∩) shows glacigenic triangular slopes. (■) shows ground moraine overlays on both valley flanks up to ca. 5000 m asl. (▲) are glacigenic abrasion-forms up to ca. 4800 m asl. (⊠) shows areas of weathering occurring after the deglaciation. (□ white) is past glaciofluvial gravel ca. 180 m above the valley bottom and subrecent fluvial gravels and mega-flood sediments on the valley bottom (□ black). (Δ) are alluvial and debris-flow cones of dislocated moraine from the side valleys. (- - - -) is the maximum High Glacial glacier level at ca. 6000 m asl. Analog photo 27/08/2013 by M. Kuhle

glacial ice levels were adjusted to that of the Shyok outlet glacier. The exemplary sample of till taken at 4500 m asl (4400 m aneroid measurement) on the orographic left side in the northern Khardung La valley at 34° 20′ 51.00″N 77° 39′ 22.89″E, 6.2 km south of the Khardung settlement (Fig. 2.31: sample 01.09.13/1 and Fig. 2.10: No. 19), as well as its downvalley elongation, proves a corresponding glacier filling of this side valley. The till sheet still reaches to near the junction into the Shyok Valley on the orographic left side at an altitude of 4400 m asl. On the orographic right side, the ice level even exceeded 5900 m asl. This assessment is supported by the typical abrasion roundings at 34° 20′ 55.29″N 77° 42′ 33.44″E (see above) reaching up to this height.

The largest side valley on the orographic right side is the Nubra Valley. It has brought to the Shyok outlet glacier a ca. 12 km wide ice stream from the NNW at the 5000–5300 m level (Fig. 2.23 between Character B and No. 9) (Fig. 2.32). At 34° 43′ 25.06″N 77° 32′ 22.39″E there are two “riegelberge” (“riegels” after Flint) on the valley bottom, 40–60 m above the thalweg (Fig. 2.23: No. 9) at about 3213 m asl (3100 m aneroid measurement). Their bedrock presumably consists of fine-grained granodiorite to quarzdiorite (Fig. 2.33).

The “riegelberge” are divided into several classic roche moutonnées, showing glacial striations and even polishes that are nearly completely freshly preserved (Fig. 2.34).

A further 1.8 km long “riegelberg” complex lies NNW of the Tiricha settlement. It is divided into three single roche moutonnées (34° 44′ 38.34″N 77° 33′ 5.88″E with a highest culmination at 3230 m asl). They run in the longitudinal direction of the valley (Fig. 2.32). The two northern ones are separated from each other by a polished, basin-shaped depression (“Schliffwanne”).

SSE of Tiricha there are further roches moutonnées, partly covered with perfectly glacigenic polishes and glacial striations that are preserved freshly and nearly intact (Figs. 2.35 and 2.36: 34° 43′ 13.18″N 77° 34′ 14.57″E, 3200 m asl (3170 m aneorid measurement); Fig. 2.37: sample 28.08.13/1). All the documented roches moutonnées are partly covered by till material (Fig. 2.38: sample 29.08.13/3 and Fig. 2.10: No. 16).

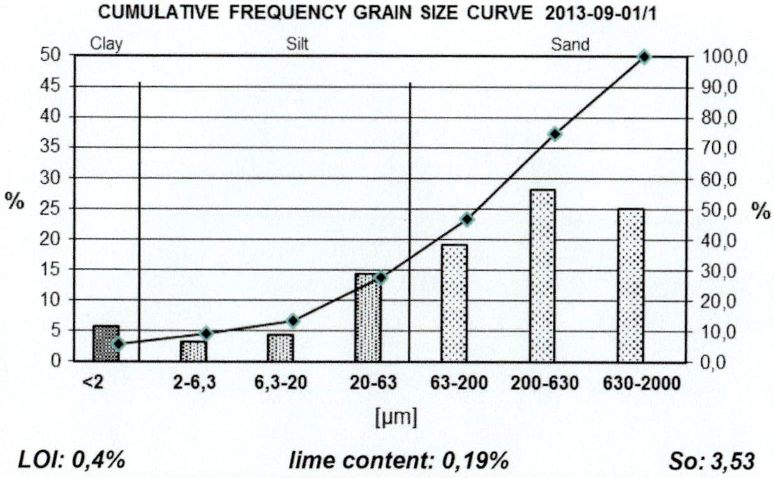

Fig. 2.31 (Grain-size diagram 2013-09-01/1): Granulometry of the fine material matrix of ground moraine at 34° 20′ 51.00″N 77° 39′ 22.89″E, 4500 (4400 aneroid-measurement) m asl in the Khardung-NE orographic left valley. This fine material matrix is situated between meter-sized boulders. Exposure and sampling depth ca. 0.5 m. The maximum occurs with the 28% peak in the middle sand, and the secondary maximum with nearly 6% in the clay fraction. The cumulative grain-size curve has a bimodal course (Dreimanis and Vagners 1971). The sorting coefficient calculated according to von Engelhardt (1973: 133) is So = 3.53 (So = $\sqrt{Q_3/Q_1}$); the lime content is 0.19%; the loss on ignition (LOI) amounts to 0.4%. Locality: Fig. 2.23 west (left) of No. 4. See Fig. 2.10a, b: No. 19; 2013-09-01/1. Sampling: M. Kuhle

Fig. 2.32 Ca. 300° panoramic photograph from 3213 (3100 aneroid measurement) m asl at 34° 43′ 23.95″N 77° 32′ 21.89″E, WSW of the Tiricha settlement, ca. 55 m above the river (Fig. 2.23: No. 9); from facing WSW into the orographic right flank of the upvalley glaciated Nubra Valley, via facing NNW upvalley, across facing ENE into the left valley flank, up to facing SSE downvalley to the Nubra Valley. (∩) indicate glacigenic triangular slopes. (■) show ground moraine overlays on both valley flanks, in places several decameters in thickness, up to 4400 m asl. (☗) mark well-preserved glacigenic abrasion-forms up to ca. 4400 m asl. (⊠) shows areas of rock fall on walls that have become exposed since the deglaciation. (Δ) shows an alluvial and debris-flow fan of dislocated moraine (see ■ above) from a side valley. (↓, ←, →) are roches moutonnées and glacigenic surfaces of ground-polish from bedrock, which possibly consists of fine-grained granodiorite up to quartz diorite (A. Vollbrecht). (- - - - -) marks the Late Glacial to High Glacial glacier level up to a height of ca. 6000 m. Analog photo 29/08/2013 by M. Kuhle

Fig. 2.33 Mineralogical rock determination of sample 29.08.13/2. Bedrock of a roche moutonnée with glacier striations that possibly consists of fine-grained granodiorite up to quartz diorite (rock determination and documentation by A. Vollbrecht). Locality: 34° 43′ 23.95″N 77° 32′ 21.89″E on the orographic right side in the lower Nubra Valley about 3213 (3100 aneroid measurement) m asl (Fig. 2.23: No. 9, Fig. 2.34). Sampling: M. Kuhle

Fig. 2.34 Roche moutonnée (see Fig. 2.33 sample 29.08.13/2) with very freshly preserved glacier striations, in part covered by ground moraine (Fig. 2.38), not far from the exit of the Nubra Valley, seen facing SSE along the orographic right valley flank looking downvalley toward the Shyok Valley. Locality: 34° 43′ 23.95″N 77° 32′ 21.89″E situated about 3213 (3100 aneroid measurement) m asl (see Fig. 2.23: No. 9). Analog photo 29/08/2013 by M. Kuhle

Fig. 2.35 Two small roches moutonnées SSE of the Tiricha settlement, 34° 43′ 13.18″N 77° 34′ 14.57″E, 3200 (3170 aneroid measurement) m asl (Fig. 2.23 on the right (east) of No. 9); photo facing SSE down the Nubra Valley. The larger roche moutonnée is covered with perfectly glacigenic polishes and glacier scratches. Despite its same age, the smaller one is already strongly weathered. This is further empirical proof of the—even on a small scale—extremely diverse erosion rates of rock surfaces. For this reason, too, TCN datings factually are impossible. (■) show overlays of ground-moraine and (▲) glacigenic abrasions on the orographic left valley side. Analog photo 28/08/2013 by M. Kuhle

Fig. 2.36 This picture shows a polish-face of the intact roche moutonnée in Fig. 2.35; even the scratches are well preserved (as to the petrography, see Fig. 2.37). The rim of the polished rocks is covered with ground moraine (■). Analog photo 28/08/2013 by M. Kuhle

On the two flanks of the same valley chamber one also finds very well preserved, fresh flank abrasions with scratches and polishes displayed continuously over a distance of ca. 100–320 m of altitude (at 34° 43′ 5.87″N 77° 34′ 38.31″E and 34° 43′ 86″N 77° 32′ 0.65″E) (Fig. 2.32).

Additionally, on the orographic left valley flank, subglacial meltwater holes with remnants of potholes were recorded at an altitude of ca. 3250 and 3500 m asl (e.g., at 34° 41′ 21.47″N 77° 35′ 33.61″E and 34° 41′ 0.95″N 77° 35′ 45.08″E) (Fig. 2.39). These subglacial hollow formations caused by meltwater were formed in the phyllite bedrock. In the area above, there is no catchment area. Therefore, they are today still partly filled with till. This material is polymictic, and partly erratic (e.g., granite boulders up to meters in size) (Fig. 2.39). The

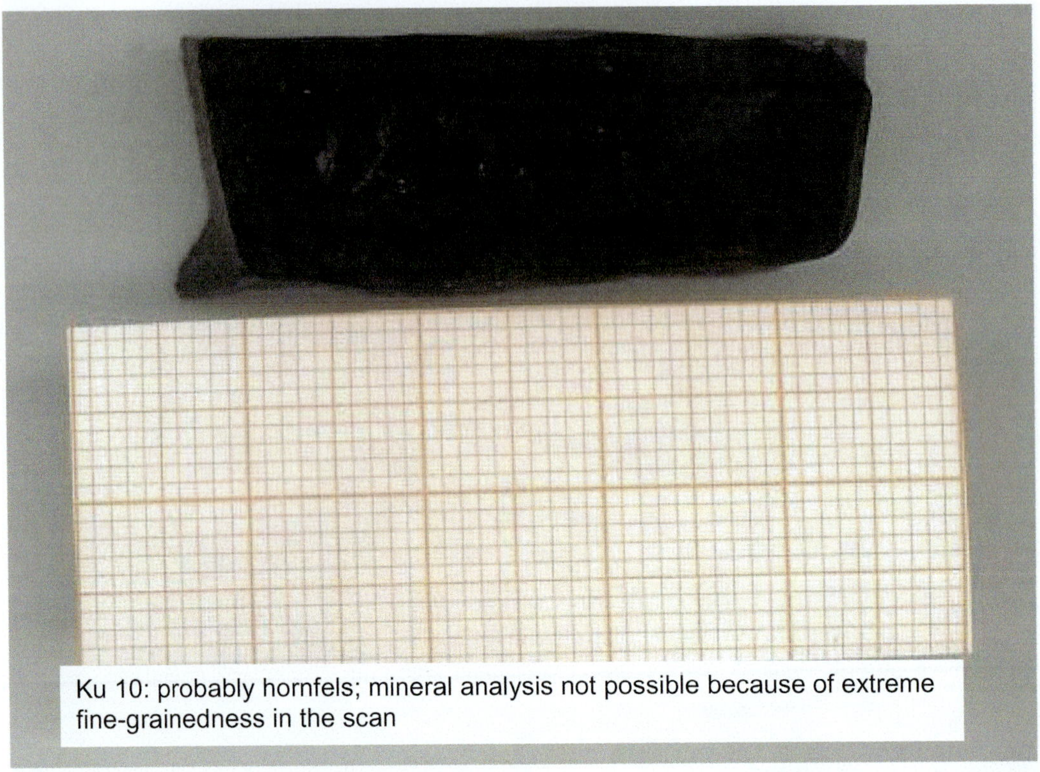

Fig. 2.37 Mineralogical rock determination of sample 28.08.13/1, 3170 m asl (aneroid): rock-fragment of the roche moutonnée with scratches on the orographic left side of the Nubra Valley (Siachen Glacier valley) SSE of the Tiricha settlement; 34° 43′ 13.18″N 77° 34′ 14.57E, 3200 (3170 aneroid measurement) m asl. Probably hornfels is concerned here. Because of the extremely fine grains in the scan, it is not possible to identify the minerals (rock determination and documentation: A. Vollbrecht). Sampling: M. Kuhle

fine-grained material matrix of this ground moraine was sampled 15 km upvalley of the junction into the Shyok Valley (Fig. 2.40: sample 28.08.13/2 and Fig. 2.10: No. 14). Within the falling trajectory of the slope where the sample was taken, the moraine cover continues in an upslope direction up to a height of 4600 m asl. There the upper end of the valley slope is located, and there too lies the top of one of the numerous late Late Glacial triangular slopes (Fig. 2.41).

Furthermore, ca. 35 km up the Nubra Valley, on the opposite, orographic right valley flank, a corresponding till cover has been sampled (Fig. 2.23, see also http://extras.springer.com: Character B: No. 10: Fig. 2.42: sample

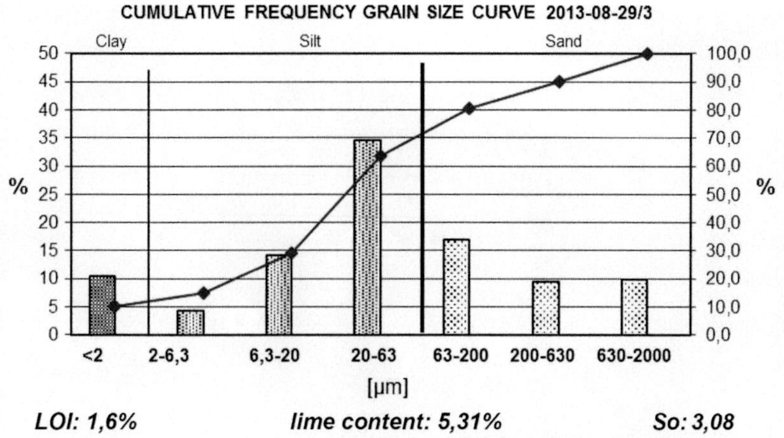

Fig. 2.38 (Grain size diagram 2013-08-29/3): Granulometry of the fine material matrix of typical ground moraine at 34° 43′ 27.59″N 77° 32′ 19.11″E about 3213 (3100 aneroid measurement) m asl WSW of the Tiricha settlement in the Nubra Valley (Fig. 2.23: No. 9), taken ca. 60 m above the river on a polished rock surface (see Fig. 2.34). This fine material matrix is situated between meter-sized boulders. Exposure and sampling depth ca. 1 m. The maximum occurs with the ca. 35% peak in the coarse silt, a secondary maximum with nearly 11% in the clay, and a tertiary maximum in coarse sand with 10%. Thus, the grain size cumulative curve has a trimodal course. The sorting coefficient calculated according to von Engelhardt (1973: 133) is So = 3.08 (So = $\sqrt{Q_3/Q_1}$); the lime content is 5.31%; the loss on ignition (LOI) amounts to 1.6%. Locality: Fig. 2.23: No. 9. See Fig. 2.10a, b: No. 16 sample 2013-08-29/3. Sampling: M. Kuhle

29.08.13/1 and Fig. 2.10: No. 15). In this area, it was documented up to an altitude of 3975 m asl (Fig. 2.41).

The highest traces of abrasion that provide proof of the High Glacial glacier level, are situated on the orographic left side at ca. 5725 and 5625 m asl (34° 40′ 53.36″N 77° 39′ 28.15″E and 34° 44′ 37.89″N 77° 37′ 22.84″E). On the orographic right side, there is a comparable mountain ridge with an abraded cupola at ca. 5560 m asl (34° 42′ 13.62″N 77° 28′ 29.95″E). At 4940–5120 m altitude similarly rounded rock cupolas are found in the valley flanks positioned closer to the valley (34° 41′ 36.36″N 77° 30′ 20.93″E and 34° 37′ 54.05″N 77° 31′ 37.41″ E). In this case, the height of the abraded cupolas merely proves that the ice level was reaching even higher. Based on the observations made in the Shyok main valley (see above), to which the glacier level of the Nubra Glacier must have been adjusted, this ice level must have been located at somewhat above 6000 m asl during the High Glacial. Here, too, the conjecture applies—in the same way as with the Shyok Glacier—that the Nubra Glacier level on both valley sides must have passed over at a low angle into extended local glacier surfaces of surrounding mountain groups. On the orographic left side, its nourishing areas were connected to the 7480 m high Saser-Kangri Group and on the orographic right side to the 6800 m high La Yongma Ri Group. At the valley head, the numerous summits exceeding 7000 m, belonging to the Rimo Kangri-, Teram-Kangri-, Singi Kangriv-, Ghent Kangri-, Sherpi-Kangri-, Saltoro Kangri-, and K12 Group, are connected to the area. They culminate in the 7684 m high Saltoro Kangri. Even today, they still nourish, through

Fig. 2.39 Facing NE onto the orographic left flank of the lower Nubra Valley from ca. 3320 m asl, at 34° 41′ 21.47″N 77° 35′ 33.61″E (Fig. 2.23, halfway below and to the right of No. 9). (■) shows up to decameter-thick ground moraine covers (see. Fig. 2.40) with up to meter-sized, polymictic, edged, round-edged, and up to more rounded, partly erratic boulders of granites and phyllites. (⏶) marks glacigenic abrasion-roundings. (◉) is one of several subglacially developed potholes on this valley slope. In the center of the picture is a person to provide a sense of scale. Analog photo 28/08/2013 by M. Kuhle

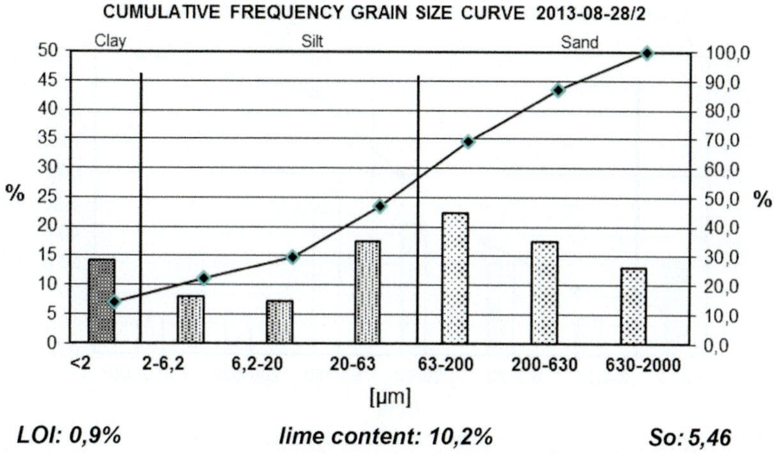

Fig. 2.40 (Grain size diagram 2013-08-28/2): Granulometry of the fine material matrix of ground moraine at 34° 41′ 23.07″N 77° 35′ 36.84″E at 3356 (3255 aneroid-measurement) m asl, taken ca. 300 m above the valley bottom east of the Charasa settlement close to Diskit. This fine material matrix is situated between meter-sized granite boulders. Exposure and sampling depth ca. 1 m. The maximum occurs with the ca. 22.5% peak in the fine sand, and the secondary maximum with nearly 14% in the clay. The grain-size cumulative-curve has a bimodal course. The clay content is rather high with ca. 14% on a steep slope. Accordingly, it is a very typical ground moraine. The sorting coefficient calculated according to von Engelhardt (1973: 133) is also high, with So = 5.46 (So = $\sqrt{Q_3/Q_1}$); the lime content is 10.2%; the loss on ignition (LOI) amounts to 0.9%. Locality: Fig. 2.23 halfway below and to the right of No. 9; Fig. 2.39 taken on the right above of a pothole. See Fig. 2.10a, b: No. 14; 2013-08-28/2. Sampling: M. Kuhle

Fig. 2.41 Ca. 220° panoramic photo from ca. 3471 (3340 aneroid) m asl at 34° 46′ 28.28″N 77° 30′ 48.46″E (see Fig. 2.23 on http://extras.springer.com: Character B: No. 10), Nubra Valley, ca. 35 km upvalley from its inflow into the Shyok Valley, SSW of the Panmik settlement, ca. 300 m above the valley bottom, on the orographic right flank above the Warisfistan monastery. From facing NNW along the orographic right valley flank upvalley via facing ENE into the left valley flank via facing SSE downvalley up to facing SSW into the right valley flank. (∩) mark glacigenic triangular slopes. (■) denote ground moraine overlays (see Fig. 2.42) on both valley flanks of partly several decameters in thickness up to ca. 4400 m asl. (☗) show well preserved roches moutonnées and glacigenic abrasion-forms up to ca. 4300 m asl. (⊠) mark areas of weathering on rock faces that have become exposed since the deglaciation. (Δ) show alluvial and debris-flow fans of moraine dislocated from the side valleys. (▼) is decameter-thick ground moraine that has slumped on the slope. (□) shows the 1 to nearly 3 km wide gravel floor of the Nubra River. (- - - -) Late Glacial to High Glacial glacier level up to a height of somewhat above 6000 m asl. The valley is shaped like a trough. Analog photo 29/08/2013 by M. Kuhle

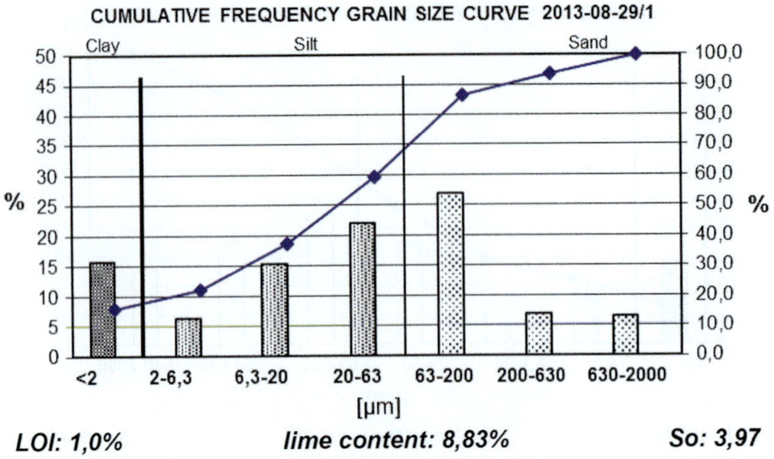

Fig. 2.42 (Grain size diagram 2013-08-29/1): Granulometry of the fine material matrix of ground moraine at 34° 46′ 28.28″N 77° 30′ 48.46″E, 3471 (3340 aneroid measurement) m asl in the Nubra Valley, ca. 35 km upvalley from the junction into the Shyok Valley, SSW of the Panmik settlement ca. 300 m above the valley bottom, on the orographic right side of the valley. The ground moraine material is situated on a slightly inclined rock face. Exposure and sampling depth ca. 0.4 m. The maximum occurs with the ca. 27% peak in the fine sand, with a secondary maximum of nearly 16% in the clay. The clay content is high. The grain-size cumulative curve has a bimodal course. The sorting coefficient calculated according to von Engelhardt (1973: 133) is So = 3.97 (So = $\sqrt{Q_3/Q_1}$); the lime content is 8.83%, the loss on ignition (LOI) amounts to 1.0%. Locality: Fig. 2.23: No. 10; Fig. 2.41 ■ foreground. See Fig. 2.10a, b: No. 15; 2013-08-29/1. Sampling: M. Kuhle

their numerous and still completely glaciated side valleys, the 70 km long Siachen Glacier that fills the upper Nubra Valley.

2.2.2.1 The High Glacial Shyok Outlet Glacier Between the Junction of the Nubra Glacier and the Gorge Section of Yagulung

The primary focus of this section is Fig. 2.43, see also http://extras.springer.com: Character C. On both flanks of this 45 km long valley chamber, till covers are preserved in many places (Figs. 2.20, 2.21, 2.25, 2.26, 2.27, 2.29, and 2.30), as is generally characteristic of the upper valley sections. Continuing downvalley, they indicate the former existence of an uninterrupted continuum on a par with the level reached by the Shyok outlet glacier (see below).

Evidence for a chronological classification can, for instance, be gathered in an orographic left-side valley named the Hundar Valley. Its glacigenic characteristics are indicative of a Late Glacial Shyok outlet glacier. The Hundar Valley has an extraordinary thick till sheet; 200–300 m high moraine accumulations that were cut into pedestal moraine terraces since the deglaciation cover the valley floor (Figs. 2.44, 2.45, and 2.46). Numerous samples have been taken from the thick till covers as far as ca. 4000 m asl (Fig. 2.47: sample 30.08.13/1 and Fig. 2.48: sample 31.08.13/1; Fig. 2.10: Nos. 17 and 18;

Fig. 2.43 Satellite image, Google Earth Pro: center 34° 39′ 49.31″N 77° 14′ 29.60″E; http://extras.springer.com: Character C: Nos. 1–10

Fig. 2.44 Ca. 200° panoramic photograph across the uppermost Hundar Valley in the valley chamber of the filial (new) settlement of Hundar Dog, from ca. 3965 (3700 aneroid measurement) m asl at 34° 31′ 15.99″N 77° 24′ 40.22″E (Fig. 2.43: No. 2); from facing SSW up-valley to the 6054 m-peak, via facing SE into the orographic right valley flank up to facing NE downvalley into the Shyok main valley. (∩) signify glacigenic triangular slopes. (■) are pedestal ground moraines on the valley bottom with a thickness exceeding a hundred meters and overlays of ground moraine (see Fig. 2.48) on both valley flanks of a decameter thickness with meter-sized, round-edged to rounded boulders (◉) up to ca. 4400 m asl. (◓) mark glacigenic flank-polishings with rock-polishes in the outcropping granite as well as abrasion-roundings, in part with a ground moraine overlay up to ca. 4440 m asl. (⊠) indicates area of weathering on rock walls developed after the deglaciation. (▼) marks cones and slopes of detritus from dislocated ground moraine. (- - - -) denotes the Late Glacial to High Glacial glacier level at a height exceeding 5000 m asl. The thickness of ground moraine has been and is further exposed by fluvial depth-erosion (⇓). The valley shows a trough-like shape. Analog photo 31/08/2013 by M. Kuhle

Fig. 2.45 Ca. 180° panoramic photograph across the middle and lower Hundar Valley, from ca. 3730 (3700 aneroid measurement) m asl at 34° 31′ 27.49″N 77° 25′ 17.34″E (Fig. 2.43 on the right above No. 2); from facing NW into the orographic left valley flank, via facing NNE downvalley into the Shyok main valley, looking (∩) into the orographic right valley flank facing SE. (∩) marks a triangular glacigenic slope, covered with ground moraine (the two ■ in the foreground) and pedestal ground moraines on the valley bottom with a thickness up to over a hundred meters and with edged to round-edged boulders several meters in diameter (⭘). (■) show ground moraine overlays on both valley flanks as well as in the Shyok main valley at a thickness of up to decameters as far as 4000–4400 m asl. (◓) indicate glacigenic flank-polishings with rock-polishes (◓ on the very left) in the granite and abrasion-roundings in the bedrock up to ca. 4900 m asl. (▼) marks debris cones and debris slopes of dislocated ground moraine. (- - - -) denotes the Late Glacial to High Glacial glacier level at a height exceeding 5100 m asl. The thickness of the ground moraine has been and still is exposed by deep fluvial erosion (■ white below). The valley has a trough-like form. Analog photo 31/08/2013 by M. Kuhle

Fig. 2.43, see also http://extras.springer.com: Character C: Nos. 1 and 2) and confirmed in the laboratory.

In many places along the valley flanks, the moderately thick, Late Glacial moraine covers reach up to 5000 m asl, partly up to 5100 m asl, and in some places even higher (Figs. 2.44 and 2.45). These heights are similar to those described for the valley of Agham (see Sect. 2.2.1). They can be explained by the back-damming of the small, only 28 km long Hundar side valley glacier against the Late Glacial Shyok main glacier. Without this back-damming of ice, the Hundar Glacier would have flowed steeply down over 2000 m in altitude until reaching the main valley bottom at 3070 m, where it would have spread out in the form of a hammer head. It would have remained flat and the reconstructed thickness of up to 1400 m would have been impossible.

As was shown above, the pedestal moraines have been cut since the deglaciation. This is mainly correct for the lowest decameters of the fluvial cutting. The sections nearer to the surface, however, were already cut before that by the subglacial meltwater underneath the glacier. The pedestal moraines were built up below the ELA. Syngenetically, meltwater at the same time discharged from the glacier surface underneath the glacier bottom. It flowed down underneath the ice as hydrostatically confined water and thus was strongly erosive. Accordingly, it subglacially cut into the loose rock of the moraine pedestals. This process involved

1. a washing-out of the till matrix with a synchronous condensation of the larger boulders and gravel components, and
2. a dislocation and downvalley accumulation of these subglacially washed and coarsely classified (sorted) fluvial sediments.

Coarse gravel bands such as these that developed in the respective small tunnel valleys below the glacier were re-sedimented within these cuts of moraine material and dovetailed

Fig. 2.46 From ca. 3600 m asl, at 34° 34′ 16.60″N 77° 26′ 35.70″E (Fig. 2.43, halfway below and to the left of No. 1) facing east along the thalweg of the lowest orographic left side valley of the Hundar Valley, looking downward to the main Shyok Valley (background in sunlight). (■) mark ground moraine. (■ background) denotes pedestal ground moraine exceeding 100 m in thickness (see Fig. 2.47). (▣) marks subglacially outwashed gravel material in the Late Glacial, eroded from ground moraine and thus losing a great part of its fine material matrix. (**O**) indicate up to several meter-sized, round-edged to rounded polymictic boulders. (- - - -) denotes the High Glacial glacier level at a height exceeding 5800 m asl. The thickness of the ground moraine and the glacio-fluvial gravels is exposed by the fluvial erosion depth (⇓) adjusted to the main valley. Analog photo 31/08/2013 by M. Kuhle

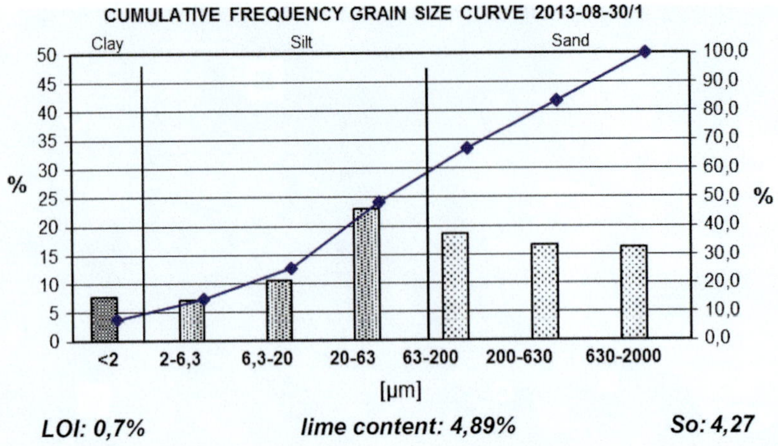

Fig. 2.47 (Grain size diagram 2013-08-30/1): Granulometry of the fine material matrix of sub-glacially and—after the deglaciation—glacio-fluvially washed (as well as dislocated by a debris flow) ground moraine at 34° 34′ 41.95″N 77° 27′ 8.65″E at ca. 3300 (3200 aneroid) m asl in the Hundar Valley next to the valley exit. The fine material matrix is situated between polymictic, meter-sized, round-edged boulders. Exposure and sampling depth ca. 1 m. The maximum occurs with the ca. 23% peak in the coarse silt, and a secondary maximum with nearly 8% in the clay. The grain-size cumulative curve has a bimodal course. The sorting coefficient calculated according to von Engelhardt (1973: 133) is So = 4.27 (So = $\sqrt{Q_3/Q_1}$); the lime content is 4.89%; the loss on ignition (LOI) amounts to 0.7%. Locality: Fig. 2.43: No. 1; cf. Fig. 2.46; see Fig. 2.10a, b: No. 17; 2013-08-30/1. Sampling: M. Kuhle

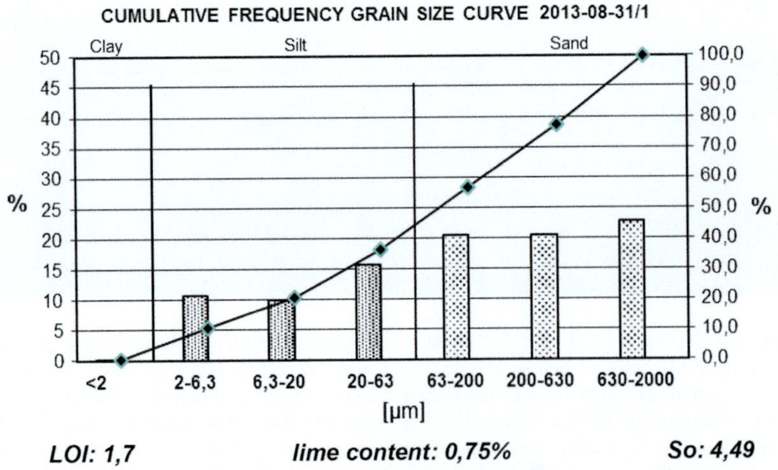

Fig. 2.48 (Grain size diagram 2013-08-31/1): Granulometry of the fine material matrix of primarily deposited ("exogenous") ground moraine at ca. 3965 (3700 aneroid measurement) m asl at 34° 31′ 15.99″N 77° 24′ 40.22″E, in the orographic left flank of the Hundar Valley in the valley chamber of the filial (new) settlement of Hundar Dog. The fine material matrix is situated between up to meter-sized, round-edged to rounded granite boulders (see Fig. 2.44). Exposure and sampling depth ca. 6 m. The maximum occurs with the ca. 23% peak in the coarse sand, and the secondary maximum with ca. 11% in the fine silt. The grain-size cumulative curve has a bimodal course. The lack in clay and the altogether coarse-grained composition of the matrix point to a local moraine from the granite area of the immediate vicinity. The sorting coefficient calculated according to von Engelhardt (1973: 133) is So = 4.49 (So = $\sqrt{Q_3/Q_1}$); the lime content is 0.75%; the loss on ignition (LOI) amounts to 1.7%. Locality: Fig. 2.43: No. 2; see Fig. 2.10a, b: No. 18; 2013-08-31/1. Sampling: M. Kuhle

Fig. 2.49 Facing WNW in the middle Hundar Valley from ca. 3730 (3700 aneroid measurement) m asl at 34° 31′ 27.49″N 77° 25′ 17.34″E (Fig. 2.43 right above No. 2) looking into the orographic left valley flank. The granite bedrock has been concavely scoured and polished by glacigenic flank-polish. In the lower half of the section until recently this polish has been preserved in sections and in an upper one has already been completely splintered by weathering since deglaciation and thus roughened. This heterogeneity proves once more the variability of erosion rates over even small areas by weathering and the empirical impossibility of TCN surface datings. Analog photo 31/08/2013 by M. Kuhle

with the pedestal moraine. This dovetailing, which can be identified in exposures by horizontally and vertically alternating deposition, can be observed repeatedly all the way down to the valley exit of the Hundar Valley (Fig. 2.46). Pedestal moraines with intercalated subglacial gravel bodies were dammed up just like the hanging glacier of the Hundar side valley against the large Shyok main valley- and outlet glacier.

Before the ELA was uplifted by about 400–600 m during the Late Glacial (cf. Fig. 2.5), no pedestal moraines worth mentioning were able to build up. During the High Glacial (Stage 0 = LGP), no till sheets could develop on the valley floor nor on the valley flanks. Accordingly, it is from this period that the glacier polishings on both valley flanks of the Hundar Valley must originate (Fig. 2.49).

We now turn again to the Shyok main valley.

Opposite the exit of the Hundar Valley, to the north, a thick Late Glacial till cover exists. It is situated on the orographic right flank of the Shyok Valley on the polished rock spur at the junction of the Nubra Valley (34° 37′ 21.85″N 77° 32′ 36.13″E). There it reaches up to ca. 4440 m asl. Towards the highest position its thickness decreases from several decameters to a few meters (Fig. 2.50). Concomitantly, it begins to show gaps.

Downvalley 6 km there is a further spacious till cover (34° 38′ 56.13″N 77° 28′ 58.23″E). It reaches up to ca. 4200 m asl. The next orographic

Fig. 2.50 Ca. 150° panoramic photograph in the middle Shyok Valley from ca. 3140 m asl, 34° 33′ 36.13″N 77° 30′ 24.48″E (Fig. 2.43 to the right of No. 1; Fig. 2.23 to the left of B); from facing ENE toward the junction of the Nubra Valley with the Shyok main valley via facing north into the orographic right flank of the Shyok Valley up to facing WNW looking downward into the Shyok Valley. (■) mark ground moraine covers up to a height of 4750 m asl (Fig. 2.43: Character C: No. 3). (∩) denote back-polished, triangular glacigenic slopes. (◓) indicate glacigenic flank-polishings and abrasions. (▣) shows the current valley bottom with a Holocene, glacio-fluvial gravel cover and dunes in the middle foreground. (◁) marks alluvial and debris-flow fans consisting of dislocated Late Glacial moraine material in the valley exit of the Hundar Valley. (- - - -) denotes the High Glacial glacier level exceeding ca. 5600 to ca. 6000 m asl. Analog photo 30/08/2013 by M. Kuhle

right-hand till remnant that is similarly extended is situated another 5 km downvalley (Fig. 2.43, see also http://extras.springer.com: Character C: No. 3: 34° 40′ 24.59″N 77° 26′ 17.68″E; Figs. 2.50 and 2.51). It reaches up to ca. 4110 m asl. Another 21 km further downvalley unmistakable complexes of till lie connected at 34° 41′ 8.62″N 77° 24′ 57.79″E, 34° 41′ 48.41″N 77° 23′ 13.54″E, 34° 41′ 38.15″N 77° 21′ 26.94″E, and 34° 44′ 39.11″N 77°13′ 55.65″E (Fig. 2.43, see also http://extras.springer.com: Character C: No. 4; Fig. 2.51). The last one reaches up to ca. 4240 m asl. Scattered remnants of till have been diagnosed on the orographic right side up to ca. 4800 m (34° 46′ 43.72″N 77° 13′ 22.31″E). These till covers rest on large, High Glacial, glacigenically formed triangular slopes. These slopes were abraded by intensive flank polishing and are situated on polished mountain spurs between hanging side valleys. They reach heights of 4600 up to ca. 5200 m asl. Due to their continued polyglacial genesis during the numerous Pleistocene ice ages, most recently during the LGP (Last Glacial Period = Stage 0), they tower above the occurrences of Late Glacial till sheets (as described above) by a vertical distance of ca. 1000 m (Figs. 2.51 and 2.52).

On the orographic left flank of the Shyok Valley, corresponding till covers are also preserved. However, due to a greater steepness of the flanks they are less thick. They are located in the following positions in a downvalley direction: 34° 41′ 17.50″N 77° 14′ 6.77″E (Fig. 2.43, see also http://extras.springer.com: Character C: No. 5; Fig. 2.51) and as far up as ca. 3900 m; 34° 41′ 37.18″N 77° 13′ 23.79″E (Fig. 2.43: Character C: No. 6) and as far up as ca. 4250 m; 34° 42′ 00.34″N 77° 12′ 6.62″E (ca. 4170 m asl) and up to ca. 4300 m; 34° 42′ 59.03″N 77° 11′ 4.94″E (ca. 3830 m asl) and up to ca. 4170 m (Fig. 2.43: No. 6: 34° 41′ 33.40″N 77° 13′ 19.63″E; Fig. 2.51). Glacigenic flank

Fig. 2.51 Ca. 270° panoramic photograph of the middle Shyok Valley downvalley from the Nubra Valley from ca. 3050 m asl, at 34° 41′ 12.36″N 77° 15′ 56.92″E (Fig. 2.43 to the right of No. 5); from facing SE from the Shyok Valley upward via facing north into the orographic right flank of the Shyok Valley via facing NW down the Shyok Valley up to facing SW into the orographic left flank of the Shyok Valley. (■) mark ground moraine covers (Fig. 2.43: Character C: Nos. 3–6) in part with decameter thicknesses. (☗) are glacigenic flank-polishings and abrasions up to ca. 5500 m asl. (⊠) denote areas of weathering of bedrock, which since the Late Glacial deglaciation are roughening the valley flanks. (▼) show active debris cones and debris slopes that are covering ground moraine cores. (Δ) is a debris-flow fan and alluvial fan from dislocated moraine, which are interlocked with the current valley bottom (◘) of glaciofluvial gravels. (◉) are round-edged and rounded boulders on an orographic left debris-flow fan. (- - - -) shows the High Glacial glacier level at ca. 5600 to nearly 6000 m asl. Analog photo 30/08/2013 by M. Kuhle

polishings are preserved in many places and sometimes are lying under covers of till. Here are some examples: at 34° 30′ 45.35″N 77° 38′ 59.37″E in 3740 m asl (Fig. 2.53); at 34° 37′ 59.80″N 77° 23′ 47.65″E; at 34° 43′ 31.01″N 77° 10′ 45.15″E up to ca. 4600 m; at 34° 46′ 13.07″N 77° 8′ 47.80″E at 3260 m asl on the one (as well as neighboring) classic-shaped roche moutonnée (240–260 m high) blocking the Shyok Valley (Fig. 2.43: Character C: No. 7); and further from here on the orographic right valley flank at 34° 48′ 7.47″N 77° 7′ 50.50″E up to a height of ca. 4400 m asl (Fig. 2.43, see also http://extras.springer.com: Character C: No. 8; Fig. 2.52).

The highest glacigenic abrasion forms that are appropriate for the determination of the maximum past ice thickness are located in the orographic left valley flank of the relevant valley chamber of the Shyok main valley at 34° 35′ 11.95″N 77° 24′ 29.58″E at ca. 5480 m asl; at 34° 35′ 31.12″N 77° 21′ 32.32″E at ca. 5500 m asl; at 34° 36′ 26.09″N 77° 19′ 4.31″E at ca. 5260 m asl, and at 34° 43′ 38.92″N 77° 6′ 42.21″E at ca. 5270 m asl (Fig. 2.43, see also http://extras.springer.com: Character C: No. 9; Fig. 2.51). This is their sequence in a downvalley direction. Among others, the last named abrasion form indicates a polish demarcation line and thus a once-reached glacier level. The highest abrasion form that is situated opposite the just-mentioned one (No. 9) on the orographic right flank is a rounded rock cupola at 5600 m asl (34° 47′ 10.38″N 77° 12′ 26.42″E; Fig. 2.43, see also http://extras.springer.com: Character C: No. 10; see also Fig. 2.52). Accordingly, this is not a polish demarcation line but only indicates a minimum glacier ice level. The difference in height of over 330 m (from No. 9 to No. 10 in Fig. 2.43) has to

Fig. 2.52 Ca. 170° panoramic photograph of the middle Shyok Valley from ca. 3050 m asl, at 34° 44′ 29.43″N 77° 11′ 14.98″E (Fig. 2.43 to the left of No. 4); from facing NW down the Shyok Valley via facing SW into the orographic left flank of the Shyok Valley seen from across the inflow of a gorge-like, orographic left tributary valley. (■) signify ground moraine remnants in some places several decameters in thickness (e.g. ■ on the right). (▲) mark glacigenic flank-polishes and flank-abrasions up to ca. 5500 m asl (see also Fig. 2.43: Nos. 7 and 10). (⊠) marks areas of weathering in the outcropping bedrock, which have roughened the valley flanks since the Late Glacial deglaciation. (▼) marks present-day debris cones and debris slopes covering cores of ground moraine. (▷) shows a debris-flow fan of dislocated moraine, which is undercut by the current valley-bottom (□) of glacio-fluvial gravels. (⇓) marks a subglacial meltwater-gorge developed during the Late Glacial. (- - - -) denotes the High Glacial glacier level at ca. 5600 m asl. Analog photo 30/08/2013 by M. Kuhle

be attributed to the higher altitude of the nourishing area of the right valley side to the north—the La Yongma Ri mountain group with numerous 6000–6800 m high summits and glaciers that are currently still larger than those found south of the Shyok Valley.

Accordingly, the Last Glacial glacier level (Stage 0 = LGP = Last Glacial Period) on the orographic right side was situated at least somewhat above 5600 m and on the orographic left side at about 5270 m asl on the ca. 8.2 km (from No. 9 to 10) wide valley cross-profile.

This is the valley cross-profile in the Shyok Valley treated in this book that is situated farthest downvalley. It had a glacier filling with an ice thickness of at least 2270 m in the center of the valley, plus the thickness of the gravel of the current valley bottom (Fig. 2.52).

2.2.2.2 Summary of Sects. 2.1 and 2.2: The High Glacial (LGP) Glaciations of the Pangong Co Area in WSW Tibet and the Connected Shyok Outlet Glacier from the Western Margin of Tibet Down to the Valleys of the Karakorum; the Classification of a Dating of Lake Terraces as Well as the Results in the Older Literature

The primary focus of this section is Fig. 2.1: No. 36. The 170 km long Pangong Co (lake) was

Fig. 2.53 Facing SW into the orographic left valley flank of the Shyok Valley opposite the junction of the Nubra Valley seen from ca. 3160 m asl at 34° 31′ 33.53″ N 77° 40′ 2.93″E (Fig. 2.23: right hand side, i.e. east of character B). (■) signify meter-thick remnants of ground moraine on the glacigenically polished rock wall with flank-polishes and ground moraines on a steep wall (e.g., at 34° 30′ 45.35″N 77° 38′ 59.37″E at 3740 m asl) and (☗) glacigenic flank-polishes and flank abrasions up to ca. 4500 m asl, more than 1300 m above the current valley bottom (□) of glacio-fluvial gravels. (- - - - -) shows the High Glacial glacier level at far more than 5000 m asl in a distorted perspective. Analog photo 27/08/2013 by M. Kuhle

filled with glacier ice from the ESE end of the lake to its NW end (Fig. 2.2, see also http://extras.springer.com: Character A). The former ice level was probably situated 2000 m above the lake level, i.e., at at least somewhat above 6000–6300 m asl. This is evidenced by the glacigenically abraded and rounded mountain ridges and cupolas NE to SE of the Kangju Kangri (6700 m), e.g., at 33° 39′ 17.97″N 78° 35′ 22.56″E, at 33° 39′ 51.30″N 78° 34′ 42.93″E, at 33°41′ 59.48″N 78° 34′ 26.32″E, and at 33° 44′ 27.84″N 78° 33′ 7.97″E.

The ice thickness in the upper Pangong Tso Valley, through which the glacier of the Pangong Co area was flowing toward the NW, amounted to ca. 2000 m.

Again downvalley, where the Pangong Co outlet glacier joined the Shyok Valley and continued as the Shyok outlet glacier, an ice thickness of 2000–2800 m was reached (Fig. 2.23, see also http://extras.springer.com: Character B); a similar thickness existed in the valley chamber downward from the settlement of Agham. The glacier level thereby lay at at least 6050 m asl in the region downvalley from the Pangong Co all the way to downvalley of Agham.

Again downvalley, the similarly thick Nubra Valley glacier joined the Shyok outlet glacier from the north (Fig. 2.23: Character B).

From the inflow of the Nubra Valley glacier into the Shyok outlet glacier as far as 49 km downvalley to the gorge section of the Shyok Valley at the Yagulung settlement, an outlet glacier thickness of ca. 2300 m asl has been preserved (Fig. 2.43, see also http://extras.

springer.com: Character C). In this lower valley chamber the ice level was located at least a little higher than 5600 m asl.

The downvalley, 89 km long Shyok Valley section continues to remain narrow like a gorge. The valley bottom descends from 3000 to ca. 2500 m asl. There lies the well over 4 km wide junction area of the Hushe Valley. The reconstruction of the continuous extension of the High Glacial Karakorum ice stream network is based on findings in the field and laboratory that have already been introduced in detail in earlier publications (Kuhle 1988d, 1997, 1998a, 2001a, 2010, 2013a).

The area located 150 km NNW of the valley chambers of the Shyok and Nubra valleys also has been investigated in detail (Kuhle 1988b, 1994a, 2005b, 2013a). Here, too, on the opposite northern edge of the current glacier faces of the Mustagh Karakorum, a corresponding ice stream network has been reconstructed. It represented the northern part of the ice stream network which together with its southern part, i.e., the Nubra and Shyok outlet glaciers, formed a continuous glaciation across the area in between. Due to the currently still thick ice cover in this region, the connection of the glaciers of the two areas during the High Glacial is evident in any event.

This northern part of the ice stream network was—similar to its southern part—connected to the western margin of the Tibetan ice (Kuhle 1999, 2013a). The equivalent to the Shyok outlet glacier here was the High Glacial Kolakoshih Ho (valley) outlet glacier flowing down the northern W-margin (Kuhle 2013a: 162–164).

To conclude the summary, the only older ^{14}C-date obtained in the area of the Tibetan Plateau has to be acknowledged. It concerns a lake terrace 70 m above the extant lake level of the Pangong Co and is dated to ~39 ka B. P. (Rhodes et al. 1996: 118; Li and Zhu 2001: 39). Originating from the "Kunlun-Karakorum Sino-French team," this dating has not been published officially (Rhodes et al. 1996: 118). The reason for this is that the Pangong Co has an overspill at 4270 m asl (or 4255 m), that is, 24 m (or 9 m) above the current lake level so that a lake level of +70 m requires an artificial damming of the lake. Because there do not exist any indications of geological causes for this damming, its cause is obvious. A Late Glacial glacier dammed up the lake up to this level. A reasonable catchment area for this glacier would be the one described above (Sect. 2.1) on the orographic left side of the NW Pangong Co (Fig. 2.2, see also http://extras.springer.com: Character A: on the left of Nos. 1–4). Here, SW of the lake and north of the Kangju Kangri (6700 m), the current glacier endings reach down to 5500–5400 m. Thus, an ΔELA of 600–700 m is necessary to dam up the Pangong Co by about 70 m or even more.

The ^{14}C-date of ca. 39 ka B.P., however, is not at all compatible with this ΔELA, as the latter belongs to the Late Glacial Sirkung stage or Dhampu stage (Stage IV or III; Fig. 2.5). If the date were correct, then lake development would have taken place during an interstadial between Würm I and II, i.e., between MIS 3 and 2. During this period, the ELA could have been uplifted by about 650 m, so that it would have been only 600–700 m lower than today. During the subsequent LGM (Würm II; i.e., MIS 2), with an ΔELA of more than 1000 (ca. 1200) m, the valley of the Pangong Co would have been completely filled with glacier ice. This glacier filling (see above) would by necessity have hindered any lake development. So, should the date be correct, then the lake terrace as posited by the above authors (Rhodes et al. 1996) would have been completely overridden by glacier ice.

Godwin-Austen (1868, cited by Ambolt and Norin 1982: 48) interpreted the Pangong Co basin as a glacigenic form, and pointed out the orographic left moraines as well as their postglacial cutting. Drew (1875: 324 note, cited by Ambolt and Norin 1982: 48) described for the orographic right side opposite the settlement called Man horizontal field-marks up to ca. 300 m above the lake level. Godwin-Austen, who did not observe these forms himself, interprets them as being indicators of a past glacier ice body in the Pangong Co basin. Between this body and the valley slopes a lake might have formed, leaving these marks as shorelines (Ambolt and Norin 1982: 48). Huntington (1905, cited by Ambolt and Norin 1982: 50) described the wide Pangong Valley—at that time named the Yaktil-Murgab Valley—as the "U-shaped Yaktil-Murgab valley." It continues downvalley of the Pangong Co.

Ambolt and Norin (1982: 56) note that during the Late Glacial ("Late Glacial Time") the Pangong Co was dammed up by the upper "Shyok ice-stream." Thus it had also occupied the entire Pangong Valley (at that time "Tankse Valley") with its lake surface (ibid., see also the accompanying map). After the reconstruction of the present author (Kuhle), this assessment is not correct, however. During the Late Glacial period, the Pangong Co basin itself was at least partly filled with glacier ice and so was the Pangong Valley in a downvalley direction. Accordingly, the lake—if present at all—must have been considerably smaller than today. This finding also compromises another assessment of the aforementioned authors, namely, that the accompanying Shyok Glacier must have already ended ca. 10 km downvalley of the inflow of the Pangong Valley (ibid.: map). At least based on the indicators presented here, this cannot be confirmed by the present author.

Yet, the geomorphological descriptions of the Nubra Valley given by Thomson (1852, pp. 195), Drew (1873, pp. 446; 1875, p. 275), Longstaff (1910, 638), and Neve (1913, 236)—all cited by Ambolt and Norin 1982: 52)—appear to be very important. They have already been summarized by Dainelli (1922, Ser. 2, Vol. 3, 415–444, cited by ibid.) and again by Ambolt and Norin (1982: 52–53). Particularly noteworthy is the Shyok Valley glacier that Norin (Ambolt and Norin 1982: 52) reconstructed to a thickness of 670 m and which, according to him, flowed from Agham ca. 85 km downvalley as far as to the gorge section. Dainelli classified this ice margin below Phastan as "the third main stage of glaciation in Baltistan," belonging to the Last Ice Age (ibid.: 53). According to the present author (Kuhle), on the other hand, it must be classified as belonging to at most the last Late Glacial Stage IV (see Fig. 2.5), although only during its recessionary phase (melting).

Thomson, Drew, and others (see above) as well as Norin (Ambolt and Norin 1982: 53) assumed that the highest glacier level of the Nubra-Shyok ice stream in the area of the junction of Nubra and Shyok glaciers reached no higher than 4400 m. The thickness of this glacier in the Nubra Valley junction, accordingly, was only ca. 1400 m. At that time of maximum glaciation this glacier, according to them, flowed down no further than near the inflow of the Shyok Valley into the Indus Valley.

By contrast, the findings of the present author prove a 1600 m thicker Nubra-Shyok ice stream, reaching an altitude of ca. 6000 m asl. This glacier actually reached the Indus Valley, and even flowed into the Indus Glacier with a thickness of more than 1450 m (Kuhle 2010, 2013a: 188).

In addition, this reconstructed glaciation featured a High Glacial outlet glacier of a Tibetan inland ice stream that flowed down from the Pangong Co—a reconstruction not shared by the other authors mentioned above.

The reconstructions of the aforementioned authors, accordingly, are classified by the present author as belonging to the late Late Glacial (Stage III, see Fig. 2.5). An alternative assessment must be ruled out. The reason is that Thomson and others refer to the thick glacigenic accumulations on the valley flanks up to 4400 m asl as the maximum glacier level. Necessarily these must have been deposited several hundred meters below the contemporary ELA (see Sect. 2.2.1), i.e., at least 250–350 m further below. Accordingly, for the deposition, an ELA located at ca. 4750–4650 m asl must be assumed. At a current ELA of ca. 5550 m asl, this means a ΔELA of 800–900 m. This is the ELA depression of the Late Glacial Stage III (Fig. 2.5).

2.3 The High Glacial Indus Glacier in the Central Indus Valley at the Western Margin of Tibet

The primary focus of this section is Fig. 2.1: Nos. 17 and 36. The upper 145 km long section of the Indus Valley located in High Tibet and also named Gar Zangbo, has been investigated geomorphologically by the present author in the field, and also sampled carefully (Kuhle 1999: Fig. 2: Nos. 119/119–131). It extends from the 25 km wide transfluence pass (4820 m asl) leading down from the Tsangpo Valley with the Manasarowa Lake in the east into the Indus Valley via Gar to the west as far as to Langmage (Langmacun). In these 20–26 km wide cross-profiles of the Indus Valley, High Glacial glacier thicknesses of at least 1500 m have been reconstructed. The ice level was located at 5700 m asl (ibid.: 37–43, 197–214, Photos 143–145), while the current valley bottom (consisting of glacio-fluvial gravels and alluvial fans) lies at 4200 m asl. This was a section of the SSW part of Tibetan inland ice (Fig. 2.1: No. 23 on the left of Kailas).

The glacier reconstructions presented now in what follows refer to the valley section that begins ca. 250 km further downvalley between Tunah and Upshi. Here, on the orographic left side, remnants of pedestal moraines at up to 3570–3630 m asl and thus ca. 200 m above the Indus Valley floor (3400 m) are preserved (e.g., at 33° 47′ 6.31″N 77° 50′ 12.62″E and at 33° 50′ 43.20″N 77° 47′ 17.75″E). This valley flank, which developed in outcropping edges of sedimentary rock, is very steep and has no flattenings. On the orographic right valley slope, however, flattenings do exist. There, on small rock terraces and sporadically preserved, round-polished remnants of valley shoulders as well as on mountain cupolas rounded by abrasion, till is located at 3750 m (33° 48′ 0.19″N 77° 51′ 14.51″E and 33° 56′ 42.65″N 77° 44′ 55.36″E) up to 4220 m asl (Fig. 2.54, see also http://extras.springer.com: Character D: No. 1). Another example of glacigenically abraded and rounded valley shoulders with till cover is also found in the locality at 33° 50′ 18.55″N 77° 50′ 32.03″E at ca. 3960 m asl.

The findings already published for the Indus Valley and its side valleys in the area between the settlements of Upshi (see above) and Khalsi (34° 19′ 4.22″N 76° 52′ 23.90″E, 2950 m asl), 125 km downvalley, need not be repeated here. They speak in favor of a continuing minimum ice thickness of 900 m (e.g., at 34° 08′–20′N 77° 25′–76° 50′E) (Kuhle 1998a: 88). The reconstruction of a High Glacial outlet glacier flowing down from the junction area of the Taglang La-North valley with the 5100–5400 m high Taglang La (pass) through the Taglang La-N-valley near Upshi into the Indus Valley has also been published, as well as the connected glaciation of the 6401 m high Ruberung Massif south of the Indus Valley (ibid.: 88). This in turn formed a continuum expanding south to SW from the Indus Valley via a beeline distance of 240 km, thus forming an ice stream network from the Indus Valley up to the exit of the Solang Nala (valley) on the southern side of the Himalaya (Lahoul [Lahul] or Pir Panjal) (ibid.: 88).

In the following, new and unpublished findings are presented and discussed. They were gathered from the field partly in 1993 and partly in 2013:

Due to their very narrow gorge morphology, the obsequent side valleys on the orographic left side contain till remnants only very sporadically. An example are moraine remnants of light, erratic material with large blocks transported along the valley on reddish bedrock sandstones (Fig. 2.54, see also http://extras.springer.com: Character D: No. 2). These are located at 33° 52′ 6.61″N 77° 39′ 46.50″E at 4050 m asl (above the settlement) and at 33° 52′ 19.39″N 77° 39′ 49.85″E at 4070 m asl; further at 33° 52′ 27.96″N 77° 39′ 55.68″E at 4130 m asl, at 33° 52′ 30.63″N 77° 39′ 51.27″E at 4130 m

Fig. 2.54 Satellite image, Google Earth Pro: center 34° 3′ 28.13″N 77° 45′ 4.29″E; http://extras.springer.com: Character D: Nos. 1–10

asl, as well as at 33° 51′ 7.46″N 77° 42′ 36.32″E at 3725 m asl. Both localities are situated in the side valley that joins the Indus Valley near the settlement of Martselang.

The valley with the Chemrey and Sakti settlements (Fig. 2.54: Character D) was connected to the local ice stream network of the Ladak (Ladakh) Range between the Shyok and Indus valleys.

On both sides of the Waring La (see Sects. 2.2 and 2.2.1) and the Chang La Baba, classic trough-shaped valleys have been developed, mantled with till sheets. The extent of the LGM glaciation of this mountain group becomes especially clear in the valley that leads down from the Chang La Baba to the Pangong Valley —referred to in Ambolt and Norin (1982) as the “Tankse valley” (Sect. 2.1; see Fig. 2.2, see also http://extras.springer.com: Character A: Nos. 9 and 10). Though only a few of the highest mountains reach scarcely more than 6000 m, the glacier of this valley joined the Pangong Valley outlet glacier. This occurred below the valley chamber of Darbuk and is evidenced by till remnants at up to 5200–5400 m asl (Fig. 2.54: Character D: No. 3).

Correspondingly, the glacier in the valley of Sakti on the south side of the Chang La Baba flowed into the Indus Glacier. Despite this, at Sakti about 3700 m downvalley, the U-shaped valley changes into a V-shaped valley. It is probable that the trough valleys on both sides of the Chang La Baba are of an Early to Late Glacial age. In contrast, the High Glacial Sakti Valley glacier was dammed by the Indus Glacier. It had become stuck nearly without a movement and therefore without erosion in the area of its lower 300 m of altitude within a V-shaped valley cross-profile. Corresponding examples are provided by the European Alps, where among others the lower Lötschental side glacier was dammed against the Rhone Glacier. This, too, has left behind a U-shaped valley above and a V-shaped valley below. Further examples can be found on the South Island of New Zealand, with the orographic right side valleys east of Jacksons between Mt. Wilson and Mt. Alexander, as well as in British Columbia west of Calgary in the orographic left side valleys west of Banff and NNW of Kananaskis.

In addition, it is a general observation that naturally the subglacial meltwater erosion from the Early to the Late Glacial increases in the sections downvalley. The altitude-dependent balance of erosion between glacigenic ground polishing and linear erosion has thus also led to more subglacial developments of V-shaped valley sections in those regions with lower absolute altitude (Kuhle 1982a: 59–61; 1983: 158–160).

In the orographic left side valley exit where the Stok settlement is situated (Stok Valley), there lies a Late Glacial glacier tongue basin at 3500–3600 m asl (aneroid measurement). It has been filled by a younger, glacio-fluvial alluvial debris fan.

A 30 m high, a remnant of an orographic left moraine or kame lies at ca. 34° 4′ 19.65″N 77° 32′ 45.16″E at 3500 m asl (Fig. 2.54, see also http://extras.springer.com: Character D). A current alluvial fan is connected to the tongue basin, spreading out on the bottom of the Indus Valley; 46% of the SiO_2 grains of its medium sand fraction are glacially crushed (Fig. 2.10: No. 1, sample 21.07.1993/1, 3305 m asl [aneroid measurement]). This points to the fact that it contains rearranged moraine material. The 30 m high moraine or kame remnant, however, also contains only 55.3% glacially crushed grains (Fig. 2.10: No. 3, sample 22.07.1993/1, 3500 m asl). At 34° 3′ 34.36″N 77° 32′ 11.24″E (Fig. 2.54, see also http://extras.springer.com: Character D) at 3655 m (aneroid measurement), a sample was taken of a remnant of a lateral moraine (i.e., kame) terrace (Fig. 2.10: No. 4, sample 22.07.1993/2). It contained 56.2% glacially crushed grains. Figure 2.54: Character D: No. 4 marks a cushion-formed remnant of till that belongs to a somewhat older stage. At that time, the Stok Valley glacier still had a thickness of more than 140 m and so reached into the Indus main valley. As seen from the 4965 m high (4800 m aneroid measurement) Stok La (pass) at the valley head down to the NNE, the Stok Valley does not exhibit the shape of a trough valley. It is marked by vertically standing, colored, quartzite-like metamorphic bedrock and formed in complete dependence on the structure. In several of its valley chambers, decameter-thick pedestal moraines are preserved (Fig. 2.54: 34° 0′ 40.65″N 77° 29′ 55.89″E). Till covers such as these have already been described above, i.e., in reference to the parallel side valleys south of here (Fig. 2.54, see also http://extras.springer.com: Character D: No. 2, as well as 33° 51′ 7.46″N 77° 42′ 36.32″E at 3725 m asl). In addition, glacigenic deposits such as these were observed in the Kongmaru-Northeast valley (Shang Valley), the source valley of the valley leading down to Martselang on the orographic right, upvalley from the Chukirmo hamlet. Here, northeast of the Kongmaru La at ca. 3800 m (3875 m aneroid measurement) m asl, pedestal moraines consisting of two decameter-thick moraine covers have been mapped. The hanging, more light-colored moraine cover develops classic earth pyramids (Kuhle 2004: 188, Fig. 34).

The valley leading southward down from the Kardung La (pass) toward Leh, named Puchu Chu, has been proved by Dainelli (1922) as once glaciated. The impressively preserved tongue basin above Leh has been classified as belonging to the LGM. The present author has classified the tongue basin (Kuhle 1996a: 153, Abb. 10) with a ΔELA of ca. 1200 m as belonging to the Late Glacial Stage I (Ghasa stage, Fig. 2.5) (Kuhle 1998a: 88). Downvalley, till covers as well as roches moutonnées prove a High Glacial (Stage 0) side-valley glaciation (Stage 0) down to ca. 3000–3200 m asl, which reached the valley bottom of the Indus Valley (ibid.) (see below).

Earlier findings on those Late Glacial end moraines fringing the tongue basin (see above) have by now been confirmed by granulometric (Fig. 2.55: sample 1.9.13/2) as well as morphometric sample analyses (Fig. 2.10: No. 20). Further investigations resulted in the identification of coarse-blocky till material on the

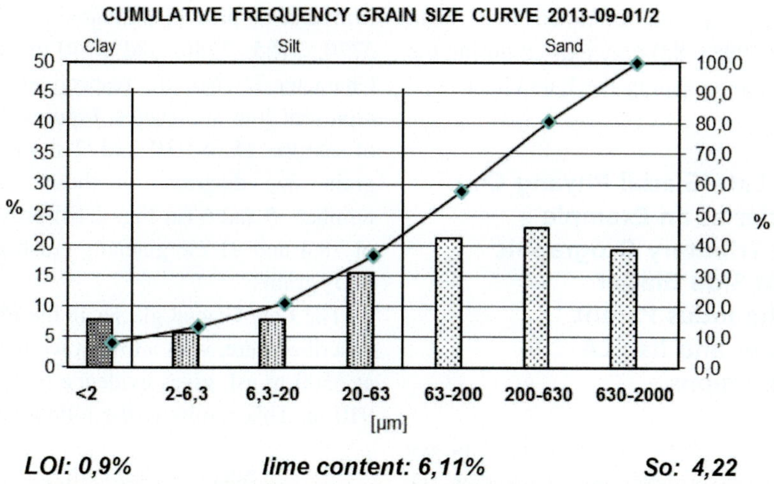

Fig. 2.55 (Grain size diagram 2013-09-01/2): Granulometry of the fine material matrix of ground moraine (i.e., lateral moraine) at 34° 12′ 5.87″N 77° 36′ 37.23″E, at 3920 (3750 aneroid) m asl, on the orographic left flank of the southern Khardung La Valley, upvalley from the Leh settlement and orographically left of the Late Glacial tongue basin (after Kuhle 1996a: 153, Abb. 10), so that the moraine material we are speaking about must be of an older, Late Glacial or High Glacial age. This fine material matrix is situated between meter-sized boulders. Exposure and sampling depth ca. 4 m. The maximum occurs with the ca. 23% peak in the middle sand, and the secondary maximum with nearly 8% in the clay. The grain-size cumulative curve has a bimodal course. The sorting coefficient calculated according to von Engelhardt (1973: 133) is So = 4.22 (So = $\sqrt{Q_3/Q_1}$); the lime content is 6.11%; the loss on ignition (LOI) amounts to 0.9%. Locality: Fig. 2.54 halfway above and to the right of No. 6. See Fig. 2.10a, b: No. 20; 2013-09-01/2. Sampling: M. Kuhle

orographic left side, 250 m above those end- or lateral moraines. They are located at an altitude of up to ca. 4260 m asl (Fig. 2.54: Character D: No. 5: 34° 13′ 10.04″N 77° 36′ 54.58″E). Possibly they, too, belong to the Late Glacial. In this case they would have to be classified as belonging to Stage I, while the aforementioned tongue basin would belong to Stage II.

The sampling of the matrix of the 3.8 km wide coarse-blocky fan in the junction area of the Puchu Chu with the Indus Valley resulted in a SiO_2 grain portion of 58.9% that is glacially crushed (Fig. 2.10: No. 2, sample 21.07.93/2, 3300 m asl). In addition to the remarkably large skeletal portion in the form of 2–3 m long boulders, this is proof of the moraine character of this material. Without doubt these are deposits that have been reshaped fluvially, i.e., by debris flows after the deglaciation.

In the immediate junction area of the Puchu Chu with the Indus main valley there can be observed in some places and at maximum up to 3950 m asl biconvex, coarse-blocky accumulations on the upper slopes (Fig. 2.54, see also http://extras.springer.com: Character D: No. 6: 34° 9′ 52.46″N 77° 31′ 30.05″E).

These accumulations are also found on the orographic right valley flank of the Indus main valley (e.g., at 34° 7′ 48.06″N 77° 36′ 50.71″E; 34° 8′ 3.23″N 77° 36′ 51.50″E; 34° 8′ 5.04″N 77° 36′ 49.54″E; 34° 8′ 9.90″N 77° 36′ 35.32″E; 34° 8′ 11.03″N 77° 36′ 14.55″E; 34° 8′ 51.41″N 77° 31′ 12.25″E). At least some of them that do not show a local catchment area upslope (Fig. 2.54: Character D: No. 6) must be considered to be meter- to decameter-thick till covers. Due to their maximum altitude only a little below the High Glacial (LGM = Stage 0) ELA, this is probably Late Glacial till cover originating from the Indus main valley glacier (on the theory of the accumulation of till sheets, see Sect. 2.2.1). In the past, the present author has argued that the complete Indus main valley glaciation of the section under discussion from the locality Upsi as far as downvalley from Khalsi (Kuhle 1998a: 88) only existed during the LGM, but no longer during the Late Glacial.

2.3.1 The Late Glacial Phyang Chu Glacier as an Example of a Tributary Orographic Right Side Glacier of the Indus Parent Glacier and Its ELA Depressions

The primary focus of this section is Fig. 2.54, see also http://extras.springer.com: Character D: Nos. 7–9. For the WNW-oriented valley of the Puchu Chu, the Phyang Valley, there has been described a lowest Late Glacial ice margin at 3500 (3400 aneroid measurement) m asl. This was assessed by means of the end moraine on which the Phyang monastery is situated (Kuhle 1988a: 88) (Fig. 2.54: Character D: No. 7). The end moraine shows polymictic, striated, rounded as well as faceted boulders more than 2 m in length (Kuhle 2004: 190, Fig. 36). The large boulders lie isolated from each other and are embedded in a clayey matrix. In his description of the locality (ibid.), the author has cited the name "Stok valley" instead of "Phyang valley," a geographical latitude located 12.5° further WNW of that given by Google Earth, as well as an altitude of 3300 m asl. The reason for this difference was the use of the ONC map 1:1 Mio (scale 1:1,000,000) in 1993. On both sides of this terminal moraine remnant a glacier outlet existed, so that the end of the glacier tongue was divided into two parts.

Upvalley the accompanying lateral moraines could be observed along the two valley flanks up to a height of 4150 m asl (Fig. 2.54: Character D: 34° 13′ 29.65″N 77° 30′ 36.43″E), i.e., 200 m above the thalweg (see also 34° 12′ 17.47″N 77° 31′ 0.41″E). On the orographic left side this moraine generation, as seen from a cross-profile, is divided into four levels at 3770, 3755, 3746, and 3740 m asl (Fig. 2.54: Character D: No. 8), respectively, which thus represent four sub-stages. Figures 2.56 and 2.57 of samples 15.08.93/1 and /2 show characteristic grain size diagrams of these moraines. The numbers 5 and 6 (in Fig. 2.10) provide evidence of 78.4 and 91.7% glacially crushed portions of SiO_2 grains.

The Late Glacial glacier in the Phyang Valley described here, with a lowest ice margin position at 3500 m asl, gives evidence of a ΔELA of ca. 910 m. This results in the following calculation:

$$\text{Sdepr} = (\text{tim} - \text{tif})/2 \tag{2.6}$$

where tim = the current terminus of the glacier tongue; tif = the former terminus of the glacier tongue; and Sdepr = the ELA depression (equilibrium-line depression).

The current glacier tongues are situated at tim = 5320 m asl; the past glacier tongue is situated at tif = 3500 m asl:

$$\text{Sdepr} = (5320 - 3500)/2 \tag{2.7}$$

$$\text{Sdepr} = 1820/2 \tag{2.8}$$

$$\text{Sdepr} = 910 \tag{2.9}$$

$$\text{ELA-depr} = 910\ \text{m} \tag{2.10}$$

The ΔELA of ca. 910 m classifies this glacier position as belonging to Stage II or III (Taglung stage up to Dhampu stage (see Fig. 2.5)).

A youngest Late Glacial ice margin of Stage III or IV (see Fig. 2.5), preserved with clear ramparts of lateral moraines (Fig. 2.54: Character D: No. 9), came to an end at 4040 m asl. Its ΔELA of ca. 640 m classifies this glacier stage with a good probability as belonging to the Sirkung stage (IV), as the youngest Late Glacial stage. This results from the following calculation (see Eq. 2.6):

The current glacier tongues are located at tim = 5320 m asl (see above); the past glacier tongue is located at tif = 4040 m asl:

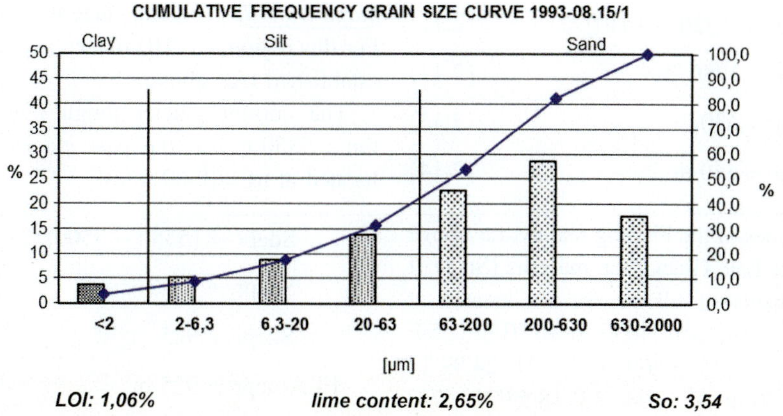

Fig. 2.56 (Grain size diagram 1993-08-15/1): Granulometry of the fine material matrix of ground moraine in the Phyang Chu at 3625 (aneroid measurement) m asl taken on the basis of the orographic left inner slope of a lateral moraine between meter-sized boulders. Exposure and sampling depth of several meters. The maximum occurs with the nearly 28% peak in the middle sand. The clay content is 4%. The sorting coefficient calculated according to von Engelhardt (1973: 133) is So = 3.54 (So = $\sqrt{Q_3/Q_1}$); the lime content is 2.65%; the loss on ignition (LOI) amounts to 1.06%. Locality: Fig. 2.54 near No. 8. See Fig. 2.10a, b: No. 5; 1993-08-15/1. Sampling: M. Kuhle

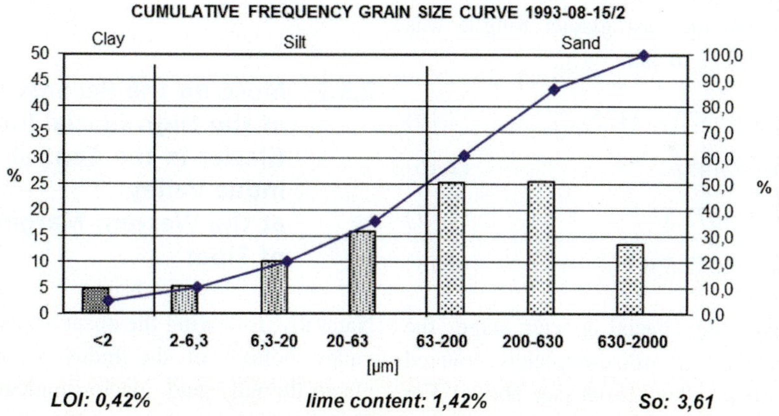

Fig. 2.57 (Grain size diagram 1993-08-15/2): Granulometry of the fine material matrix of moraine on the orographic left in the Phyang Chu at 3550 (aneroid measurement) m asl taken between meter-sized boulders. Exposure and sampling depth is several meters. The maximum occurs with little more than 25% dual peaks in the fine and middle sand. The clay content is 5%. The sorting coefficient calculated according to von Engelhardt (1973: 133) is So = 3.61 (So = $\sqrt{Q_3/Q_1}$); the lime-content is 1.42%; the loss on ignition (LOI) amounts to 0.42%. Locality: Fig. 2.54 near No. 7. See Fig. 2.10a, b: No. 6; 1993-08-15/2. Sampling: M. Kuhle

$$\text{Sdepr} = (5320 - 4040)/2 \quad (2.11)$$

$$\text{Sdepr} = 1280/2 \quad (2.12)$$

$$\text{Sdepr} = 640 \quad (2.13)$$

$$\text{ELA-depr} = 640\ \text{m} \quad (2.14)$$

On the slopes of the Phyang Valley, far above of those older Late Glacial ice margins (Stage II or III), remnants of till covers are located at between ca. 4220 and 4650 m asl (Fig. 2.54: Character D: 34° 13′ 32.69″N 77° 31′ 24.99″E, up to 4220 m asl; 34° 13′ 18.53″N 77° 29′ 50.65″E, up to 4300 m asl; 34° 14′ 26.86″N 77° 30′ 21.56″E, up to 4650 m asl). They prove that the older LGM stage, the Ghasa stage (I, Fig. 2.5) or stage II, must have at least reached the Indus main valley bottom at ca. 3175 m asl. This means that the local ΔELA in the respective southern exposure amounted to at least ca. 1070 m:

The current glacier tongues are located at tim = 5320 m asl; the past glacier tongue was located at tif = 3175 m asl:

$$\text{Sdepr} = (5320 - 3175)/2 \quad (2.15)$$

$$\text{Sdepr} = 2145/2 \quad (2.16)$$

$$\text{Sdepr} = 1072.5 \quad (2.17)$$

$$\text{ELA-depr} = 1072.5\ \text{m} \quad (2.18)$$

During this Late Glacial glacier stage, the Indus main valley was still completely covered by the Indus main valley glacier (see above). For this reason it is not possible that end moraines of the contemporary (synchronous) Phyang side valley glacier developed in the exit area of the Phyang Valley.

For comparison, the Late Glacial Stok Valley glacier in the NNE-oriented Zanskar–(Zansgar) Himalaya and in the opposite orographic left flank of the Indus Valley reached down to ca. 3500 m asl (Sect. 2.3). The current glaciers in the catchment area of the Stok Valley come to an end at ca. 5350 m asl. This is proved by a ΔELA of ca. 925 m and confirms the findings from the Phyang Chu for the medium Late Glacial Stage II or III (Taglung stage to Dhampu stage). For this region, ca. 910 m (see Fig. 2.5) could be established (see above):

The current glacier tongues are located at tim = 5350 m asl; the past glacier tongue was located at tif = 3500 m asl:

$$\text{Sdepr} = (5350 - 3500)/2 \quad (2.19)$$

$$\text{Sdepr} = 1850/2 \quad (2.20)$$

$$\text{Sdepr} = 925 \quad (2.21)$$

$$\text{ELA-depr} = 925\ \text{m} \quad (2.22)$$

These results strongly contrast with those of Burbank and Fort (1985). Those authors determine in the Zanskar Himalaya—even for the LGM (Last Glacial Maximum)—an ELA depression of only 500–600 m. This difference is far more than could possibly be explained with reference to the use of different ELA calculation methods (see Kuhle 1988a).

2.3.2 More on the Reconstruction of the High Glacial Indus Glacier in the Central Indus Valley at the Western Margin of Tibet

Naturally, following the onset of deglaciation the valley bottom of the Indus Valley has been glacio-fluvially and glacio-limnically reshaped. This is evidenced by analysis of an exemplary sample (Fig. 2.58: sample 15.08.93/3; Fig. 2.10: No. 7). Protruding above more recent Holocene accumulations are moraine ridges, such as for instance the highlighted 4 km long and up to ca. 80 m high till complex (Fig. 2.54, see also http://extras.springer.com: Character D: No. 10).

The till sheets located further downvalley, which in many places also have been developed as pedestal moraines, are depicted in Fig. 2.59 (see also http://extras.springer.com: Character E) at No. 1 (34° 11′ 23.54″N 77° 22′ 32.17″E, 3570 m asl; Fig. 2.60), at No. 2 (34° 9′ 43.03″N

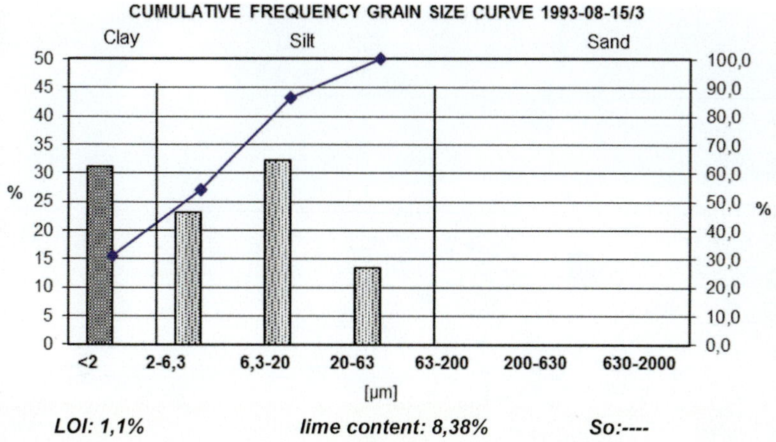

Fig. 2.58 (Grain size diagram 1993-08-15/3): Granulometry of the fine material matrix of limnic (i.e., glacio-limnic) sediments on a mountain spur on the orographic right in the valley exit of the Lee Valley at its inflow into the Indus Valley at 3190 (3200 aneroid measurement) m asl on the valley bottom at ca. 34° 8′ 17.06″N 77° 29′ 53.11″E. Exposure and sampling depth ca. 0.5 m. The maximum occurs with a ca. 32% peak in the middle silt. The clay content is 31%; the lime content is 8.38%; the loss on ignition (LOI) amounts to 1.1%. Locality: Fig. 2.54: No. 6. See Fig. 2.10a, b: No. 7; 1993-08-15/3. Sampling: M. Kuhle

Fig. 2.59 Satellite image, Google Earth Pro: center 34° 16′ 53.47″N77° 3′ 32.04″E; http://extras.springer.com: Character E: Nos. 1–10

Fig. 2.60 Facing south toward a ca. 60 m high pedestal of ground moraine (■) at 34° 11′ 23.54″N 77° 22′ 32.17″E on the Indus Valley floor that reaches ca. 3570 m asl (Fig. 2.59: No. 1). It is situated in the middle of the Indus Valley and consists of ca. 5 m sized, round-edged to rounded granite-boulders. They are isolated from each other in many places, and embedded in a fine material matrix. Analog photo 25/07/1993 by M. Kuhle

77° 21′ 54.60″E, 3375 m asl), at No. 3 (34° 10′ 1.96″N 77° 20′ 14.04″E, 3225 m asl), at Character E (34° 12′ 13.68″N 77° 15′ 21.16″E, 3400 m asl), at No. 4 (34° 15′ 18.43″N 77° 13′ 26.15″E, 3556 m asl, Figs. 2.61, 2.62, and 2.63; cf. Fig. 2.64), at No. 5 (34° 15′ 25.21″N 77° 12′ 24.03″E, 3465–3485 m asl), at No. 6 (34° 13′ 49.52″N 77° 11′ 7.12″E, 3360 m asl), at No. 7 (34° 14′ 17.20″N 77° 9′ 46.51″E, 3100–3180 m asl), at No. 8 (34° 14′ 59.46″N 77° 2′ 22.79″E, 4060–4150 m asl, Fig. 2.65; cf. Figs 2.66, 2.67, and 2.68), as well as at No. 9 (34° 19′ 28.06″N 77° 14′ 10.64″E, 4115 m asl). No. 7 marks a classic glacial band polishing of outcropping edges of the stratum between 3100 and 3180 m asl, in part covered with till and lake sediments. No. 8 also marks flank polishing on the orographic left side, with a moderately thick till cover (Fig. 2.65).

The three samples taken at locality No. 4 in Fig. 2.59 (see also http://extras.springer.com: Character E) deserve particular mention. Here meter-thick till material covers the ground. The material was subjected to granulometric and morphoscopic analyses (see sample 20.08.13/3, Fig. 2.62; Fig. 2.10: No. 10). As the petrographic analysis shows, this is erratic till material with meter-sized granite boulders (Fig. 2.63a: sample 20.08.13/2a leucogranite) lying on outcropping fine-grained and strongly "greened" granodiorites (Fig. 2.63b: sample 20.08.13/2b; analysis of the rock and documentation by A. Vollbrecht; Fig. 2.61).

Further evidence of till material is provided by Figs. 2.65, 2.66, and 2.67, with erratic granite boulders on the orographic right valley flank both up- and downvalley of the Uletokba (Uletokop) settlement.

Fig. 2.61 A meter-thick ground moraine (■ black) in the Indus Valley at 34° 15′ 18.43″N 77° 13′ 26.15″E, 3556 m asl (see Fig. 2.59 at http://extras.springer.com: Character E: No. 4) facing south onto the orographic left opposite valley flank. (■) mark up to decameter-thick ground moraine pedestals and covers that are preserved on the valley flank up to 4200 m asl. (∩) is a glacigenically back-abraded truncated spur, i.e., triangular glacigenic slope. (◓) are rounded mountains, i.e., glacial knobs and glacigenic flank-polishes and flank-abrasions. (**O**) are round-edged to rounded, up to 6 m long isolated erratic boulders (Fig. 2.63a, b), embedded in the fine material matrix of the ground moraine (Fig. 2.62). Analog photo 20/08/2013 by M. Kuhle

It is certain that all of these findings allow the reconstruction of at least two glacial Indus Glacier levels:

1. a younger Late Glacial level (Stage II or I, Fig. 2.5) demarcated by till sheets preserved in the highest sections of the valley slopes, and which are massive enough to cover the outcropping rock on a large scale;
2. an older High Glacial level (LGP, LGM, MIS 3-2, Stage 0, Fig. 2.5) that due to abrasion lines and traces of flank polishing on the valley flanks can be delimitated in the uppermost slope regions.

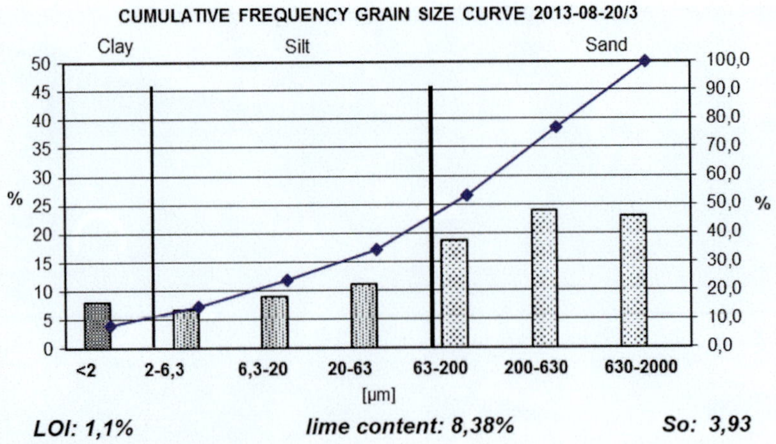

Fig. 2.62 (Grain size diagram 2013-08-20/3): Granulometry of the fine material matrix of ground moraine orographically right in the Indus Valley at 34° 15′ 18.43″N 77° 13′ 26.15″E, 3556 (3500 aneroid measurement) m asl. This fine material matrix is situated between meter-sized boulders (see Fig. 2.61). Exposure and sampling depth ca. 1.5 m. The maximum occurs with the ca. 24% peak in the middle sand, and a secondary maximum with nearly 8% in the clay. The grain-size cumulative curve has a bimodal course. The sorting coefficient calculated according to von Engelhardt (1973: 133) is So = 3.93 (So = $\sqrt{Q_3/Q_1}$); the lime content is 8.38%; the loss on ignition (LOI) amounts to 1.1%. Locality: Fig. 2.59 under No. 4. See Fig. 2.10a, b: No. 10; 2013-08-20/3. Sampling: M. Kuhle

As shown above (Sect. 2.3) the uppermost findings of till sheets between Upsi and Leh are located between 200 and 780 m above the current valley bottom of the Indus Valley. Thus a *Late Glacial* glacier thickness of ca. 780 m—plus the accumulation of the valley bottom since the deglaciation—would be realistic.

From the junction of the Puchu Chu over a distance of ca. 71 km downvalley down to the junction of the orographic left Wanla Chu side valley, downvalley of Khalsi, two levels of the upper margins of the till sheets are observed:

1. *A higher level* forming a continuation of the one at the exit of the Puchu Chu at ca. 3950 m asl (Sect. 2.3; Fig. 2.54: Character D: at No. 6). Examples of this continuation are found at 34° 11′ 29.80″N 77° 26′ 10.83″E at 3720 m asl (Fig. 2.54: Character D: above No. 10) near the settlement of Taroo, as well as between Fig. 2.59 No. 4 and 34° 15′ 48.96″N 77° 14′ 2.23″E, i.e., between 3550 and 3600 m asl (Fig. 2.61), and at 34° 11′ 59.39″N 77° 11′ 35.86″E at 3600 m asl on the same valley cross-profile of Fig. 2.59: No. 4. However, on the orographic left side this continuation occurs up to even 4200 m asl (Fig. 2.64), and at Fig. 2.59: No. 8 at 3950 m asl; finally also on the orographic left side at 3650 m asl (Fig. 2.59: No. 10) (see below);
2. *A lower level* whose uppermost margins demarcated by till sheets lie at 3390–3440 m asl. This is, for example, a 240–275 m high level above the thalweg in the pertinent valley cross-profiles. The differences in height apply, for instance, to the Indus Valley cross-profile of Fig. 2.54: Character D: at No. 10 as well as Fig. 2.59: Nos. 1 and 2. A difference in height of 300 m is reached between the upper margin of till material and the thalweg on the valley cross-profile of Fig. 2.59: No. 3 and on the valley cross-profile of Fig. 2.59: Character E. On the valley cross-profile of Fig. 2.59: No. 6, differences in height of 330 m are reached. These differences are a measure of the pertinent Late Glacial glacier thickness of

Fig. 2.63 **a**, **b** Mineralogic rock determination of sample 20.08.13/2a/b. No. 4a: Erratic boulder: leuco-granite ("Leukogranit"). No. 4b: Bedrock: fine-grained granodiorite ("Granodiorit") strongly greened ("vergrünt"). Mineralogic rock determination and documentation by A. Vollbrecht. Locality: 34° 15′ 18.43″N 77° 13′ 26.15″E, 3556 (3500 aneroid measurement) m asl; orographically right in the Indus Valley on saddle of ground moraine 10 km upvalley from the Alchi settlement; see Fig. 2.59 at http://extras.springer.com: Character E: No. 4; see also Fig. 2.61. Sampling: M. Kuhle

Fig. 2.64 Facing south into the orographic left valley flank of the Indus Valley. (■) show up to decameter-thick pedestals and covers of ground moraines, the remnants of which reach up the valley flank as high as ca. 4200 m asl (34° 10′ 16.23″N 77° 15′ 6.62″E). (◓) marks a glacigenically rounded hill towering up to ca. 3700 m (34° 13′ 5.82″N 77° 14′ 26.28″E; Fig. 2.59 between No. 6 and character E) that consists of vertically standing sedimentary rocks. Analog photo 20/08/2013 by M. Kuhle

the Indus Glacier. Further downvalley the highest position of the upper margin of the till sheet decreases from more than 260 m above the thalweg at 34° 15′ 11.52″N 77° 6′ 54.30″ E (Fig. 2.59, see also http://extras.springer.com: Character E) (cf. Fig. 2.67) to ca. 175 m above the thalweg at 34° 15′ 35.30″N 77° 4′ 44.58″E (Figs. 2.65 and 2.66).

Figure 2.68 (sample 19.08.13/2) concerns analyses of till material lying up to 260 m above the thalweg. It contains rounded components of coarse pebbles of complex petrography that stem from a neighboring bedrock conglomerate. The granulometric characteristics confirm the moraine character, as also does the 97.6% SiO_2 grain size portion of the medium sand fraction of the

Fig. 2.65 Ca. 180° panoramic photograph from ca. 3300 m asl at 34° 15′ 46.80″N 77° 4′ 36.85″E from the orographic right flank of the Indus Valley (Fig. 2.59 on the right somewhat above No. 8). From facing ESE upvalley via south into the left valley flank up to facing WSW looking diagonally downvalley. (◓) mark glacigenic flank-polishings, reaching up to ca. 4300 m asl at 34° 13′ 36.91″N 77° 5′ 33.90″E (up to - - - - white). (■) show meter- to decameter-thick ground moraine pedestals and covers with erratic granite boulders (■ below, see Fig. 2.66) that reach up to ca. 4150 m asl. (Δ) shows a debris-flow fan/ alluvial fan, the material of which consists of dislocated ground moraine. (⊠) mark areas of weathering that have developed since the deglaciation, which have roughened the flank-polishings. (◀) shows a post-glacial detrital (talus) slope developed by weathering (see. Fig. 2.66). (- - - -) denotes the Glacial minimum level of the Indus valley glacier at about 4300 m asl. Analog photo 20/08/2013 by M. Kuhle

fine material matrix, as it is glacially crushed (Fig. 2.10: No. 8).

Thus the pertinent glacier terminus must have been situated several km downvalley at ca. 3000 m asl. It reached as far as into the valley chamber of the Khaltse settlement. This glacier stage has to be classified as belonging to the Late Glacial Stage II (Taglung stage, Fig. 2.5).

In the same, aforementioned valley cross-profile (Fig. 2.59, see also http://extras.springer.com: Character E: No. 8; Fig. 2.65), the higher level of till cover (see Eq. 2.6) confirms an ice thickness of ca. 910 m. It belongs to the Late Glacial Stage I (Ghasa stage, Fig. 2.5).

Based on the present author's (Kuhle's) findings, the Lamayuru lake has to be identified as a Late Glacial catchment lake of the Indus Glacier. During Stage I (see above) the Indus main valley glacier still had a thickness of at least 700–800 m. This is proved by the analysis of the aforementioned till cover at 3650 m asl. The till is meters to decameters in thickness and consists of erratic material (Fig. 2.59, see also http://extras.springer.com: Character E: No. 10: 34° 17′ 4.07″N 76° 48′ 16.06″E). It is situated on an abraded mountain cupola and thus affords proof only of the minimum altitude of the Indus Glacier level. The actual level (upper limit) of the I was stage probably was situated at least 100 m higher (see above) at ca. 3750 m asl. The orographic left ice margin of the Indus Glacier extended down into the Wanla Chu side valley and there was attached to the preserved lake sediments of the Lamayuru lake. On both sides of the lake outlet, that is, in the contact area, remnants of lateral moraine (i.e., till material) are interlocked with the lake sediments. The latter prove a lake level of at least 3540 (3480 aneroid

Fig. 2.66 Looking NE into the orographic right valley flank of the Indus Valley onto pedestal ground moraine material at ca. 3160 m asl at 34° 15′ 32.37″N 77° 4′ 35.72″E, 130 m above the thalweg (Fig. 2.59 to the right of No. 8; see Fig. 2.65). (⭘ white and ◘) signify edged, larger and smaller polymictic boulders of metamorphic sedimentary rocks and phyllites outcropping at this locality. (⭘ black) is one of the meter-sized, round-edged to rounded erratic granite boulders (the matrix between, i.e., under the boulders is clayey). (⏶) shows a glacigenic truncated spur, i.e., a triangular glacigenic slope that up to its culmination at 3630 m asl has been rounded. (⊠) marks post-glacial weathering at the upper margin of a debris slope. Analog photo 20/08/2013 by M. Kuhle

measurement) m asl. The results have been obtained on the basis of field work carried out in 1993; a more detailed description shall not be presented here.

This is no isolated incident: probably somewhat younger, Late Glacial lake sediments have also been observed in the orographic right side valleys at 34° 16′ 50.48″N 77° 3′ 24.70″E, 34° 19′ 51.84″N 76° 54′ 53.65″E, 34° 20′ 3.60″ N 76° 52′ 33.83″E (Fig. 2.59, see also http://extras.springer.com: Character E). These typical catchment lakes in an ice margin position were

Fig. 2.67 Facing NE into the orographic right valley flank of the Indus Valley, onto several meter-sized, rounded, erratic granite boulders (**O**) at ca. 3080 (3170 aneroid measurement) m asl at 34° 14′ 41.84″N 77° 6′ 47.71″E, seen from ca. 40 m above the Indus River (Fig. 2.59 between Nos. 7 and 8). (**□**) indicates an edged boulder from local phyllite, which outcrops in this valley cross-profile. (◓) mark glacigenic flank-abrasions and flank-roundings ("Schichtkopfstreifenschliff") on the outcropping rocks of that phyllite. (■) show pedestal ground moraine, the breaking-edge of which is visible here at ca. 3200 m asl; erratic boulders rolled down (**O**) from it (post-depositionally). Analog photo 20/08/2013 by M. Kuhle

once dammed by the Indus Glacier during the late Stage I at the levels of ca. 3365, 3375 and 3285 m asl.

In some places of the Indus Valley, several hundred meters above the highest remnants of till cover, there are preserved polish or abrasion lines. These are indications of the maximum glacier infilling and ice thickness. A highest flank polishing of this sort has been observed on the orographic left side up to ca. 4300 m asl (Fig. 2.59, see also http://extras.springer.com: Character E: 34° 13′ 36.91″N 77° 5′ 33.90″E; Fig. 2.65). It reaches a still about 200 m greater height than the glacigenic abrasion of the layer

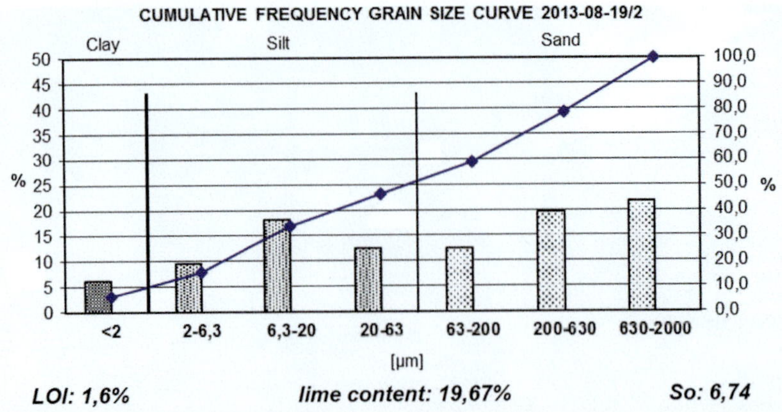

Fig. 2.68 (Grain-size diagram 2013-08-19/2): Granulometry of the fine material matrix of ground moraine orographically right in the Indus Valley at 34° 15′ 3.52″N 77° 6′ 54.37″E, 3280 (3160 aneroid measurement) m asl, ca. 250 m above the Indus River above the Uletokbo settlement, above the locality of Fig. 2.67. Exposure and sampling depth ca. 1 m. The maximum occurs with the ca. 23% peak in the coarse sand, and a secondary maximum with 18.5% in the middle silt. The grain-size cumulative curve has a bimodal course. The clay content is ca. 6%. The sorting coefficient calculated according to von Engelhardt (1973: 133) is rather high with So = 6.74 (So = $\sqrt{Q_3/Q_1}$); the lime-content is 19.67%; the loss on ignition (LOI) amounts to 1.6%. Locality: Fig. 2.59 nearly in the middle between Nos. 8 and 7. See Fig. 2.10a, b: No. 8; 2013-08-19/2. Sampling M. Kuhle

edges ("Schichtkopfstreifenschliff") that can be diagnosed at the valley ridge 5–6 km further downvalley (Fig. 2.65; Fig. 2.59: No. 8). The fact that no till cover is found here must be explained by the fact that the position exceeds that of the High Glacial ELA that was located at ca. 4000 m asl (see Kuhle 2001a: 115, Fig. 2/2).

With the abrasion line at 4300 m asl, a High Glacial (LGP, MIS 3-2, Stadium 0, Fig. 2.5) ice thickness of ca. 1260 m of the Indus glacier in this mid-valley section is likely.

Finally, a glacial-geomorphological consequence of this ice level must be pointed out. The Likir Chu side valley (Fig. 2.59, see also http://extras.springer.com: Character E: No. 9) shall be discussed as an example. A main valley glacier level such as this necessarily must have led to the damming of the side-valley glaciers. This continued well into the Late Glacial. The resulting reduction of the flow dynamics put a stop to the High Glacial development of trough valleys. Only when in the late Late Glacial (from Stage III onwards, Fig. 2.5) the Indus main valley glacier did not dam the side valleys anymore could trough valley development set in. In the cross-profile of the Likir Chu it was limited to the lower 100–200 m. This means that when the side-valley glacier was still thicker than 100–200 m, its discharge was dammed and trough valley erosion stopped. Therefore, during the High Glacial, the preglacial V-shaped valleys were largely preserved. Trough profiles only emerged later during the Early and Late Glacial period, and only close to the valley floor (Fig. 2.69). Such reduced glacial erosion in side valleys is also proved by an enormous thickness of till sheets of several decameters. The composition of the moraine material, of predominantly very large boulders with sizes of several meters is especially striking and by inference implies that the trituration under the glacier was insignificant. The fine material matrix yet shows the characteristic features of till (see Fig. 2.70, sample 20.08.13/1: 34° 18′ 52.57″N 77° 13′ 55.03″E, 3989 m asl; Fig. 2.10: No. 9).

Fig. 2.69 Ca. 135° panoramic photograph from ca. 4110 m asl at 34° 19′ 28.26″N 77° 14′ 11.38″E from the valley bottom of the Likir Chu—a hanging orographic right side valley of the Indus Valley (Fig. 2.59 on http://extras.springer.com: Character E: No. 9)—from facing NW into the orographic right valley flank, facing NNE upvalley up to facing NE, looking into the left valley flank. (▲) mark glacigenic abrasion-roundings. (⊠) are areas of weathering that have developed after deglaciation and are roughening the abrasion-roundings in the granite bedrock. (◄) marks a post-glacial debris slope developed by rock weathering and surficially covered by ground moraine. (∪) shows a clearly preserved trough-, i.e., a U-shaped valley cross-profile. (■) mark meter-thick ground moraine with edged, meter-sized granite boulders (●). (- - - -) denotes the Glacial minimum level of the ice stream network of ca. 5250–5600 m asl. Analog photo 20/08/2013 by M. Kuhle

2.3.3 Summary of Sect. 2.3: The High Glacial Indus Glacier in the Central Indus Valley

The primary focus of this section are Figs. 2.54 (see also http://extras.springer.com: Character D) and 2.59 (see also http://extras.springer.com: Character E). The detailed field observations from 1993, which were complemented by the results of the expedition in 2013, confirm a connected Indus main valley glacier with a thickness of at least 900 m (Kuhle 1998a: 88). The maximum High Glacial thickness was ca. 1260 m. With nearly the same thickness, this glacier passed the junction of the orographic left Wanla Chu side valley. In this same location, the Indus Glacier dammed the Lamayuru lake during the Late Glacial.

In the area of the inflow of the Wanla Chu, the surface of the glacial Indus Glacier was located at ca. 4250 m asl, that is, still above the High Glacial ELA (snowline). In addition to numerous smaller side-valley glaciers such as the Kongmaru-Northeast valley (Shang Chu or Martselang Valley), Stok Chu (valley), Puchu Chu, Phyang Chu, and Likir Chu, the ice stream networks of Sakti Chu (valley of Sakti) and the Taglang La-North valley also belonged to the large-scale connected nourishing area of the Indus Glacier. Taken together these ice stream networks yielded the connecting glacier junction between the Pangong Shyok outlet glacier system (Sects. 2.1 and 2.2) in the northeast (Sakti Chu via Chang La

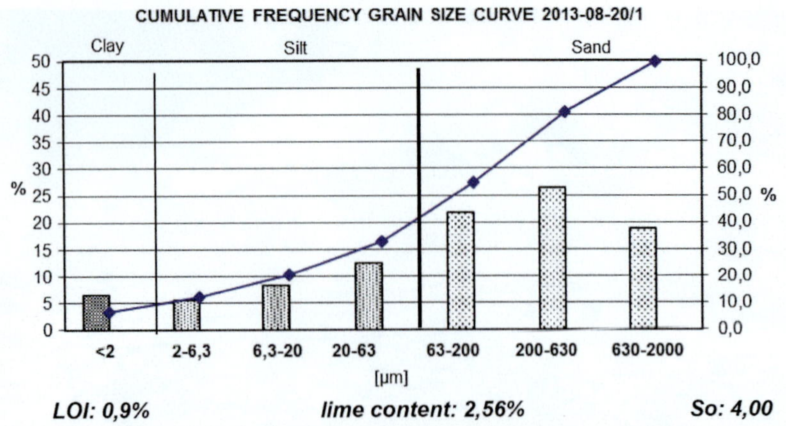

Fig. 2.70 (Grain size diagram 2013-08-20/1): Granulometry of the fine material matrix of middle ground moraine on the orographic left side in the Likir Chu, upvalley from the Likir monastery, ca. 17 m above the thalweg at 34° 18′ 52.57″N 77° 13′ 55.03″E at 3989 (3845 aneroid measurement) m asl. Exposure and sampling depth ca. 1.5 m. The maximum occurs with the ca. 26% peak in the middle sand, and a secondary maximum with ca. 6.5% in the clay. The grain-size cumulative curve has a bimodal course. The sorting coefficient calculated according to von Engelhardt (1973: 133) is So = 4.00 ($So = \sqrt{Q_3/Q_1}$); the lime content is 2.56%; the loss on ignition (LOI) amounts to 0.9%. Locality: Fig. 2.59 between Nos. 5 and 9. See Fig. 2.10a, b: No. 9; 2013-08-20/1. Sampling: M. Kuhle

Baba) and the 240 km distant southern slope of the Himalaya (Lahoul, Pir Panjal) (ibid.: 88) on the south to SW (Taglang La-North valley).

In the Indus main valley, the highest occurrences of till sheets demarcate the two Late Glacial Stages I (Ghasa stage, Fig. 2.5) and II (Taglung stage, Fig. 2.5); their glacier levels were comparatively lower than during the LGM. The terminus of the Indus Glacier during Stage II was situated at ca. 3000 m asl in the valley chamber of the Khaltse settlement. During Stage I, the Indus Glacier still passed the orographic left Wanla Chu side valley with a thickness of 800–900 m.

On the basis of new altitude values, correcting those from 1993, the Late Glacial glacier stages of Stage I, II or III, and IV (Fig. 2.5) were reconstructed for the side valleys Stok Chu, Puchu Chu, and Phyang Chu. The chronological classification was achieved via the calculated ΔELAs (snowline depressions); these amounted to at least 1070 m (Stage I), 910–925 m (Stage II or III), and 640 m (Stage IV), respectively.

The glacial Indus Glacier will not be described in this book further downvalley from the junction of the Wanla Chu side valley (Fig. 2.59, see also http://extras.springer.com: Character E: above No. 10). From here on it flowed through the ice stream network of the Nun Kun group in the south and the Karakoram in the north. For this specific flow discharge running parallel to the Shyok outlet glacier (Sect. 2.2) up to the junction of the Indus and Shyok glaciers, the interested reader is referred to previous published work (Kuhle and Kuhle 1997: 121, Photo 3; Kuhle 1998a: 88/89; Kuhle 2013a: 177).

This work needs only be summarized here as follows: 83 km down the Indus Valley from the junction of the Wanla Chu side valley, in the valley chamber of the Marol settlement, the Karcha or Suru valley glacier flowed from the south into the Indus Glacier. It came from the 128 km distant, 7135 (7315) m high Nun Kun Massif, i.e., the Nun Kun ice stream network. Under the northern flank of the Nun Kun it had a verifiable thickness of 1000 m and in the 8 km wide valley chamber of Kargil, a thickness of 600–700 m. 73 km down-valley, the Indus main valley glacier reached its confluence with the Shyok Glacier (Sects. 2.2.1, 2.2.2, and 2.2.2.2; Kuhle 2011a: Glacier Maps SJ32, TJ12, TJ11).

2.4 The Current, Historical, Neoglacial, and Late Glacial to High Glacial Glaciation of Central Tibet

The primary focus of this section is Fig. 2.1: Nos. 5, 10, 12, 19, and 35. After having started with the SW margin of Central Tibet, we now move on to the NW margin. There, however, we do not turn to the outer margin of Tibet, but to the southern slope of the peripheral mountains leaning towards the Aksai Chin, that is, towards western Central Tibet (Fig. 2.1: No. 23).

2.4.1 The Current, Historical to Neoglacial Glaciations of the Chongce Shan I-South Slope

The ca. 115 by 30 km large, interconnected current glacier face of the ca. 6850 m high Chongce Shan I Massif has ca. 30 glacier tongues that belong to the south slope of the massif. We choose to focus on the largest glacier tongue because naturally it reacts the most slowly to climate oscillations. It terminates at 5340 m asl (Fig. 2.71, see also http://extras.springer.com: Character F: No. 1). Three historical glacier stages and a neoglacial one are connected in the forefield: they are situated at 35° 10′ 37.93″N 81° 11′ 20.66″E at 5270 m asl, at 35° 9′ 22.30″N 81° 12′ 38.57″E at 5230 m asl, at 35° 8′ 5.20″N 81° 12′ 58.96″E at 5190 m asl, and at 35° 4′ 26.28″N 81° 7′ 48.36″E at ca. 5080 m asl.

They confirm a ΔELA of 35 m:

The current glacier tongues are located at tim = 5340 m asl; the former glacier tongue is located at tif = 5270 m asl:

$$\text{Sdepr} = (5340 - 5270)/2 \tag{2.23}$$

$$\text{Sdepr} = 70/2 \tag{2.24}$$

$$\text{Sdepr} = 35 \tag{2.25}$$

$$\text{ELA-depr} = 35\ \text{m} \tag{2.26}$$

as well as a ΔELA of 55 ([5340 − 5230]/2), 75 ([5340 − 5190]/2), and 260 m ([5340 − 5080]/2).

These snowline depressions have to be classified as belonging to their corresponding Stages X, VIII, VII or ‘VII, and V (Fig. 2.5), or maybe to a slightly older stage between the late Late Glacial Stage IV and Stage V in the middle or early Holocene.

The relatively large size of the present glacier tongue—which today is still 4 km wide with a central glacier length of ca. 28 km—is the reason why further glacier advances, i.e., glacier stages that with regard to time lie in between, have not been manifest in ice margin positions of their own. The positive mass balance of this glacier tongue obviously had its consequences only in the increases of the ice thickness. These increases in thickness became delayed, i.e., they were realized as glacier advances only during a later stage. This delay was also further enforced by the —during the Neoglacial period—especially large-scale glacier tongue in the form of a hammer head with a lateral extension of 15 km. This is made especially clear by the large time leap from Stage V to Stage VII.

During the earliest advance of Stage V—or maybe during an older stage between the late Late Glacial Stage IV and Stage V—the glacier tongue filled the entire lake basin of Gozha Co and reached up to No. 2 in Fig. 2.71. This would necessitate a ΔELA of more than 260 m (see

Fig. 2.71 Satellite image, Google Earth Pro: center 34° 42′ 49.30″N 80° 22′ 45.95″E; http://extras.springer.com: Character F; Nos. 1–10

above), ca. 300 m (Fig. 2.5). At the time of this Neoglacial (middle or early Holocene) advance (Stage V or a little older), the ENE-adjacent glacier tongues of the accompanying valley head flowed together with the large glacier tongue and formed a common piedmont glacier lobe.

Below (south) of No. 1 (Fig. 2.71, see also http://extras.springer.com: Character F: No. 1) up to nearly the Gozha Co, till sheets are preserved that developed within the last ca. 150–2000 years (ca. Stage X to 'VII). They show the typical pattern created by dead-ice holes and the pertinent small meltwater rills. Light-colored, up to more than 2 m long, moraine boulders transported over a long distance also can be observed on the till cover as far down as to the bank of Gozha Co.

Corresponding covers of till continue to the west as far as SW of Gozha Co (left of No. 2 in Fig. 2.71).

At a ΔELA of 300 m the glaciers of the 6240 m high massif, which flowed down to the north and were located south of the Gozha Co, also reached the glacier filling of Gozha Co.

This was the case during Neoglacial Stage V or maybe during an older stage between the late Late Glacial Stage IV and Stage V. Currently, the extant glacier tongue ends ca. 375 m above the lake. Therefore, the ends of the glacier tongues must have reached the lake at a ΔELA of ca. 188 m ([5460 − 5085]/2).

In a similar fashion, the enlargement of the glacier faces and the confluence of the glaciers between all adjacent, currently glaciated mountain massifs in the entire study area of the eastern Aksai Chin must have occurred. These are the mountain massifs between the Chongce Shan I Massif and Pangong Co.

2.4.2 The Late Glacial to High Glacial Glaciation of West-Central Tibet Between the Chongce Shan I-South Slope and the Himalaya SW of Pangong Co

During the youngest Late Glacial stage, Stage IV (Sirkung stage or maybe during an older stage [III] between the late Late Glacial and Stage V; Fig. 2.5), at a ΔELA of ca. 700 m, the valleys and depressions between the still currently glaciated mountain groups were filled by glacier ice.

This can be inferred by the calculation of the ΔELA. It is immediately proved by the identified covers of till. These are located in all valleys and depressions between those mountain massifs. In these regions, they have only been partly covered and reshaped by younger glaciofluvial activity, alluvial fans, and lake sediments.

Even in the area of the current ice margins, the burial of the still fresh till material by glaciofluvial deposits and alluvial fans can be observed in many places (see the anthracite-grey alluvial fans below and on the right of No. 1 in Fig. 2.71).

In the following, further examples of till covers in the discussed area of western Central Tibet will be given. Without exception, these are the lowest areas among those mountain massifs. They are the till sheets at No. 3 (Fig. 2.71, see also http://extras.springer.com: Character F) at 5050 m asl, at 34° 46′ 28.62″N 81° 16′ 43.15″E at 5010 m asl, at 34° 46′ 25.94″N 82° 1′ 51.36″E at 5120–5170 m asl, at No. 8 (Fig. 2.71, see also http://extras.springer.com: Character F) at 5230–5260 m asl, at No. 10 at 5190–5270 m asl, at Character F (Fig. 2.71, see also http://extras.springer.com: Character F) at 5180–5220 m asl, at 34° 37′ 28.36″N 80° 42′ 39.03″E at 5200 m asl, at No. 4 (Fig. 2.71) at 5010 m asl, at No. 5 (Fig. 2.71) at 5150–5180 m asl, at No. 6 (Figs. 2.71 and 2.72) at 5350–5400 m asl, at 34° 32′ 53.89″N 80° 25′ 23.26″E consisting of erratic material on a 100 m high rock hill at 5290 m asl, at 34° 43′ 16.44″N 80° 10′ 37.58″E at 5150–5300 m asl (Fig. 2.73), at No. 9 (Fig. 2.71) at 5450 m asl, and at No. 7 (Fig. 2.71) at 5280–5300 m asl (see also Kuhle 1999: 266–271, Photo 198–200).

The ice filling of these depressions between the mountain massifs in the respective area, where currently there are lakes in many places (Fig. 2.71), led to the till covers discussed here. These very freshly preserved deposits of till were possible because an ice cover still existed during the late Late Glacial (Stage IV). This necessarily implies a ΔELA of ca. 700 m (Fig. 2.5). During this period, 100 m high rock hills at 5290 m asl (see above) as well as rock ledges and flat transfluence passes at ca. 5300–5420 m asl were also covered with till (Kuhle 2013a: 159–160, Photo 100). In the middle to early Late Glacial of Stages III–I (Dhampu to Ghasa stage), a ΔELA of 800–1100 m (Fig. 2.5) existed. This ELA nearly reached the level of the lowest valley bottoms, i.e., of the bottoms of the lowest field depressions, or perhaps reached even beyond them.

Accordingly, during that period, processes of glacial erosion, i.e., ground polishing of the inland ice, were most active and persistent on those lowest relief surfaces of the Tibetan area discussed here. At the same time, there existed a greater ice thickness that must have led to the removal of the till sheet of the Early Glacial. Due to the greater ice load, this process must have been even more intensive during the High Glacial. During this period of the LGM (LGP, MIS 3-2, Stage 0) there existed a still larger ΔELA of up to 1300 m (Fig. 2.5).

During this same period, hill chains, intermediate valley ridges with transfluence passes (-saddles), and mountain ridges from ca. 5500–5800 m asl (e.g. at 34° 49′ 1.92″N 80° 4′ 56.67″E, at 34° 52′ 54.83″N 80° 37′ 46.79″E, at 34° 49′ 55.15″N 80° 34′ 51.24″E, at 34° 49′ 21.13″N 80° 39′ 52.30″E, at 34° 48′ 53.37″N 80° 40′ 50.45″E, at 34° 45′ 53.50″N 80° 53′ 14.64″E, at 34° 42′ 36.81″N 80° 59′ 20.94″E, at 34° 49′ 8.38″N 80° 6′ 25.84″E, and 33° 53′ 51.57″N 79° 27′ 33.73″E) up to a level of ca. 6300 m (e.g., at 34° 47′ 24.58″N 80° 39′ 0.85″E, at 34° 47′ 0.05″N 80°36′ 17.44″

Fig. 2.72 Facing NE into the orographic left valley flank of a WNW-ESE oriented high valley in central western Tibet from 5350 to 5400 m asl, at 34° 32′ 53.89″ N 80° 25′ 23.26″E, showing erratic ground moraine material (■ foreground; Fig. 2.71 on http://extras.springer.com: Character F: No. 6) on two ca. 100 m high rock hills (◓). (■) mark meter- to decameter-thick, very strongly clayey ground moraine covers. (◓) are glacigenic abrasions and polish-forms, such as glacial knobs up to a height of 5570 m asl. (- - - -) indicates the High Glacial glacier level exceeding 6000 m asl in a distorted perspective. Analog photo 09/09/1996 by M. Kuhle

E, at 34° 27′ 5.34″N 79° 7′ 12.28″E, at 34° 28′ 18.45″N 79° 31′ 5.85″E, at 34°21′ 42.47″N 81°1′ 27.16″E, at 34° 19′ 32.06″N 81° 1′ 1.74″ E, at 34° 13′ 33.74″N 81° 4′ 59.52″E, at 34° 11′ 42.97″N 81° 2′ 31.96″E, at 34° 0′ 27.35″N 80° 4′ 40.43″E, at 33°58′ 15.39″N 79°48′56.23″ E, at 34°10′ 50.44″N 78°52′ 15.58″E, at 33° 53′ 50.89″N 79°54′ 59.58″E, at 33° 53′ 19.60″N 79° 54′ 50.40″E, at 33°46′ 29.18″N 79° 51′ 10.35″E, at 33° 41′ 30.42″N 79° 46′ 48.75″ E, as well as at 33° 21′ 44.59″N 78° 55′ 40.05″ E), and maximally up to 6730 m asl with the clearly rounded summit of Zangqun Kangri (34° 5′ 42.74″N 79° 50′ 25.08″E) were reworked by glacigenic abrasion. The introduced SSW continuation of the cross-profile through western Central Tibet with its highest abrasion forms, has already been described in Sect. 2.2.2.2. Here, only the rounded mountains SW of the NW end of Pangong Co are taken into consideration. The ice level was located there at more than 6000–6300 m asl. According to the findings introduced above, in the intervening area—that is, in the area between the Chongce Shan I Massif (Liushi Shan) and Nako Tso, or in other words the ESE end of Pangong Co lake (Kuhle 1999)—the inland ice surface also thickened up (in the shape of a flat cupola) and reached altitudes of more than 6300 m asl or at maximum even more than 6730 m asl.

At 34° 17′ 9.16″N 79° 59′ 39.85″E (ca. 6480 m asl), at 33° 53′ 19.16″N 79° 54′ 47.64″ E (ca. 6340 m), at 33° 17′ 43.91″N 79°46′ 0.19″ E (ca. 6200 m), and at 33° 28′ 59.69″N 79° 20′ 13.52″E (ca. 6250 m), further rounded summit cupolas exist in the region. They

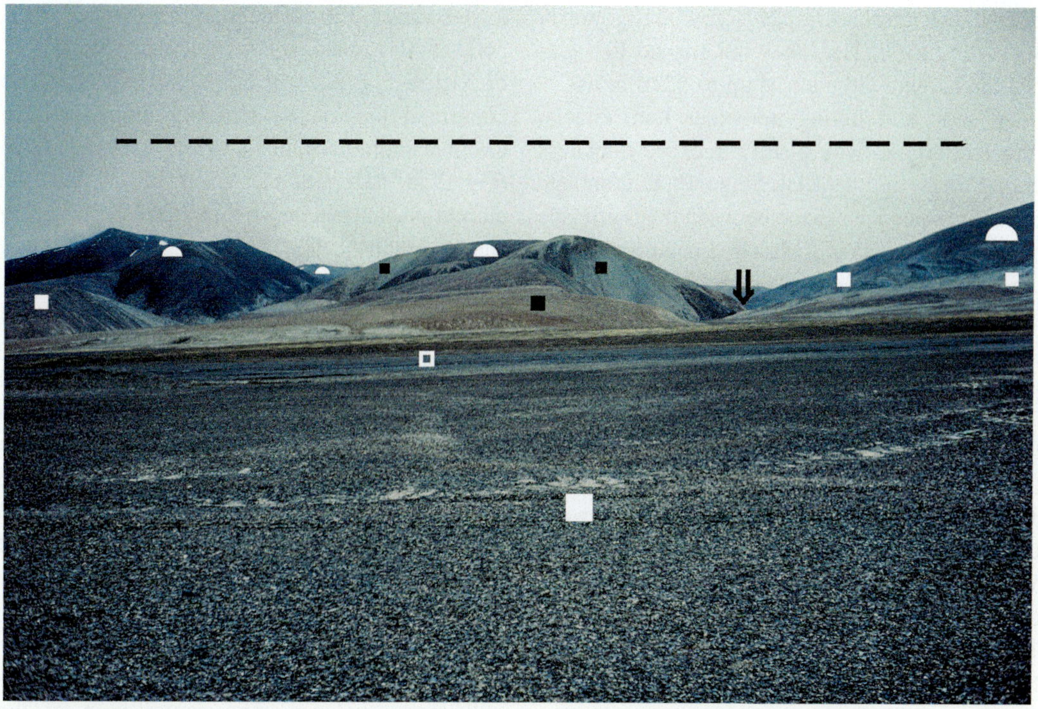

Fig. 2.73 Erratic ground moraine-material seen facing WSW in central western Tibet from 34° 43′ 16.44″N 80° 10′ 37.58″E at 5150–5300 m asl (■ foreground; Fig. 2.71: Character F, left of No. 5) overlooking a hill landscape. (■) signify meter- to decameter-thick very clayey ground moraine covers. (⏶) mark glacigenic abrasions and ground-polish forms such as glacially streamlined hills up to a height of 5530 m asl. (□) shows a streambed from which the clay and silt portions have been flushed from the ground moraine surface. (⇓) marks a V-shaped form, laid out by a subglacial meltwater spillway. (- - - -) indicates the High Glacial glacier level exceeding 6000 m asl in a distorted perspective. Analog photo 09/09/1996 by M. Kuhle

make clear that the current (historical) and former insignificant interglacial glaciations time and again sharpened the summit cupolas, which during the glacial periods had been glacigenically rounded. This sharpening has taken place regressively, i.e., from below. It can be diagnosed by the fresh working edges, which are the abrupt steepenings setting in below the summits and running from there downward. On this basis the summit cupolas—that are only rounded at the top—are recognizable as being forms abraded by the inland ice. In this context, the 6730 m high Zangqun Kangri (see above) also shows interglacial steepenings on all four flanks of a rounded summit cupola. Quite generally, this interglacial sharpening of glacial forms can be observed in all landscapes once formed by inland ice, e.g., in Rondane and in the Dovre Fjell of Scandinavia.

The highest glacigenic abrasion of the layer edges ("Schichtkopfstreifenschliff") occurs at 5960 m asl in an E-W direction (34° 35′ 36.59″ N 78° 58′ 40.93″E). All these abrasions took place under cold-based conditions. Under the influence of "cold-based" ice, coarse abrasion occurs by means of exaration (detersion), etc. Such ground abrasions are preserved at No. 10 (Fig. 2.71, see also http://extras.springer.com: Character F) at 5200 asl between some veils of till. They also occur on a large scale in dark rock above No. 8 (Fig. 2.71), running from east to west, and also west of No. 6 (Fig. 2.71,

2.72 background) at 34° 41′ 13.68″N 80° 14′ 52.61″E. Here they are shaped like glacial knobs with altitudes of up to 5570 m asl.

At least still during the older Late Glacial (Stages II-I) and with a still larger ice thickness during the High Glacial Stage 0, a complete cover of inland ice existed between the north side of the Chongce Shan I Massif (Liushi Shan) and the Pangong Co with its outlet glacier reaching down into the Shyok and Indus valleys (Sects. 2.1 and 2.2). Towards the west the inland ice covered the entire Aksai Chin area, including the Lingzi Thang (Kuhle 1999: 43–48, 260–272), up to the outlet glacier of the Kolakoshih Ho (valley) (Kuhle 2004 and 2011a: Glacier Maps TJ12, TJ11; 2013a: 157–164), so that between the Kuenlun and the Karakorum as well as the Western Himalayas (see Sect. 2.3) a widely connected glacier cover existed (Kuhle 1988b, 1994a). In places, it must have reached an ice thickness of ca. 2000 m.

2.4.3 The Present, Historical, Neoglacial (Holocene), and Late to High Glacial Glaciation of the Muggargoibo Massif in Central Tibet as a Further Representative Example

The focus of this section is on Fig. 2.74 (see also http://extras.springer.com: Character G: 32° 16′ 10.73″N 87° 23′ 44.76″E). Unlike the forefield of the large glacier tongue of the Chongce Shan I Massif, in which only three Historical glacier stages and one Neoglacial glacier stage are developed due to topographic reasons (see Sect. 2.4.1), a glacier has been chosen here as the object of study that has preserved all eight Historical to Neoglacial (Holocene) stages.

The southern glacier of the 6247 m high Muggargoibo shows eight past glacier stages as far as to the valley of the next-higher order. The lowest current (2012) ice margin is located at ca. 5600 (5604) m asl. It corresponds to Stage XIV or XIII in Fig. 2.5. All the eight past stages are confirmed by concise terminal moraines. The terminal moraine rampart of the youngest stage (Fig. 2.74: ①) lies ca. 25 m lower than the current glacier tongue, so that a ΔELA of 12.5 m can be established (Stage XII or XIII, see below). The lowest end moraine rampart lies at ca. 4860 m asl (Fig. 2.74: ⑧). Thus it proves a ΔELA of 372 m. This ΔELA classifies the ice margin as belonging to the oldest Neoglacial Stage V (Nauri stage) that, accordingly, is from the Holocene (see Fig. 2.5). The stages located in between are therefore Stages XI–VI. Only one single stage does not show a very prominent end moraine rampart. Because of the overall good state of preservation of the moraines, their classification as belonging into the Holocene to Historical period can already be clearly confirmed geomorphologically. However, there is still more evidence. The demarcation of the small, rampart-shaped terminal moraines of this (at maximum 14 km long) valley glacier from the till cover of the closer and more distant surroundings of the Muggargoibo Massif (Fig. 2.74: 5 times Character G) also is sufficient evidence for a chronological classification into the Holocene. These till sheets prove a glaciation that completely covered the entire region. For this a ΔELA of more than 400 m was necessary. Accordingly, the covering glaciation has to be classified as belonging to the late Late Glacial Stage IV (Sirkung stage, Fig. 2.5).

This furthermore means that in the earlier Late Glacial Stages III, II, and I, the ELA must have been located ca. 150, 300, and 400 m lower (Fig. 2.5), thereby yielding a correspondingly thicker ice cover. During the LGP (Last Glacial Period, Stage 0), the Muggargoibo Massif with its neighboring, still glaciated mountains must have been entirely covered by the inland ice. The ice cover here reached a thickness of more than 1650 m.

The Muggargoibo Massif is surrounded on a large scale by extensive High to Late Glacial

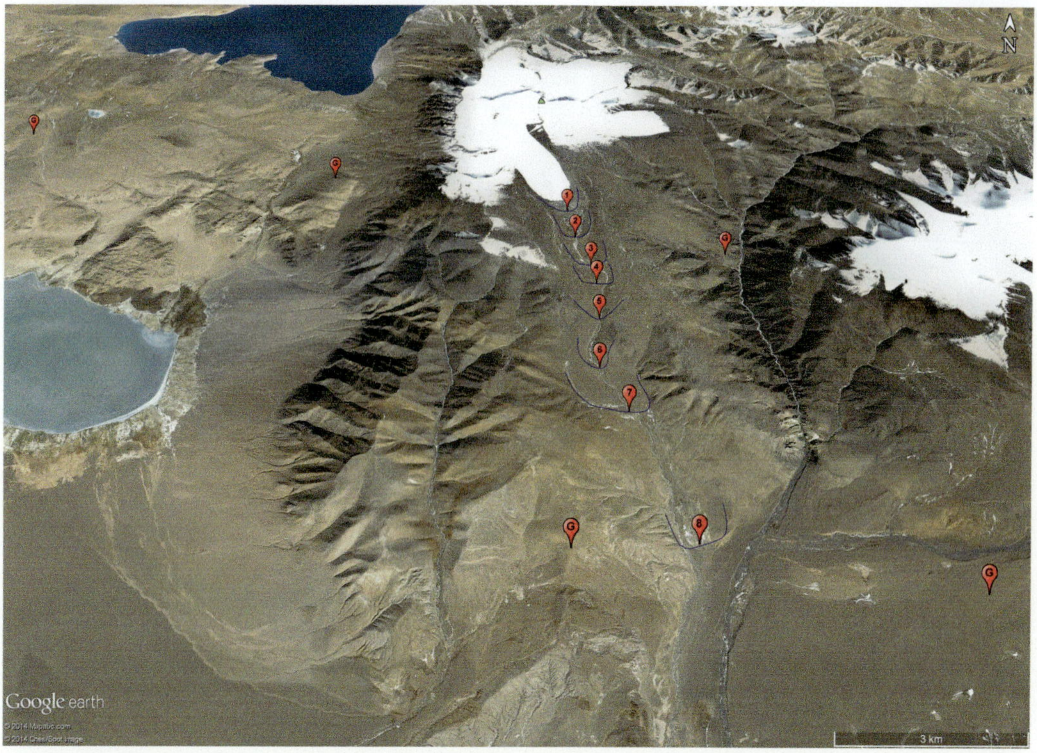

Fig. 2.74 Satellite image, Google Earth Pro: center 32° 12′ 48.39″N 87° 22′ 51.71″E; http://extras.springer.com: Character G: Nos. 1–8. On the south slope of the Muggargoibo massif (6247 m asl; 32° 18′N 87° 23′E) in Central Tibet, the well-differentiated stadial sequence 1–8 corresponds with ELA depressions (ΔELA) of between 12.5 and 372 m. Only during the older stages has there been a confluence with the valley glacier of another catchment area

covers of till, such as for example ca. 60 km SSW at 31° 49′ 35″N 87° 07′ 04″E at about 4630 m asl.

The present author gathered ample evidence of the concentric extension of a High Glacial till cover and the characteristics of a large-scale ice sheet that completely covered the terrain. This evidence was documented at both higher as well as minor absolute altitudes (Kuhle 2013a) and at various locations across the entire area: thus, for example, ca. 230 km to the W (Kuhle 1999: 22–27, 131–140; 2014a: 133 Photo 26), 380 km to the ENE (Kuhle 1991: 134–140, 148–150; 2003: 79–81; 2014a: 128, 129, Photos 15–18), 375 km to the E (Kuhle 1991: 140–143, 151–153; 2003: 81–85; 2014a: 131–133, Photo 25), 350 km to the SE (Kuhle 1988c: 460–462; 1991: 166–167, 179–181), and 340 km to the S (Kuhle 1999: 15–22, 84–92).

2.4.4 The Current, Historical to Neoglacial (Holocene) Glaciation of the Nyainqentanglha Shan in Central Tibet

The focus of this section is Fig. 2.1: No. 5. In 1989, on the orographic right flank of the 16 km long valley SW below the 7162 m high Nyainqentanglha main summit, the hanging glacier tongue terminated at 5630 m asl (Fig. 2.75, see also http://extras.springer.com: Character H: No. 1; Kuhle 1991: 158–159, Photo 33). On 12 November 2007, the satellite image (Google Earth) shows the end of the same glacier tongue at 5655 m asl. Accordingly, in the year 1989, the ΔELA for Stage XIII (Fig. 2.5) was 12.5 m lower than the ΔELA for Stage IV (Fig. 2.5).

Fig. 2.75 Satellite image, Google Earth Pro: center 30° 19′ 22.79″N 90° 35′ 37.51″E; http://extras.springer.com: Character H: Nos. 1–10

A corresponding ΔELA was observed at the Muggargoibo southern glacier (Sect. 2.4.3).

In 1989, a second glacier tongue terminated at 5550 m asl (Fig. 2.75: Character H: No. 2; Fig. 2.76) while in 2007 its terminus lay at 5720 m asl. Here, a ΔELA of 85 m of Stage XIII in comparison with Stage XIV can be calculated.

A third glacier tongue terminated at 5500 m asl in 1989 (Fig. 2.75: Character H: No. 3; Fig. 2.76) and at 5574 m asl in 2007, thus yielding a ΔELA of 37 m of Stage XIII over Stage XIV.

A fourth glacier tongue with NE exposure terminated at 5540 m asl in 1989 (Fig. 2.75: Character H: No. 4) and at 5570 m asl in 2007, yielding a ΔELA difference of 15 m for Stage XIII over Stage XIV.

Yet another glacier tongue hanging in the valley flank of the opposite orographic left side of the valley with a SW exposure had its terminus at 5540 m asl in 1989 (Fig. 2.77). In 2013, it still terminated in nearly the same position (30° 21′ 58.11″N 90° 33′ 15.69″E). In this case, the ΔELA difference of Stage XIII over the current glacier stage lies at ca. 0 m. This glacier tongue flows down from the most northwesterly of the four more than 7000 m high main summits of the Nyainquentanglha Massif.

The glacier tongue of the main valley glacier terminated at 5340 m asl in 1989 (Fig. 2.75: Character H: No. 5; Fig. 2.78). On 10 April 2013, the satellite image (Google Earth) shows the glacier terminus at 5360 m asl (30° 22′ 4.55″ N 90° 32′ 22.62″E). This amounts to a ΔELA of 10 m in comparison with the current glacier stage (Stage XV?). Over the past 24 years, the current glacier terminus has retreated ca. 500 m.

An end moraine rampart, the inner slope of which is situated at 30° 21′ 19.07″N 90° 33′ 22.88″E at 5320 m asl, gives evidence of the last advance of the main valley glacier tongue 40 m lower than the current position of

Fig. 2.76 Facing SW in the Nyainqentanglha Shan in Central Tibet from 30° 20′ 11.52″N 90° 35′ 12.46″E at 5430 m asl from the orographic left valley flank looking to the 6000 m summit and its NE slope (Fig. 2.75: Character H: Nos. 2 and 3). (↙, ↘) marks the lowest tongue-end of two hanging glaciers on 14/08/1989. (■ higher) show historical end moraines of the hanging glaciers. (■ lower) show Late Glacial ground moraines on the main valley flank. (▲) marks glacigenic abrasion-roundings of the Glacial main valley glaciation. (⊠) shows Holocene to Historical areas of weathering on walls that have been undercut and steepened by the hanging glaciers. Analog photo 14/08/1989 by M. Kuhle

the glacier terminus. Accordingly, the ΔELA amounted to ca. 20 m, which makes this advance belong to Stage XII (Fig. 2.5). During this period (since 60–33 years B.P.), a horizontal retreat of the main valley glacier of ca. 1.6 km has taken place.

At 5285 m asl, a further historical ice margin existed (Fig. 2.75: Character H: No. 6). An end moraine serves as an indicator of a past advance. It proves a ΔELA of ca. 38 (37.5) m, and therefore belonged to Stage X (Fig. 2.5). This ice margin was located ca. 4 km from the current glacier terminus.

The second-largest glacier of the valley is the 3.8 km-long “7110 and 7040 m south-summit glacier” (Fig. 2.79). In 1989, at 30° 20′ 19.90″N 90° 35′ 14.00″E, its lowest ice margin was at 5400 m asl, and remained so in 2007 (Fig. 2.75, see also http://extras.springer.com: Character H). Accordingly, over a period of ca. 18 years, i.e. from Stage XIII to Stage XIV (Fig. 2.5), the ELA hardly changed. However, the ice complex at the glacier terminus that is covered by surface moraine became somewhat narrower compared with its condition in 1989. But there was no significant glacier retreat.

At the exit of the 16 km long valley SW below the Nyainqentanglha Massif, which consists of four summits with altitudes of more than 7000 m, the next very clearly indicated terminal moraine can be found at 4800 m asl (Fig. 2.75: Character H: No. 7). More specifically, it is an end moraine series consisting of two to three ramparts situated densely next to each other. On 10 April 2013, the actual glacier terminus was at 5360 m asl (see above), resulting in a ΔELA of

Fig. 2.77 Facing NE in the Nyainqentanglha Shan in Central Tibet in the orographic left main valley flank onto a glacier tongue at 30° 21′ 58.11″N 90° 33′ 15.69″E (Fig. 2.75 to the right of No. 5) that has flowed down on 13/08/1989 to 5540 m asl. (■) shows the border between Neoglacial (Holocene) and Historical ground moraine (below ■) and Late Glacial ground moraine (above ■) on the main valley flank. (▲) show young, past glacier-polishings of the hanging glacier. (⊠) marks Historical areas of weathering of surfaces undercut and steepened by the hanging glacier. Analog photo 13/08/1989 by M. Kuhle

Fig. 2.78 Facing NW in the Nyainqentanglha Shan in Central Tibet looking toward the glacier tongue of the main valley, which has come to an end on 13/08/1989 at 30° 21′ 54.14″N 90° 32′ 39.98″E at 5340 m asl (Fig. 2.75: Character H: No. 5). Analog photo 13/08/1989 by M. Kuhle

280 m (5360 − 4800/2 = 280). Applied to the current glacier terminus of the "7110 and 7040 m south-summit glacier" (Fig. 2.79) at 5400 m asl, which also belongs to the catchment area, a ΔELA of 300 m (5400 − 4800/2 = 300) can be calculated. Both of these ΔELAs classify the ice margin as belonging to the Nauri stage (V), ca. 5500–4000 years B.P. (Fig. 2.5).

Likewise during the Neoglacial, 0.9 km ESE of that ice margin, a NE and parallel-running valley glacier terminated at 4765 m asl (Fig. 2.75: Character H: No. 8). In 2007, its active glacier tongue had its terminus at 5240 m asl (30° 19′ 52.19″N 90° 38′ 16.57″E). Thus, a ΔELA of ca. 237.5 m (5240 − 4765/2 = 237.5) exists in comparison with Stage XIV. Accordingly, this ice margin at 4765 m asl also has to be classified as belonging to Stage V.

The matching Holocene age between the two last-mentioned ice margin indicators is further confirmed by the equally well preserved geomorphological state of the terminal moraine ramparts.

The next-younger ice margin of the aforementioned NE and parallel-running valley is situated at 30° 17′ 59.44″N 90° 39′ 13.62″E at 4950 m asl. It indicates a ΔELA of 145 m (5240 − 4950/2 = 145). Accordingly, the ice margin has to be classified as belonging to the older Dhaulagiri stage (VI) from before ca. 4000–2000 years B.P. (henceforth YBP).

A geomorphologically clearly younger ice margin in comparison to Stage V is situated upvalley at 30° 18′ 38.04″N 90° 39′ 24.37″E at 5050 m asl. It indicates a ΔELA of 95 m (5240 − 5050/2 = 95). Accordingly, the ice margin can be classified as belonging to the middle Dhaulagiri stage (‘VII) from before ca. 2000–1700 YBP.

Again upvalley, another seven end moraine ramparts and eight glacier stages that belong to the Historical glacier stages of the past ca. 1700 YBP (see Fig. 2.5) are evident: (1) at

Fig. 2.79 In the Nyainqentanglha Shan in Central Tibet seen from 5430 m asl facing NNE up the "7110 and 7040 m south summit glacier," looking toward the 7040 m peak. The deepest tongue end of the glacier at the time of the picture lay at 30° 20′ 19.90″N 90° 35′ 14.00″E, 5400 m asl (Fig. 2.75: Character H: to the right above No. 2). Analog photo 14/08/1989 by M. Kuhle

30° 18′ 43.26″N 90° 39′ 18.45″E at 5090 m asl, with a ΔELA of 75 m (5240 − 5090/2 = 75), that is, the younger Dhaulagiri stage (VII); (2) at 30° 18′ 50.29″N 90° 39′ 13.66″E at 5110 m asl, with a ΔELA of 65 m (5240 − 5110/2 = 65), that is, Stage VIII; (3) at 30° 19′ 4.00″N 90° 39′ 4.81″E at 5150 m asl, with a ΔELA of 45 m (5240 − 5150/2 = 45), that is, Stage IX; (4) at 30° 19′ 14.59″N 90° 38′ 53.88″E at 5165 m asl, with a ΔELA of 38 m (5240 − 5165/2 = 37.5), that is, Stage X; (5) at 30° 19′ 18.56″N 90° 38′ 49.39″E at 5170 m asl, with a ΔELA of 35 m (5240 − 5170/2 = 35), that is, Stage XI; (6) at 30° 19′ 22.52″N 90° 38′ 44.42″E at 5172 m asl, with a ΔELA of 34 m (5240 − 5172/2 = 34), that is, Stage XII; (7) at 30° 19′ 25.90″N 90° 38′ 41.58″E at 5180 m asl, with a ΔELA of 30 m (5240 − 5180/2 = 30), that is, Stage XIII; and (8) dead ice at 30° 19′ 43.19″N 90° 38′ 26.14″E at 5200 m asl with a ΔELA of 20 m (5240 − 5200/2 = 20), that is, Stage XIV (end of 2007). In 2007, the still active ice margin is located at ca. 5240 m asl at 30° 19′ 52.19″N 90° 38′ 16.57″E (see above).

2.4.4.1 Summary of the Present, Historical to Neoglacial (Holocene) Glaciation of the Nyainqentanglha Shan in Central Tibet

The focus of this section is Fig. 2.1: No. 5. Due to the previous detailed analysis of the Present to Neoglacial glaciation of the SE side of the Nyainquentanglha Shan, the sequence of the glacier stages in High Asia (see Fig. 2.5) has once again been reconstructed and confirmed. All the way up to the exits of the 16 km long valley SW below the ca. 7162 m high Nyainqentanglha Massif, and its 11 km long, NE and parallel-running valley, the glacier stages are partly (in the longer valley) and completely (in the shorter valley) represented. It is not possible to explain the gaps in the longer valley due to the climate. An explanation is not possible because of the complete sequence in the NE-, parallel-running valley. Hence, instead, this incompleteness has to be attributed to the 10 side-valley and hanging glaciers that must have reached down from either side. These side glaciers have very different sizes. Because of that, they have been added to the main valley glacier without any regularity. Due to this topographic variety, an orderly sequence of glacier advances consisting of small steps could not be realized in the longer valley.

In the shorter valley, by contrast, there are only four hanging glaciers. Their influx could only modify their snowline depressions (ΔELAs) during the Historical Stages VIII to XIV if seen in comparison to Fig. 2.5. But they were not able to bring about the complete disruption of stages.

By comparison, the complete omission of stages observed in the longer valley was caused by those 20 tributary glaciers on either side.

Both valley glaciers still reached the valley exit during the Holocene. Accordingly, their tongue termini reached the large, very broad main valley running from NE to SW. This valley forms the boundary of the Nyainquentanglha Shan in the SE. The 16 km long valley glacier reached the valley exit during Stage V. The 11 km long valley glacier already had done so during the earlier Neoglacial Stage VI at ca. 4000–2000 YBP (Fig. 2.5). In either case, their terminal moraines formed during the Holocene and are interlocked with the orographic right Late Glacial till sheet of the large main valley glacier.

2.4.5 The Late Glacial to High Glacial Glaciations of the Nyainquentanglha Shan in Central Tibet

The focus of this section is Fig. 2.1: No. 5. Immediately at the valley exit of the SW 16 km long valley below the over 7000 m high Nyainquentanglha summits, Late Glacial till covers are found on the orographic right flank of the ca. 14–18 km wide main valley (Fig. 2.80). They are found not only there, but also are present at large scale over a distance of more than 100 km in many places (Kuhle 1991: 155–161, 164–166; Photos 24, 29–32, 35; 2013a: 52–55, Photo 24).

Fig. 2.80 Facing WSW at the SE margin of the Nyainqentanglha Shan (viewpoint: Fig. 2.75 to the right of No. 8; 30° 17′ 21.92″N 90° 39′ 53.85″E, 4760 m asl) into the orographic right flank of the very large and broad main valley seen SE of the mountain group downvalley. (■) marks Neoglacial (Holocene) (■ below) and Late Glacial (■ above) ground moraine covers in the side valley exit and at the main valley flank. (☗) shows glacigenic flank-abrasions and glacigenic triangular slopes or truncated spurs (∩), respectively, as these are characteristic keyforms for inland ice areas (e.g., Scandinavia in Europe). (- - - - -) denotes the highest provable glacier level at about 6000 m asl. (□) marks the glacier stream and its glacio-fluvial gravel floor on the ground moraine and underneath. Analog photo 10/08/1989 by M. Kuhle

For instance, Character H in Fig. 2.75 (see also http://extras.springer.com) marks one of three small lakes. They were formed in that orographic right till cover during the process of meltdown by melting blocks of dead ice (Kuhle 1991: 157–159: Photo 31). The lake lies at ca. 5165 m asl and thus ca. 575 m above the thalweg (4590 m asl) of the large main valley. Such obviously young, freshly preserved forms in the loose rock material that can very easily be reshaped solifluidally (i.e., by solifluction), are proof of the minor age of these till covers. Therefore, the decameter-thick till sheet has to be classified as belonging to the late Late Glacial Stage IV or III. WNW above this very thick till the author found a till cover with rounded, erratic boulders at 5270 m asl (Fig. 2.75, see also http://extras.springer.com: Character H: 30° 18′ 31.07″N 90° 37′ 15.17″E). They prove a minimum ice thickness of the main valley glacier of ca. 725 m (Kuhle 1991: 157–158, 164, 229–239, Photo 30; Fig. 43: No. 20).

A further extensive till remnant is marked by No. 9 in Fig. 2.75 (see also http://extras.springer.com: Character H) at 5340 m asl (Fig. 2.80). It was undercut by two orographic right side-valley glaciers and modified into a longitudinal moraine.

The highest glacigenic abrasion forms here reach ca. 5800 m asl (Fig. 2.80; Fig. 2.75, see also http://extras.springer.com: Character H: No. 10: 30° 17′ 7.50″N 90° 33′ 33.19″E and 30° 15′ 9.45″N 90° 33′ 30.27″E). These are classic truncated spurs that have been abraded by flank polishing into glacigenically triangular slopes (Kuhle 1991: 228–230; Fig. 43 between Nos. 13 and 19). As has been shown by the author in detail and on a large scale, High Glacial polish lines between 5800 and 6200 m can be detected. From these ice thicknesses of 1200–1800 m can be derived (ibid.: 155–157, 159, 160–165: Photos 24, 27–29, 39). They must be minimum thicknesses because the main valley bottom shows Late Glacial and Holocene loose rock material across a breadth of 7 km. These are decameter-thick till sheets (Kuhle 1988c: 459; 1991: 154–156, 159, 164, 228–230, Photos 19, 20, 27, 28, 39, 40; Fig. 43: No. 19; 1996a: 148, Abb. 2; 2001b: 5, Fig. 7; 2003: Figs. 3, 78, 82, 84; Sample 17.8.91/1, Fig. 11) as well as post-glacial gravel floors and debris fans (Kuhle 1991: 157, 158, 165: Photos 30 and 31: ×). The latter have evolved since the deglaciation of the main valley (Fig. 2.80).

The reconstructed ice thickness proves an inland-ice-like glacier cover that completely mantled and surrounded the massif of the Nyainqentanglha Shan. This has been confirmed by findings of erratics at the SW spur of the Nyainquentanglha Shan, that is, the mountain slope down to the Tsangpo Valley. The Chalamba La (Fig. 2.81a, see also http://extras.springer.com: Character I: arrow) is located there. This pass (also Shüke or Tschü Tschü La; 5451 m asl; 29° 53′N 90° 07′E) is of special glacio-geological interest. Here, and on the flanking mountain slopes, 100–200 m beyond the pass depression, erratic granite boulders (Fig. 2.81c) are lying on bedrock rhyolithe (Fig. 2.81b) (29° 53′ 48.46″N 90° 8′ 0.05″E). These large rounded boulders that can only be seen on the surrounding hill cupolas and remnants of old faces at corresponding sea levels, prove an Ice Age glacier transfluence across the pass. During two research expeditions the glacial-geomorphological situation of these granite erratics, transported long distances, was investigated in detail, samples were taken, and the results published (Kuhle 1987a: 414–417: Figs. 28–30; 1988c: 458–460: Figs. 2, 4, 5; 1988d: 582, 583; 1988e: 12–14; 1991: 179–181, 228–230; Fig. 43: No. 27, Photo 42; Heydemann and Kuhle 1988: 615–617).[1]

According to the detailed geological map of Wu and Xiao (1991), granites outcrop only 30–40 km further to the S and SW. There the 'Lhasa pluton' (Gansser 1964, Plate 2.1) is located, which has already been referred to in Kuhle (1988c, 1988d and 1991) as the next occurrence of granite. Should the erratic boulders on the Chalamba La come from this area—which is possible but not likely—they must have been transported upvalley. In any case this would prove that it would necessarily have been caused by a relief-overriding glaciation. More probable is that the erratics were transported from the area of the NE Nyainqentanglha Shan, from a distance of ~130 km, by the large main valley glacier situated E to SE of the Nyainqentanglha Shan (Fig. 2.82), coming from Central Tibet (Kuhle 1991: 167). However, in order to

[1]The claim made by Lehmkuhl et al. (2002) that these boulders are granites weathering in situ, which outcrop only a few meters above the pass road, so that the interpretation provided by Kuhle (e.g. 1987a, 1988c, 1991) becomes disproved (Lehmkuhl et al. 2002: 204), is a mistake. The crystalline compact rock outcropping on the pass (see Fig. 2.81a) is not granite—as considered by Lehmkuhl et al.—but rhyolite (see Fig. 2.81b), whereas the boulders deposited on it consist of granite (see Fig. 2.81c). The unambiguously allochthonous origin of these granite boulders in comparison to the bedrock rhyolite has been determined by mineralogical analysis (microsections and radiography) of the two rock types and published (Heydemann and Kuhle 1988: 615, 617; for radiography see Figs. 3 and 4; Kuhle 1987a: 416–417, Figs. 29 and 30; 1988c: 459, 460; see microsections in Figs. 4 and 5). This has been indicated in Kuhle (1991: 167 and Photo 42). Accordingly, erratics transported long distances are the matter of concern here. In addition, the well-rounded shapes of these meter-sized granite boulders proves the long-distance transport (ibid.). In the meantime, this interpretation has been supported by the geological map of Wu and Xiao (1991). Lehmkuhl et al. (2002: 204) maintain that the granite—which in their opinion outcrops on the pass—also occurs in the map of Wu and Xiao (1991). This is problematic for two reasons: first the Chalamba La (see Fig. 2.81a) is situated at 29° 54′ N 90° 10′E and thus ca. 30 km west beyond the section of the geological map compiled by Wu and Xiao. Secondly, the map of Wu and Xiao—as far as it reaches the region of the pass at all—shows a large amount of "volcanic" or "pyroclastic" rocks—as would be the case for rhyolite—but no granite at all.

Fig. 2.81 **a** Satellite image, Google Earth Pro: center 29° 54′ 37.66″N 90° 6′ 14.41″E; http://extras.springer.com: Character I. View is facing east toward the 5451 m asl high Chalamba La (pass); see arrow (29° 53′N 90° 07′E; more exactly, 29° 54′N 90° 10′E). The erratic boulders (granite) on the rhyolite that outcrops across the entire pass area prove a glaciation overriding the entire terrain (see Footnote 1). The small character J, in the background to the left is position J on Fig. 2.83. **b** Thin section (A. Heydemann) of rhyolite bedrock of the Chalambala (Fig. 2.81a arrow) between 5200 and 5600 m asl (sample taken by M. Kuhle, 24/8/1984) on which the erratics of Fig. 2.81c lie. This is a rhyolitic rock with crystals of quartz, albite-oligoclase, mica, and chlorite, which is hydrothermally altered. Its eruptive character is shown by its macro-crystalline features (xenocrysts of twinned feldspar) embedded in an amorphous matrix. Laboratory photograph: A. Heydemann and M. Kuhle; (see Footnote 1). **c** Thin section (A. Heydemann) of one of the numerous, up to meter-sized, rounded, erratic granite boulders (biotite-granite) on the Chalamba La (Fig. 2.81a: arrow) between 5200 and 5450 m asl; samples taken by M. Kuhle 24.8.1984; erratic block that lies on the solid rhyolitic rock of Fig. 2.81b. This is a biotite-granite, i.e., an acidic plutonic rock (light block) of quartz (dark, wavy extinction), biotite, albite-oligoclase, and potash feldspar crystals with a macro-crystalline structure. This structure and its diverse mineral composition (granite with potash feldspar, solid rhyolite rock with chlorite) show that these blocks are glacially transported erratics. Laboratory photograph: A. Heydemann and M. Kuhle (see Footnote 1)

overcome the counter slope of the pass (Fig. 2.81a, see also http://extras.springer.com: Character I: arrow), this would have required an ice thickness of at least ca. 1200 m (Kuhle 1987a, 1988c, d, e, 1991). Further erratics were mapped in locations up to ca. 12 km SW below of the transfluence pass (Kuhle 1991: Fig. 43: Nos. 27–28; Kuhle 1988c: Fig. 2: No. 4).

The High Glacial valley filling with glacier ice immediately east of the Chalamba La transfluence pass and its Late Glacial regression with kame formations, etc. has already been discussed elsewhere (Kuhle 1988c: 458, 459: Fig. 2: No. 3; Kuhle 1991: 161–162, 167, 228–230: Fig. 43: No. 25, Photo 40; 2013a: 56) (Fig. 2.83, see also http://extras.springer.com: Character J).

Fig. 2.81 (continued)

Fig. 2.82 At 30° 24′N 90° 57′E at 4220 (4190 aneroid measurement) m asl on the valley bottom of the ca. 8–20 km wide main valley SE of the Nyainquentanglha Shan, which leads out of Central Tibet (Fig. 2.83: upvalley to the right of B), near the Nyingzhong settlement. (■) mark ground moraine covers on the slopes and on the bottom of the main valley. (⬢) are round-edged and (●) rounded, meter-sized, erratic granite boulders (see person on the left for scale) of this ground moraine cover under which there is metamorphic sedimentary bedrock. (⏶) are glacigenically abraded and thus rounded hills and mountain slopes. (⇓) denotes a Late Glacial subglacial meltwater notch into a glacigenically abraded rock threshold. Analog photo 17/08/1991 by M. Kuhle

Further south, following the large main valley glacier downward 64 km to the Tsangpo River, further glacial glacier traces are evident. First, reconstruction of the maximum extension of ice in the area of the Tibetan Plateau is tied to the indicators of till covers and erratic boulders. With the clearly improving quality of satellite imagery, new possibilities are available to explain still-open questions in a detailed and exact way. One of these questions is the connection of the Central Tibetan glaciation to the Tsangpo Valley by the large main valley under discussion here, located E to SE of the Nyainquentanglha Shan. In 1984 and 1989, the present author reconstructed a relief-overriding glaciation south of the Nyainquentanglha Shan (Fig. 2.83, see also http://extras.springer.com: Character J: No. B), which in the SW flowed down across the Chalamba La pass (5451 m asl; Fig. 2.83: Character J: No. A). This is evidenced by erratic granites (see above). As shown by Fig. 2.83, further deposits of till from this glaciation have been found with a practically still fully intact structure of the surface. They were deposited on and beyond the pass near Nimaguo (4730 m asl; 29° 37′N 90° 21′E) in a SSW direction, and possess a thickness of at least 140 m (at 29° 45′N 90° 22′E) (Fig. 2.83: Character J: No. C). The connected valley that leads down toward the Tsangpo is on both sides mantled up to a height of 5300 m asl by covers of till. They are proof of ice thicknesses of at least 1200 m (Fig. 2.83: Character J: signatures of lines). Here, in exposures on the orographic

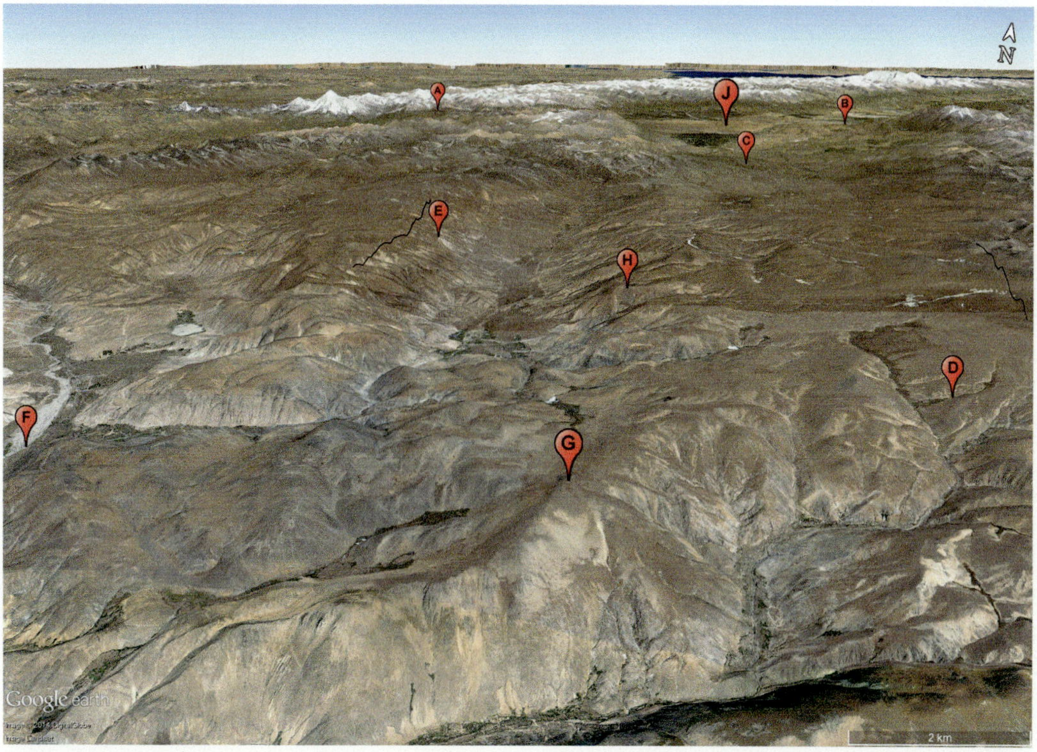

Fig. 2.83 Satellite image, Google Earth Pro: center 29° 26′ 32.13″N 90° 16′ 42.85″E; http://extras.springer.com: Character J. Research area between the southern slope of the Nyainquentanglha Shan and the Tsangpo Valley. In position A (Chalamba La; 5451 m asl; 29° 53′ N 90° 07′E, more exactly 29° 54′N 90° 10′E) erratic granite boulders prove a relief-subordinated glaciation (Kuhle, 1988c, 1991). In position B (southern slope of the Nyainquentanglha Shan), a main valley-filling glaciation of at least 725 m thickness is proved by the presence of moraine deposits with erratic detritus (Kuhle 1991, 2013a: Photo 24). In position C this glaciation continues—provable by ground moraine remnants, (dead-ice holes, and erratic granite boulders at 29° 46′ 00″N 90° 21′ 49″E)—and crosses the pass height in a southerly direction. Here, on both sides of the thalweg, moraine covers are visible as far as above the flank summits (5300 m asl) (black lines; e.g., 29° 31′ 58.00″N 90° 14′ 41.00″E and 29° 37′ 16.00″N 90° 23′ 42.00″E in the Nimaguo La area, with granite erratics). Exposures can be found, for example, in position D (see detail in Fig. 2.84), where light-colored, large boulders, embedded in ground matrix, can be seen. These boulders, too, can be seen in position G. The ground moraine covers are preserved over the entire area and only the latest are eroded by slides (positions E and H: 29° 29′ 41″N 90° 17′ 58″E). Position F marks an immediate side-valley junction with the Tsangpo Valley, 40 m above the Tsangpo River

left valley flank at 4836 m asl, light-colored, up to 4.5 m sized boulders are shown. Probably they consist of the same granite as on the Chalamba La. In addition, however, they are embedded here in a yellow fine-material matrix (Fig. 2.84; Fig. 2.83: Character J: No. D). This ice stream E to SE past the Nyainqentanglha Shan must have reached the upper Tsangpo River. It proves a ΔELA of ~1200 m. The mantling with till is almost completely preserved. Only in some places is it weakened by slides whose materials still have sharp fracture edges (Fig. 2.83: Character J: Nos. E and H), i.e., they are very young. Accordingly, there is no other geomorphological possibility than to classify loose rock that is rich in fine-material matrix—and which has hardly been reworked—as belonging to the Last Ice Age (LGP = Stage 0 in Fig. 2.5 = MIS 2) or even to the Late Ice Age (Stage I).

From the high summits of the Nyainqentanglha Shan the local glaciers flowed down to the inland ice on both sides of the mountain crest.

Fig. 2.84 Satellite image, Google Earth Pro: Center 29° 26′ 1.82″N 90° 22′ 26.05″E. Ground moraine exposure at 4836 m asl. The exposure is situated in Fig. 2.83 in position D. Visible here are light, erratic granite boulders, probably of the same sort that have been found at the Chalamba La (see Fig. 2.81c). Here they are embedded in a yellowish-grey fine material matrix, underlain by dark bedrock. This is exposed in the thalweg below

For the SE side, the local glaciation of the 16 km long valley and its NE, parallel-running valley within the catchment area of the 7162 m high Nyainqentanglha main summit has been chosen as an example (see Sects. 2.4.4 and 2.4.4.1; Fig. 2.75, see also http://extras.springer.com: Character H: Nos. 1–8).

In addition, however, a continuous ice sheet surface also existed between the SE and NW slopes of the Nyainqentanglha. A continuous ice surface without steps is the glacier-typological characteristic of an ice stream network. This applied, for instance, to the region between the large main valley situated E to SE of the Nyainquentanglha Shan and the lake basin of Nam Co (Tengri Nor) in the NW. Thus, a continuous glacier-ice surface like this existed between the till locality of Fig. 2.82 at ca. 30° 24′N 90° 57′E and the Nam Co basin. Here, the LGP glacier from the NW overflowed a transfluence pass and followed a trough-shaped valley down to the SE (Fig. 2.85). The abraded transfluence pass is situated at 5200–5300 (aneroid measurement) m asl at 30° 40′ 52.79″N 91° 5′ 49.90″E (Fig. 2.86). Across this transfluence pass the glacier overflowed the watershed between the Nam Co basin and the SE side of the Nyainqentanglha Shan (Figs. 2.206 and 2.207). This must have happened in a direction from the 500 m higher area in the NW toward the SE (Kuhle 1991: 155–159, 163, 164, Photos 25 and 26).

Further proof of the ice flow can be demonstrated by erratic boulders that are situated on the

Fig. 2.85 View from 5350 (5500 aneroid measurement) m asl from a mountain cupola in the Nyainquentanglha Shan at 30° 40′ 44.38″N 91° 5′ 32.90″E (see Fig. 2.83 to the right above B) of a classic glacigenic trough valley (∪), facing SSE downwards to the ca. 8–20 km wide main valley SSE of the Nyainquentanglha Shan, leading down out of Central Tibet. (▲) marks glacigenically abraded and thus rounded mountain ridges up to at least 5560 m in height. (**O**) show NNW-facing cirques on mountains up to 5450 m in elevation. (■) indicate Neoglacial (Holocene: Stage V, Fig. 2.5) ground- and end moraines at the bottoms of these cirques. (- - - -) denotes the High to Late Glacial glacier level of the Tibetan inland ice that has flowed down to the right. Analog photo 03/11/1989 by M. Kuhle

transfluence pass (Laken La). Up to 3 m long, erratic, light-colored gneiss or granite boulders on dark bedrock crystalline schists (phyllites) can be found here. The slopes above the pass depression change from those phyllites to reddish sandstones (ibid.: Photo 26 at the far right) without outcropping gneiss or granite being present. Accordingly, the gneiss/granite boulders are erratic. They were transported here from great distance. Numerous of these boulders are rounded. The medium to smaller boulders are mainly round-edged.

Similar to this valley glacier, there also existed additional, parallel valley glaciers that flowed diagonally to the orientation of the mountain group. They produced a glacier contact between the two inland ice complexes NW and SE of the Nyainquentanglha. The mountain ridges, here about 5400–5700 m in height, were abraded glacigenically. This happened during the glacial period (LGP, Stage 0, Fig. 2.5) when the ice level was located at 5800–6000 m asl or higher. Naturally, during the Late Glacial, due to the lowering of the ice level, a certain glacigenic

Fig. 2.86 From 5350 (5500 aneroid measurement) m asl from a mountain cupola in the Nyainquentanglha Shan at 30° 40′ 44.38″N 91° 5′ 32.90″E facing ESE to the eastern trough-valley flank across the transfluence pass from the trough valley in the south (Fig. 2.85) to the Nam Tso lake basin (see Fig. 2.83 on the right above B) in the north. The abraded transfluence pass is situated at 5200 (5300 aneroid measurement) m in elevation at 30° 40′ 52.79″N 91° 5′ 49.90″E. (▲) marks a glacigenically abraded valley flank, which up to its culmination at an elevation of ca. 5520 m is dissected by exaration rills. (■) indicates a Neoglacial (Holocene: Stages V and VI, Fig. 2.5) end moraine ("Satzendmoräne") at the bottom of a small, interglacial cirque. (- - - -) denotes the High to Late Glacial glacier level of the Tibetan inland ice sheet. Analog photo 03/11/1989 by M. Kuhle

sharpening occurred due to the increasing dominance of glacigenic flank polishing in the valleys. In accordance with this, the ice thickness in the Nam Co basin amounted to at least 1000 m.

2.4.5.1 The Late to High Glacial Glaciation of the Nyainqentanglha Shan and the Nam Co Basin in Central Tibet

The focus of this section is Fig. 2.1: No. 5. Thus, the ice cover of the Nam Co area, too, bears evidence of glaciations in the form of abraded hills and ground moraines (Kuhle 1991: 157–159, Photo 26). Till sheets are found not only on the southern shore of the lake, i.e., on the north slope of the Nyainqentanglha Shan. Rather they also continue in the glacigenic landscapes north and east of the Nam Co. Similar to the southern shore, where Holocene glaciers from the Nyainqentanglha Shan built up till sheets such as these, Holocene glaciers flowed down from the Samdain Kangsam Massif located east of the lake (30° 50′ 11.29″N 91° 29′ 21.36″E) to nearly the lake level. Thus the large NW glacier of this massif in the Holocene (Neoglacial) flowed down to ca. 4800 m asl. There it left behind three geomorphologically very fresh terminal moraine arches at ca. 31° 3′ 1.08″N 91° 21′ 51.21″E, 31° 1′ 1.96″N 91° 21′ 47.84″E, and 31° 0′ 34.68″N 91° 22′ 17.65″E. The nearest current glacier terminus is at 5400 m asl. From

Fig. 2.87 From the Nyainquentanglha Shan at ca. 5500 (aneroid measurement) m asl at ca. 30° 40′ 44.38″ N 91° 5′ 32.90″E facing WNW looking into the Nam Tso lake basin (see Fig. 2.83 on the right above B; Fig. 2.88: Character K). (▲) denote glacigenically abraded mountain ridges and hills (glacial knobs) on sedimentary bedrock with vertically standing layers this side of the lake in the middle foreground, towering up to 5400 m asl, and 20 km long glacially streamlined hills beyond the lake. (- - - -) denotes the High to Late Glacial glacier level of the Tibetan inland ice sheet. Analog photo 03/11/1989 by M. Kuhle

this follows a ΔELA of 300 m for this ice margin at 4800 m ((5400 − 4800)/2). This ΔELA classifies the ice margin as belonging to the Nauri stage (V, see Fig. 2.5). The previous ice filling, i.e., of the whole Nam Co basin, was not only nourished by the Nyainquentanglha but also the 6530 m high Samdain Kangsam Massif with its still currently strongly glaciated catchment areas.

North and NE of the Nam Co are rock ridges rounded up to 5400 m by glacigenic ground polishing. They tower ca. 700 m above the current lake level, which is located at ca. 4730 (4718) m asl. These round hills in part are covered by till sheets. The 52 occurrences of till, enumerated below, have been traced 100 km away from the northern lake shore farther to the north. These till covers are located, for example, at: 30° 56′ 9.42″N 90° 58′ 16.88″E at 4960 m asl (Figs. 2.87 and 2.205); at 30° 56′ 27.17″N 90° 57′ 38.19″E at 4870 m asl (Fig. 2.87); at 31° 0′ 10.88″N 91° 1′ 55.57″E at 4860 m asl; at 31° 0′ 40.21″N 91° 7′ 45.52″E at ca. 5060–5130 m asl; at 30° 59′ 13.14″N 91° 10′ 24.38″E at 5910 m asl; at 31° 1′ 43.82″N 91° 0′ 30.42″E at 4880 m asl; at 31° 1′ 35.93″N 91° 0′ 11.63″E at 4930 m asl; at 31° 4′ 25.78″N 91° 1′ 7.86″E at 5130 m asl; at 31° 5′ 59.99″N 91° 2′ 48.04″E at 4970 m asl; at 31° 6′ 49.50″N 91° 3′ 26.22″E at 4900 m asl; at 31° 7′ 46.45″N 91° 4′ 15.71″E at 4900 m asl; at 31° 5′ 29.57″N 91° 5′ 14.83″E at 4930 m asl; at 31° 7′ 46.34″N 91° 4′ 15.50″E at 4880 m asl; at 30° 59′ 12.94″N 91°10′ 24.24″E at 4920 m asl; at 30°58′ 24.08″N 91° 10′ 14.3″E at 5140 m asl; at 31°2′ 20.07″N

91° 15′ 40.68″E at 5000 m asl (Fig. 2.207); at 31° 4′ 12.31″N 91° 23′ 3.02″E at 4910 m asl (Fig. 2.207); at 31° 4′ 10.88″N 91° 24′ 12.25″E at 5130 m asl (Fig. 2.207); at 30° 49′ 51.85″N 90° 34′ 44.60″E at 4900 m asl (Fig. 2.87); at 30° 51′ 1.53″N 90° 36′ 57.19″E at 4870 m asl; at 30° 56′ 42.66″N 90° 39′ 40.02″E at 4970 m asl (Fig. 2.87); at 30° 58′ 18.48″N 90°39′ 21.01″ E at 4900 m asl; at 30° 58′ 12.22″N 90° 42′ 17.25″E at 5040 m asl (Fig. 2.87); at 30° 59′ 26.33″N 90° 41′ 13.04″E at 5220 m asl (Fig. 2.87); at 31° 2′ 24.17″N 90° 43′ 31.78″E at 5070 m asl (see Sect. 2.9.2); at 31° 2′ 46.84″ N 90° 43′ 56.24″E at 5010 m asl (see Sect. 2.9.2); at 31° 15′ 48.86″N 90° 49′ 36.43″E at 4950 m asl (Fig. 2.203); at 31° 35′ 10.59″N 90° 33′ 58.64″E at 4900 m asl (see Sect. 2.9.2; Fig. 2.204); at 31° 3′ 34.94″N 90° 32′ 42.22″E at 4930 m asl (see Sect. 2.9.2; Figs. 2.202 and 2.203); at 31° 1′ 38.92″N 90° 16′ 38.13″E at 5170 m asl (see Sect. 2.9.2); at 30° 55′ 1.32″N 90° 20′ 39.99″E at 4820 m asl; at 30°56′ 27.23″ N 90° 16′ 48.57″E at 4820 m asl (see Sect. 2.9.2); at 30° 51′ 56.39″N 90°14′ 14.69″E at 4840 m asl; at 30° 52′ 45.78″N 90°13′ 56.12″ E at 4890 m asl (see Sect. 2.9.2); at 31° 6′ 24.80″N 90° 7′ 45.30″E at 5120 m asl (Figs. 2.202 and 2.203); at 31° 13′ 48.87″N 90° 7′ 50.24″E at 5220 m asl (Fig. 2.202); at 31° 21′ 10.18″N 90° 11.50.67″E at 4690 m asl (Figs. 2.200 and 2.201; Sects. 2.9.1 and 2.9.2); at 31° 23′ 11.38″N 90° 12′ 20.58″E at 4800 m asl (Fig. 2.201; Sect. 2.9.2); at 31° 28′ 46.56″N 90° 11′ 32.89″E at 4740 m asl (Sects. 2.9.1 and 2.9.2); at 31° 29′ 25.19″N 90° 11′ 29.91″E at 4770 m asl (Sects. 2.9.1 and 2.9.2); at 31° 31′ 4.52″N 90° 9′ 51.88″E at 4800 m asl (Sects. 2.9.1 and 2.9.2); at 31° 31′ 2.13″N 90° 9′ 15.36″E at 4750 m asl (Sects. 2.9.1 and 2.9.2); at 31° 34′ 29.41″N 90° 3′ 28.02″E at 4790 m asl (Sect. 2.9.1); at 31° 35′ 36.27″N 90° 4′ 11.75″E at 4810 m asl (Sects. 2.9.1 and 2.9.2); at 31° 38′ 29.07″N 90° 3′ 48.05″E at 4910 m asl (Sects. 2.9.1 and 2.9.2); at 31° 40′ 4.39″N 90° 3′ 27.93″E at 4800 m asl (Sects. 2.9.1 and 2.9.2); at 31° 40′ 25.97″N 90° 3′ 23.96″E at 4820 m asl (Sects. 2.9.1 and 2.9.2); at 31° 41′ 31.08″N 90° 3′ 13.36″E at 4900 m asl (Sect. 2.9.1; Fig. 2.200); at 31° 42′ 39.57″N 90° 2′ 23.18″E in 4780 m asl (Sects. 2.9.1 and 2.9.2); at 31° 45′ 35.47″N 89° 58′ 55.64″E in 4660 m asl (Sect. 2.9.1; Fig. 2.200); and finally, for instance here again with erratic material, at 31° 46′ 12.19″N 89° 58′ 58.38″E (Sect. 2.9.1) at 4760 m asl. The numerous further occurrences of till covers will be discussed later (see Sect. 2.9.2).

At several locations (e.g., at 30° 55′ 41.01″N 90° 59′ 53.48″E and 30° 55′ 50.56″N 90° 58′ 28.63″E) the shoreline of the Nam Co undercuts this hilly landscape, which is characterized by soft rounded forms and flat slopes, forming fresh, steep cliffs 10–80 m in height (Figs. 2.87 and 2.206). These cliffs provide proof of the recent modification of the past glacigenic landscape due to infilling of the lake (see below).

This finding is confirmed by a ΔELA of only ca. 500 m that is necessary to fill the lake basin of nearly four times the size of the Bodensee with glacier ice (Figs. 2.87 and 2.206). The lowest actual glacier margins of the Nyainquentanglha-north slope and Samdain Kangsam-west slope are situated at about 5400–5540 m asl. In order to reach the actual lake level of the Nam Co, accordingly, a decline of the ELA of no more than 341–411 m is necessary ($(5400 - 4718)/2 = 341$; i.e., $(5540 - 4718)/2 = 411$). At a maximum depth of the lake of 125 m, a maximum ΔELA of another 63 ($473.5 - 411 = 62.5$) m ($(5540 - 4593)/2 = 473.5$) is necessary to glaciate the bottom of the Nam Co lake basin. Accordingly, the aforementioned ΔELA of maximally 500 m corresponds well with the empirical conditions. As a tendency, the ΔELA of 500 m is probably overestimated, because in regard to the baseline for a piedmont glaciation in the Nam Co basin one would more realistically have to calculate based on the *medium* lake depth, not the maximum depth of the lake.

This therefore rather generously calculated ΔELA must be classified as belonging to the last stage of the Late Glacial, Stage IV (Sirkung stage, Fig. 2.5), but no later than that.

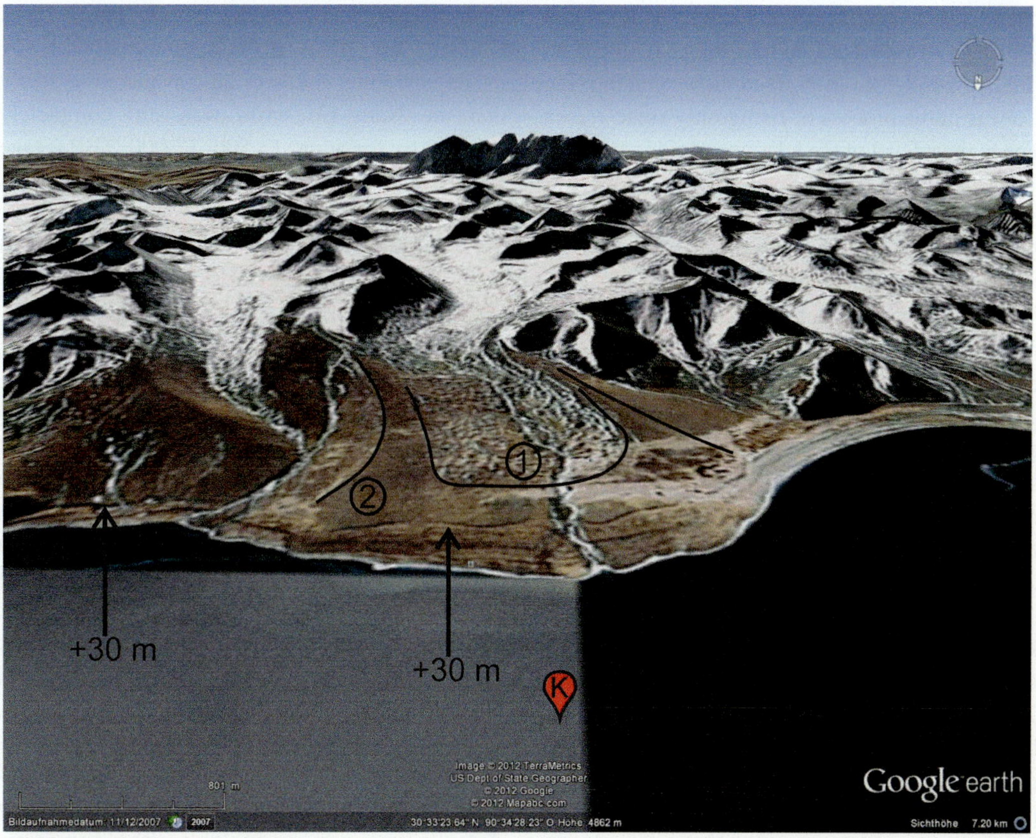

Fig. 2.88 Satellite image, Google Earth Pro: center 30° 34′ 46.94″N 90° 34′ 14.78″E; http://extras.springer.com: Character K. North slope of the Nyainquentanglha Shan on the south shore of Nam Co; in the background rise the four main summits of the massif, each exceeding 7000 m in elevation (max. 7120 m) (cf. Fig. 2.75: Satellite image Google Earth Pro: center 30° 19′ 22.79″ N 90° 35′ 37.51″E; http://extras.springer.com: Character H; Nos. 1–10). The younger glacier stage ① (older Dhaulagiri Stage VI; Holocene, Fig. 2.5) corresponds with a ΔELA of 380 m, the older Stage ② (Nauri Stage V; Holocene, Fig. 2.5) to a ΔELA of ca. 410–420 m. The Holocene high level of the lake of +30 m had been reached, according to the datings of Mügler et al. (2010), at <7.2 ka

Nevertheless, this ΔELA of 500 m was still surpassed by the one from Stage IV by ca. 200 m. This means that a ΔELA of at most 500 m cannot be dated back further than into the late Late Glacial. Therefore the present author reasons that even still during this late Late Glacial stage (IV), the Nam Co basin must have been covered completely by glacier ice.

Interestingly, this glacier reconstruction and its chronological classification receives further confirmation from a neighboring discipline (paleolimnology): the ^{14}C-dating of a drill core obtained at a water depth of 30 m by Mügler et al. (2010) established a maximum age of the Nam Co (lake) of 7.2 ka (BP). This shows that the Nam Co was in the process of rising only since 7.2 ka. The highest level of the lake on record is at +30 m above the current lake level (Fig. 2.88, see also http://extras.springer.com: Character K). Accordingly, at 7.2 ka BP, its level was still 60 m below the highest level. This can only have occurred with some delay, i.e., clearly later. The highest shoreline has been reshaped by the glaciofluvial gravel floor (glacier mouth gravel floor, gravel field, sandr No. 1; see Fig. 2.5) due to the glacier advance from Stage VI ("older Dhaulagiri stage"). The lowest ice margin of this stage reached 760 m lower (Fig. 2.88, see also http://extras.springer.com: Character K ①) than the current ice margin (ΔELA = 380 m).

According to the glacier chronology by Kuhle, this probably occurred before 2000–4000 YBP; definitely, however, clearly later than 7200 YBP (see above). The dating of the lake bottom thus confirms the very young age of the glacier advance being discussed. In a geomorphologically similarly well preserved state as the forms of this middle to young Holocene advance (Stage VI, Fig. 2.5) are the next-older traces of ice margin positions in the same glacier forefield (Fig. 2.88, see also http://extras.springer.com: Character K ②). They belong to a ΔELA of ca. 410–420 m and are undercut by the +30 m highest position of the lake level (Fig. 2.88). Accordingly, they are older than the highest position of the Nam Co; due to their state of preservation, however, they must be younger than 7200 YBP. This means that the stage referred to here (② in Fig. 2.88) is the Neoglacial (middle Holocene) Stage V (see Fig. 2.5). Please compare the similarly fresh Holocene advances of the NW glacier of the Samdain Kangsam Massif described above, down to ca. 4800 m asl, which is located close nearby.

At the same time this ^{14}C-dating shows that the Nam Co is a young Holocene lake that still did not exist during the late Late Glacial (Stage IV). Thus its existence does not contradict the empirical results of a preceding complete Tibetan inland glaciation that covered the entire Nam Co basin (Kuhle 1987a, 1988c, d, e, 1991, 2004: Glacier Map UH13 and UJ11).[2]

To conclude, it has to be considered as likely that the Nam Co basin was still completely covered by ice at ca. 13.500–13.000 YBP (Stage IV, see Fig. 2.5).

2.4.5.1.1 Summary of the Historical to High Glacial Glaciations of the Nyainquentanglha Shan and the Nam Co Basin in Central Tibet

This section focuses on Fig. 2.1: No. 5. The corresponding valley-filling glacier discharge through the large valley concerned, situated E to SE of the Nyainqentanglha Shan, is again evidenced by till sheets and erratics. On the ca. 7 km wide valley bottom (30° 19′ 36.70″N 90° 50′ 7.67″E, 4190–4290 m asl) (Kuhle 1988c, 1991, 1995, 1996a) as well as on the orographic right valley slope (30° 18′ 31.07″N 90° 37′ 15.17″E, 5270–5330 m asl), ca. 650–1000 m above the valley bottom, erratic boulders have been found. These meter-sized, partly rounded, boulders, definitely transported over long distances, prove—in addition to other key forms such as till covers with polymictic or striated glacial drift (Kuhle 1988c, 1991), flank abrasion- and trough valley forms, and relicts of a Late Glacial

[2]Lehmkuhl et al. (2002) contradict Kuhle's position and maintain that the Nam Co (lake) basin was free of ice during the last glacial maximum (LGM) and pre-last Ice Age (Mid-Pleistocene/Riss), i.e., since over 132 ka BP (ibid.: Table 4). They date the age of the advance of the small extant hanging valley glacier under discussion, down to about 760 m asl (ibid.: M1a) and thus the ΔELA of ca. 380 m (see above in the main text), to an age of 32–13 ka BP (Würm, LGM). This is refuted by the ^{14}C-lake dating of 7.2 ka BP. The next-older advance of this small glacier, which reaches the lake only marginally and points to an ELA depression of 410–420 m, is classified by these authors as being older than 132 ka BP and the +30 m stage of the lake level as having an age of 40–32 ka BP (ibid.: Table 4). Both of these assessments have been disproven. According to Kuhle's glacial chronology for High Asia, both stages (M1a and M2) are younger than 5500 YBP, i.e., they are Neoglacial advances of Stage V. Like other small glaciers they marginally reached the lake basin. Yet, they still were at least 320 m, i.e., 290–280 m of ELA depressions away from its Late Glacial ice cover (see above in the main text).

ablation landscape such as kames (Kuhle 1988c, 1991, 2013a)—a complete glaciation of the main valley (Kuhle 1991) as well as an inevitably complete ice filling of all its side valleys (ibid.). During the LGM the ice level lay at ca. 5800–6200 m asl (ibid.), so that the cross-valleys between the SE main valley concerned, situated at that time in the direction of the also ice-filled current Nam Co (Tengri Nor) lake basin NW of the Nyainqentanglha, were continuously filled with glacier ice across transfluence passes and the local water divide (Kuhle 2004: Glacier Map UH13 and UJ11). The large, ca. 14–18 km wide main valley glacier lying E to SE of the Nyainqentanglha Shan and coming from Central Tibet from the Ice Sheet center I2 (Kuhle 1985a: Figs. 2 and 3; 1985c; 1986b; 1987a: Fig. 27; 1988d: Fig. 2; 1988e: 13, Fig. 27; 1991: 167) reached the Tsangpo River in the south at ca. 3700 (3690) asl.

The local, and by comparison small, extant glaciers of the cross-valleys of all slopes of the Nyainqentanglha Shan show Holocene (Neoglacial) descents of glacier tongue termini, which lie ca. 760–840 m above them (Kuhle 1991). The accompanying fresh moraine forms and ELA depressions (ΔELA) of ca. 380–420 m categorize the advances as belonging to the Neoglacial Stages V–VI (Nauri- to “older Dhaulagiri stage”: ca. 2000–5500 YBP (Kuhle 1986a: 454, Table 4, 455). All minor advances (Stages 'VII–XII; see Fig. 2.5), also proven by till covers, occurred during the late Neoglacial and Historical periods.

2.4.6 The Present, Historical, Neoglacial, and Late to High Glacial Glaciations in the Triangle Between Chongce Shan in the West and Kukushili as Well as Ulugh Muztagh in the East to NE as Well as Zangsar Kangri in the ESE (Central Tibet)

The currently still glaciated massif of the Kukushili (6340 m) lies 360 km east of the Chongce Shan in West-Central Tibet (Sect. 2.4.1). In its eastern foreland, there exists a Neoglacial till cover (at 35° 41′ 24″N 85° 58′ 02″E as well as north and east of it; 35° 38′ 52″N 85° 59′ 19″E and S) (Kuhle 2013a: 155, 156, Fig. 48). It is located between 5200 and 5080 m asl and has the same appearance as the cover in the southern foreland of Chongce Shan I. In both cases, we are dealing with the same ΔELA of 150–210 m and till sheets belonging to Stages VI–V (Dhaulagiri and Nauri stages, Fig. 2.5). Their surfaces are covered by a pattern of a few decameters- up to more than kilometers-long dead ice kettles. They exhibit a filling with water or they are already dry. The till sheets are muddy in the active layer of the permafrost. In places they also show tundra polygons and, moreover, are compressed and striped by glacigenic molding. During this period of the middle to younger Holocene, the glaciation of the Kukushili still had a large-scale extension onto the Tibetan highland, far into the forelands of this massif.

At a ΔELA of 300 m, which has to be classified as belonging to the maximum extension of the Holocene (Neoglacial) Nauri stage (V), the ice must have even united with adjacent glacier faces. These had been built up by neighboring massifs that also rise above the plateau region. This union can be explained by two aspects:

1. by means of the equally positioned and well-preserved geomorphological condition of the till covers between the two neighboring massifs, and
2. due to a corresponding ΔELA that led to their Neoglacial ice cover.

Point 1 can be confirmed, among others, between Kukushili and the Ulugh Muztagh (6945 m) located 150 km to the ENE, namely at 35° 55′ 46.86″N 86° 33′ 0.97″E (5000 m asl, Fig. 2.89, see also http://extras.springer.com: Character L: No. 1) and 35° 57′ 7.18″N 87° 3′ 37.87E (4850 m asl). In between these two locations, the massifs are connected to one another by a continuous till cover.

Fig. 2.89 Satellite image, Google Earth Pro: center 35° 50′ 25.87″N 86° 27′ 17.83″E; http://extras.springer.com: Character L: No. 1. Character L shows at 35° 46′ 49.37″N 86° 22′ 2.12″E a ground moraine plateau at ca. 5230 m asl; it consists of ca. 80 m thick ground moraine. No. 1 marks at 35° 55′ 46.86″N 86° 33′ 0.97″E Holocene ground moraine of the Nauri stage (Stage V, see Fig. 2.5) at ca. 5000 m asl

Point 2 is confirmed by a ΔELA of ca. 300 m, necessary for the development of this connected glacier cover. Thus, the lowest area between these two massifs is situated at 4800 m asl and the nearer, current ice margin on the south side of Ulugh Muztagh at 5300 m asl. From this one can calculate a necessary ΔELA of 250 m ((5300 − 4800)/2). On the NE side of the opposite-lying Kukushili, the difference in height between 4800 m asl and the lowest current glacier margin is 640 m, so that an ΔELA of 320 m can be calculated ((5440 − 4800)/2). If one takes both values into consideration, one derives an average ΔELA of 285 m ((250 + 320)/2). Accordingly, this glacier cover has to be classified as belonging to the middle Pleistocene Stage V (Nauri stage, Fig. 2.5).

These ΔELA calculations with regard to the actual past glaciation are only very cautious approximate values that by far understate the real conditions. Two factors support the build-up of glaciers between those two massifs:

1. the self-enhancing rise of the ice surface when it flowed together from two sides in the field depression in between, and
2. the high ridges north and south of that flat basin. They show altitudes of 5200–5800 m asl and even at insignificant ΔELAs they already reached above the snowline. Thus the nourishing area would suddenly increase dramatically.

Between the Neoglacial till covers under discussion, naturally also older, not completely modified till sheets (Stage IV, Fig. 2.5) are recognizable. An example is provided by the up to 80 m thick till cover in Fig. 2.89 (see also http://extras.springer.com: Character L; 5230 m asl). Its margins are displaced by spring-line erosion in the shape of a cuesta (Kuhle 2013a: 155–156, Fig. 48).

A further example and 120 km wide study area is located between Kukushili and the Zangsar Kangri Massif (6496 m, 34° 23′N 85° 50′E), which is adjacent to the south. This area, too, is covered by till sheets of the type defined above (35° 20′ 5.93″N 85° 56′ 37.57″E, 4950 m asl; 34° 54′ 50.37″N 86° 0′ 16.80″E, 4980 m asl; 34° 52′ 1.54″N 85° 54′ 32.53″E, 5050 m asl). It has already been described in its lowest region at between 4800 and 4950 m asl, between the Ma'er Gaichaka lake and locations to the west of it, as well as the Zangsar Kangri, with their kilometer-long and up to 100 m high longitudinal moraines (Kuhle 2013a: 150–152, Figs. 44 and 45). If the Ma'er Gaichaka is identified as a tongue basin lake of a Zangsar Kangri glacier tongue, then a ΔELA of ca. 330 m is calculated for this ice margin at 4800 (4798) m asl (ibid.: Fig. 44). This glacier tongue would therefore also have to be classified as belonging to Stage V (Nauri stage, Fig. 2.5). Due to this ΔELA, too, the accompanying till cover expanding further to the west cannot be classified as older than middle Holocene in age.

At this point, the typical features of till surfaces on the western section of this N-S profile shall be discussed in more detail (Fig. 2.90, see also http://extras.springer.com: Character M).

In the same classical way, the surface of the till sheets here is also interspersed with dead-ice holes. The latter, in turn, are textured by small rills that have developed since the deglaciation. Here, too, is realized the usual picture of nearly unreshaped, Historical (e.g., Chongce Shan I-south side) to Neoglacial (Holocene) surfaces of till covers as they were analyzed and identified above by the method of comparison.

On the immediate slopes of Zangsar Kangri and Purog II Kangri (6418 m), between the mountain foreland (i.e., the Tibetan Plateau) and the current glacier termini, the same Late Neoglacial (Late Holocene) and Historical ice margins are preserved as described in detail for the massifs of Chongce Shan (Sect. 2.4.1), Muggargoibo (Sect. 2.4.3), and Nyainqentanglha (Sect. 2.4.4). Their correspondence with lake level stages of the Holocene Buruocuo lake confirms the chronological classification of the Holocene reconstructed by means of ΔELAs (Kuhle 2013a: 146–150). Today the Zangsar-West glacier reaches down to ca. 5352 m asl (ibid.: Figs. 39 and 40: A). Its Historical and Neoglacial stages with ΔELAs of 14.5 m (Stage XII; cf. Fig. 2.5), 53 m (Stage VIII), and 90 m (Stage 'VII), respectively, are evident (ibid.: Fig. 40: Nos. 1–3). Lying below these levels, there was and still is the Buruocuo lake. The next older glacier advance (ibid.: Fig. 40: No. 4) corresponded with Stage VI ("older Dhaulagiri stage" according to Fig. 2.5). It calved with a ca. 2 km wide glacier front into the lake, which was during and/or after the Holocene maximum lake level of +100 m (Kuhle 2013a: Fig. 43) in comparison with the current lake level.[3]

With a Neoglacial ice margin position, the glacier tongue flowing down from the Purog II Kangri to the SE (ibid.: 150 and Figs. 39 and 43: E) had realized an ΔELA of 180 m (Stage VI). The outlet glacier on the SE slope of the Zangsar (ibid.: Figs. 39 and 41: C), below the current glacier tongue, also shows the four stages mentioned and in addition a further fifth ice margin position, i.e., moraine generation (ibid.: Fig. 41). Stage 'VII of this glacier already reached the foreland and thus the level of the main plateau region. During this stage, the adjacent glacier termini of the Zangsar Kangri formed a piedmont glaciation with a joint ice margin (ibid.: Figs. 39 and 41: No. 3). This ice margin confirms

[3]In the meantime, a rather coherent picture has emerged of an early Holocene period of high lake levels that is probable for the entire plateau area (see Mügler et al. 2010: 626, 643). The authors cite as reliable data, obtained from complete profiles, 9.5 ka BP from Pangong Co in West Tibet (Gasse et al. 1996), 6–5.5 ka BP from Sumxi Co (Avouac et al. 1996), and 7.2 ka BP from Nam Co in middle Central Tibet (Mügler et al. 2010) (see also Sect. 2.4.5.1). Other datings have led to incorrect results with ages exaggerated by up to sevenfold, because the hard water effect was not taken into consideration; see e.g. Li and Zhu (2001).

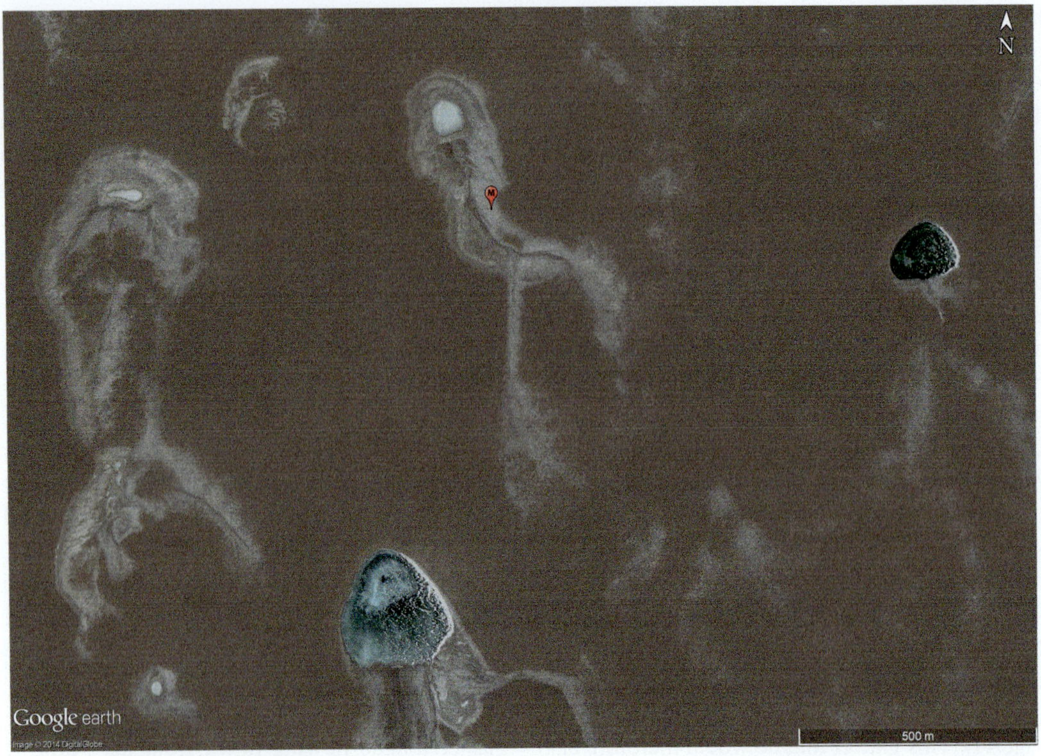

Fig. 2.90 Satellite image, Google Earth Pro: center 35° 6′ 52.48″N 85° 17′ 40.57″E; http://extras.springer.com: Character M denotes dead-ice holes with muddy ponds of past dead-ice blocks

a ΔELA of 93 (ca. 90) m in comparison with the lowest current ice margin. On this side of the mountain foreland, too, there follows an older Stage VI (ibid.: Fig. 41: No. 4). In the still more distant SSE foreland of the Zangsar there is yet another intact area of a till sheet (see 34° 03′ 00″ N 86° 12′ 00″E: till surfaces on a large scale with dead-ice holes, and 34° 02′ 49″N 85° 50′ 18″E; as well as ibid.: 147–149, Fig. 39G; Fig. 41: No. 5; Fig. 42).

It attests to a ΔELA of at least 260 m. Therefore the till cover of this stage has to be classified as belonging to the middle Holocene Stage V (Nauri stage, Fig. 2.5). Here, too, the state of preservation indicates its belonging to Stage V: it corresponds with the young to youngest till sheets in the forefields of the current glacier and also with the Chongtse Shan I south slope (Sect. 2.4.1), Kukushili east slope, and Ulugh Muztagh south slope (see above; cf. ibid.: 149).

Conditions in further support of this claim are found immediately ESE of the Zangsar in the southern foreland of the also currently glaciated 6090 m high massif (34° 8′ 26.54″N 86° 36′ 58.36″E). Here, in the depressions at about 5000 m asl, a corresponding manifestation of the till covers has been preserved. These till sheets enclose a 5300 m high mountain ridge. On it there are deposits of till lying up to 400 m above the Garkung Caka lake. They contain light-colored, meter-sized boulders (e.g. at 33° 54′ 00″N 86° 32′E, 33° 54′ 36″N 86° 34′ 13″E and 33° 55′ 34″N 86° 32′ 54″E).

On the north flank of the Purog II Kangri the current glaciers flow down to 5480–5470 m asl. At the ΔELA of 180 m (Stage VI) ascertained for this massif, the ice margin would have reached down to ca. 5120 m asl (= 5480 − (2 × 180) = 360). The level of 5120 m asl is only reached ca. 41 km away from the current glaciers, i.e., far beyond the northern mountain foreland. Up to that point the ice margin was shifted at the aforementioned ΔELA of 180 m.

Due to this side consideration, it is probable—once again by the calculation of snowline depressions—that the 120 km wide plateau area between Kukushili and Zangsar Purog-II-Kangri was covered by glaciers at a ΔELA of ca. 300 m. The ΔELA of ca. 300 m belongs to the middle Holocene Stage V (Nauri stage, Fig. 2.5).

2.4.6.1 The Late Glacial to High Glacial Glaciations in the Triangle Between Chongce Shan in the West and Kukushili as Well as Ulugh Muztagh in the East to NE, as Well as Zangsar in the West, Kukushili and Ulugh Muztagh in the East to NE, and Zangsar Kangri in the ESE

In this section we focus on Glacier Maps TJ22, TJ32, TJ21, TJ31 in Kuhle (2004). As indicated in the previous section, the triangle-shaped region under discussion was covered by glacier ice even in the postglacial Neoglacial, i.e., still during the middle Holocene. Therefore, the Late Glacial tills were reshaped by this youngest ice cover and put in their geomorphologically very fresh state. The fact of a lower ELA during the Late Glacial than during the postglacial Holocene leads to the conclusion that this region of Central Tibet must have had a larger ice thickness then than afterwards during the Neoglacial period (Stage V) (Kuhle 2013a: 146). At an extant ELA at ca. 5730 m asl (e.g. in the centrally located Kukushili Massif [see below]), and a late Late Glacial ΔELA of ca. 700 m in the Sirkung stage (IV, Fig. 2.5), this late Late Glacial snowline (ELA), which lay at ca. 5030 m asl, had already reached below the absolute altitude of the mountain forelands.

The extant ELA has been calculated according to the combination of the methods after Höfer (1879) and Kuhle (1988a). The Höfer method is a contradiction in terms because it requires an initial estimation of the ELA in order to calculate the mean altitude of the catchment area. Therefore, before the calculation of the medium height of the catchment area following Höfer can take place, one must calculate the base value after Kuhle (see the literature discussed in Kuhle 2014c). Only by determining the base value does one avoid the vicious circle implied by Höfer's method.

Accordingly, the following applies:

$$\begin{aligned}\text{Base value} = {} & 0.5(\text{height of the highest point of the catchment area} \\ & - \text{height of the current terminus}) \\ & + \text{height of the current terminus}\end{aligned} \quad (2.27)$$

$$\text{Base value} = 0.5 \times (6200 - 5450) + 5450 \quad (2.28)$$

$$\text{Base value} = 0.5 \times 750 + 5450 \quad (2.29)$$

$$\text{Base value} = 375 + 5450 \quad (2.30)$$

$$\text{Base value} = 5825 \quad (2.31)$$

$$\text{ELA} = (\text{pha} - \text{tim})/2 + \text{tim} \quad (2.32)$$

where pha = height of the accumulation area, i.e., medium altitude of the crests limiting the catchment area above the base value; tim = current terminus of the glacier tongue; ELA = equilibrium-line altitude (glacier snowline).

$$\text{ELA} = (6010 - 5450)/2 + 5450 \quad (2.33)$$

$$\text{ELA} = 560/2 + 5450 \quad (2.34)$$

$$\text{ELA} = 280 + 5450 \quad (2.35)$$

$$\text{ELA} = 5730\,\text{m asl} \quad (2.36)$$

This coincides also with the forelands of the three other massifs—the Chongce Shan, the Ulugh Muztagh, and the Zangsar Purog-II-Kangri. Accordingly, the Tibetan Plateau of the central triangle between these mountain groups must have hosted still during the late Late Glacial an inland ice–like glacier cover. This was nourished and heightened by primary snow precipitation (Kuhle 1985c, 1986b, 1987a, 1997, 2004, 2011a, 2012a). Due to the flat, nearly horizontal condition of the plateau surface, there was only a minor discharge of ice. In addition, it was slowed by the underlying permafrost. It even must have been brought to a complete standstill until an adequate ice thickness had accumulated. Up to the point when this minimum thickness was reached, the glacier cover was frozen to the ground below, as is generally true of cold-based glacier ice. With the ice thickness growing there occurred a jerky movement of the ice, i.e., a sudden release of the tension. The jerky movement of very cold glaciers (colder than −10 °C average annual temperature at the ELA level) has long been described as a so-called "block-floe movement" ("Blockschollenbewegung") (Finsterwalder 1937; Pillewitzer 1958).

During a necessarily increasing thickness of the ice cover, the discharge must have occurred by shearing. In some places, it shifted upward from the glacier bottom into the glacier body. On the one hand, this happened in places with movement-decreasing unevennesses in the field, such as depressions and basins in the ground. At the same time, a large-scale glacio-isostatic lowering must have occurred and added its effect (see below). At a sufficient glacier thickness and the resulting temporary velocity of the ice movement, a water film developed underneath the glacier, i.e., between the permafrost below and the glacier bottom. This process of melting that took place in the sub-moraine reduced the friction of the ground. Only when that pressure melting point ("Druckschmelzpunkt") was reached did the inland ice cover react with an increasing ice discharge. Accordingly, until this stage primarily only an increase in ice thickness had occurred.

In reverse chronological sequence, namely during the earlier three Late Glacial stages (Stages III, II, and I), the respective ice thicknesses must have been still larger. The cause of this was an ELA that had dropped about 800–900 m in Stage III, ca. 1000 m in Stage II, and ca. 1100 m in the oldest Stage I. The lowest temperatures must be assumed for the maximum ΔELA of ca. 1200–1300 m during the Ice Age (Stage 0 = LGP, LGM, MIS 3-2; Fig. 2.5). Accordingly, the largest inland ice thickness was probably reached during Stage 0 (Kuhle 2013a: 146).

During the High Glacial (LGP, Stage 0) the ice thickness must have reached or exceeded 2000 m. This is probable because of the lack of nunatak-like undercut mountain summits. Even the highest summits reaching nearly 6950 (6945) m asl have not been strongly sharpened. They rather have rounded forms (Kuhle 2013a: Fig. 43), which point to former glacigenic abrasions. This is mainly valid for Purog II (6418 m), Kukushili (6337 m), as well as the 6335 m high Toze Kangri (34° 46′ 25.97″N 82° 21′ 49.83″E), the abraded summit form of which corresponds to the Scandinavian "fjell" ridges. These high summits have preserved their rounded glacial form to a large extent or even completely.

During the interglacial period and also again today, several high summits were and still are sharpened by the strongly lowered level of their local ice cover and by their small hanging and cirque glaciers, but they are no longer rounded (Kuhle 1991: 136, 145–147, 149–150, Photos 1, 2, 4, 6; 2013a: Figs. 40 and 41). The Chongce Shan, the Chung Muztagh situated farther to the east (6926 m asl; 35° 38′ 55.49″N 82° 20′ 13.98″E), the Zangsar Massif (Kuhle 2013a: Figs. 40 and 41: A), and the Ulugh Muztagh provide examples of this interglacial sharpening due to their current glacier cover. Only the ca. 6945 m high Ulugh Muztagh shows a steep rock face measuring 950 m in height. Its remaining expositions have firn flanks that only descend 300–400 m until they reach the surrounding firn fields. The latter is also true of Chongce (Chongtse) Shan I and Chung Muztagh. In a genetic sense these firn-field levels have to be addressed as remnants of pre-Quaternary old faces and valley bottoms. This would correspond with the European Alps and Scandes. In connection with the triangular region of Central Tibet discussed here, this only applies to those slopes of Chongce Shan and Chung Muztagh that face toward Central Tibet. They are adjusted to the plateau surface that is more than 5000 m high.

The north sides of these massifs, however, falling away even more than 1000 m, have undergone a stronger sharpening. It is the result of a topographically caused, intensive retrogressive erosion, and results from the large relief of the plateau margin.

Due to suspected abrasion forms east and west of the Ulugh Muztagh main summit, a glacigenic undercutting can be diagnosed at ca. 6210 m asl (36° 24′ 57.79″N 87° 23′ 36.30″E and 36° 25′ 16.75″N 87° 20′ 42.79″E). It is located ca. 740 m below the summit, which during the High Glacial was covered by inland ice. East of the summit there is a rounded cavetto in the mountain ridge and west of it is the upper angle of a glacigenic triangle-shaped slope. The interpretation of the geomorphology requires the reconstruction of a High Glacial (Stage 0) glacier level at ca. 7000 m and a Late Glacial (Stage II–III, Fig. 2.5) level at about 6210 m asl.

Thus, the greatest former ice thickness made probable by the findings amounts to 2000 m. To this one has to add the thickness of the tills, which since the LGM have heightened the plateau surface.

It can be assumed that during the repeated Pleistocene glaciations an accumulative heightening took place by polyglacial till covers. At present it is not known how strongly the repeated maximum ice age (pleni-glacial) erosion removed the older tills. However, it is probable that from the last High Glacial (LGM) through the Late Glacial (Stages I–IV) up to the Holocene (Neoglacial) Stage V or VI, the till thickness in the triangle discussed here had grown larger and larger in layers. This is probable because the ice thickness and thus the ground polishing must have decreased, while the subglacial accumulation increased during a rising ELA.

With an ice burden larger than 2000 m, however, there follows a glacio-isostatic lowering of several hundred meters (Kuhle 1988d, 1993, 1995, 1999: 273–274, Fig. 38). This large-scale, basin-like depression again reduced the ice discharge close to the bottom. As a result, the High Glacial ground polishing or till removal, respectively, must have been diminished.

It is at any rate certain that the older till covers were reworked, i.e., even under the persistent influence of permafrost they were slightly removed and reshuffled on the surface. The Late Glacial or late Late Glacial to Holocene tills show their character of stacked covers and layers by means of slight differences in color and insignificant variations in texture (see Fig. 2.89, see also http://extras.springer.com: Character L and Fig. 2.90, see also http://extras.springer.com: Character M).

2.4.7 The Late Glacial to High Glacial Glaciation and the Neoglacial, Historical, and Present Glaciation in Central Tibet from the Foreland of the Zangsar Kangri in the North Across the Gandise Shan up to the Tsangpo in the South

The focus of this section lies on Fig. 2.1: No. 23 and Glacier Maps TJ21 and TJ31 in Kuhle (2004).

2.4.7.1 The North Section of the Profile: From the Zangsar Purog-II-Kangri in the North to the Nganclong Kangri (Northern Gandise Shan) in the South

As described in Sect. 2.4.6, the SSE foreland of the Zangsar Kangri at a ΔELA of ca. 260–300 m, i.e., during the middle Holocene Nauri stage (V, Fig. 2.5), was covered by glacier ice (Kuhle 2013a: Fig. 39G; Fig. 41: No. 5; Fig. 42). From the ca. 278 km long profile connected to the SW, indicators of past glacigenic ground polishing and also covers of till have been described (ibid.: 152–154).

Here, the surface of the Tibetan Plateau slants from S to SSW from ca. 5020 m to 32° 15′ 40.48″N 83° 52′ 01.00″E down to ca. 4450 (aneroid measurement 4256) m, or else to ca. 4375 (aneroid measurement 4245) m asl. The incline of this slope, however, is interrupted by intercalated counter-inclinations. They extend across the higher plateau remnants and high-mountain massifs situated in between; these are the seven mountain groups from the Chagdo Kangri in the WNW up to the Qomo Ri in the ESE, which tower above the current ELA. The strongest glaciated group of mountains in this area is the 6250 m high Mayer Kangri.

In the SSW part of the N-S profile under discussion, Late Glacial till has been pushed upwards on a counterslope at 32° 15′ 32.59″N 83° 51′ 46.35″E up to ca. 4600 m asl (Kuhle 1998b: 9, 14, Figs. 9 and 9j; 1999: 27; 2014a: 133: Photo 26). Tills such as these, which have been pushed upslope, have been observed several times in Tibet (see Fig. 2.144: 33° 48′ 19″N 88° 00′ 59″E 5450 m asl, large-scale banded and glacially scoured tills [arrow in the foreground]: 33° 48′ 17.72″N 88° 1′ 30.65″E, Detail J in Fig. 2.133, see also http://extras.springer.com: Character AI; see also Figs. 2.141, 2.142, 2.156, 2.193, and 2.190).

On this counterslope two generations of tills have to be differentiated: (1) at the foot of the slope the till cover of the Late Glacial Stage IV (at most Stage III; see Sect. 2.4.7.2.2); and (2) above it, up to the culmination of the slope, the Late Glacial Stage III (at most II) (Fig. 2.5). Adjacent to the west three generations of till have even been differentiated (Fig. 2.91): (1) at the lower slope the youngest till cover; (2) a medium-old one above it; and (3) further upslope, up to the culmination of the slope, at 4825 m asl, the oldest, i.e., third one. From bottom to top they are all classified as belonging to the Late Glacial Stages IV (at most Stage III, see Sect. 2.4.7.2.2), III (at most II), and II (at most I) (Fig. 2.5). This finding is supported by tills of this Stage II with three up to meter-sized boulders. They are located on the narrow neighboring mountain ridge at 32° 14′ 56.08″N 83° 48′ 44.64″E at 4730 m asl in continuation of that culmination.

From the point of view of this current slope of the Tibetan Plateau, in a SSW to S direction from the at present relatively strongly glaciated Zangsar Purog-II-Kangri, one must infer the former existence of an older than middle Holocene glacier cover in the lower southern areas. This rather strongly generalized and schematic approach leads to the following conclusion: those portions of older till must increase down towards the south. In addition to Figs. 2.91 and 2.92 (31° 52′ 19.36″N 84° 57′ 32.29″E; 4900 m asl) provides a further example of the condition of such older tills.

In order to examine this working hypothesis, we now examine the succession of tills from the north, i.e., from the Zangsar Purog-II-Kangri, to the south, up to those localities (Figs. 2.91 and 2.92) that are situated west and east of the Gertse settlement (Lumaringbo) in the Gandise Shan (Nganclong Shan). From their geomorphological state of preservation it is possible to draw conclusions regarding their age. In a further step, we then also consider the share of older till in comparison to the younger, medium Holocene till of Stage V on this slope.

These tills cannot be classified chronologically by TCN datings. Therefore, due to numerous unsuccessful TCN datings in Tibet, a Quaternary-geological, well-grounded fact (Kuhle 2011a, 2012a: 193–197) shall again be established here: A young Late Glacial to Neoglacial till is an older till that was again overridden and reworked by a glacier. Therefore, it consists of material that is older than the renewed

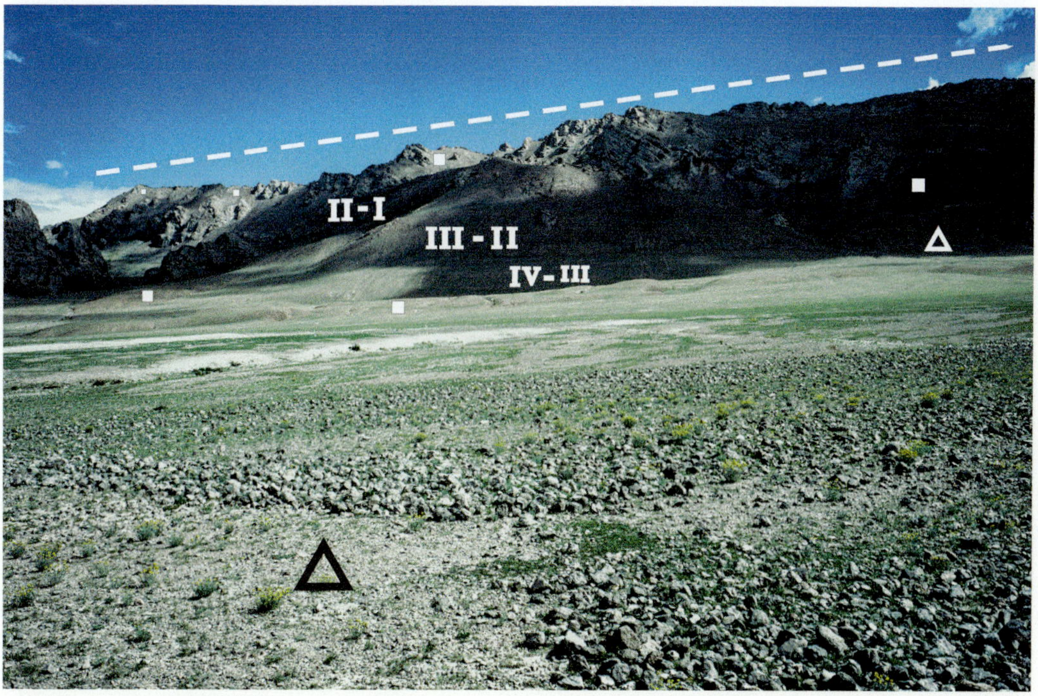

Fig. 2.91 View from the floor of a large W-E trending valley of the Gertse settlement (Lumaringbo) at 32° 15′ 38.47″N 83° 50′ 0.71″E from 4430 m asl, seen from ca. 22 km WSW of this settlement facing SSE onto the valley slope. (■) signifies Late Glacial to High Glacial (Stage 0) remnants of ground moraine covers—(II–I), (III–II), and (IV–III)—which during the Late Glacial glacier stages from I to IV (Fig. 2.5) have shifted less and less far upslope. (Δ) debris-flow fans, built up from dislocated ground moraine. (- - - -) denotes the High Glacial to early Late Glacial (Stages 0 to I) glacier level of the Tibetan inland ice. Analog photo 30/08/1996 by M. Kuhle

glacier advance. Its reshaping by the glacier advance remains minimal in the permafrost as it exists in Tibet (Kuhle 2014a: 127–143). Its reshaping takes place only superficially in the active layer. Here, mainly the fine matrix is concerned. The large boulders remain frozen into the permafrost. Accordingly, they have the TCN ages (^{10}Be years) of numerous earlier periods of radiation to which they were exposed during the absence of a glacier cover. Thus they are much older (i.e., up to several hundreds of thousands of years older) than their latest ice cover. In permafrost regions such as High Asia and Tibet, therefore, the age of tills cannot be properly determined by the TCN method.

A till that can be diagnosed by traces of dead ice as having been newly overridden and which is therefore freshly diagnosable is located south of the Zangsar at 34° 02′ 49″N 85° 50′ 18″E. It confirms the reference till for the middle Holocene Stage V (34° 01′N 86° 06′E, 5020 m asl; Kuhle 2013a: 146–149: Fig. 39G, Fig. 41: No. 5, Fig. 42) in the area ca. 21 km farther west cited at the beginning of this chapter. Accordingly, here too, the young piedmont glaciation reached down to at least 5000 m asl.

The till cover at 34° 15′ 4.19″N 84° 49′ 13.21″E at about 5000 (5036) m asl is situated along half the extent between the currently glaciated massifs Chagdo Kangri and Purog II Kangri (Kuhle 2013a: Fig. 47: No. 7). Due to the remaining dead-ice holes, this and its southern connection have to be classified as being from Stage V (Fig. 2.5). To this complex

Fig. 2.92 At 31° 52′ 19.36″N 84° 57′ 32.29″E from 4840 (4900 aneroid measurement) m asl facing west, looking onto 5130–5250 m high hills on which ground moraine covers are evident, e.g., at 31° 51′ 53.13″N 84° 56′ 10.73″E. (■) signify ground moraines such as these, which quite probably must be classified as belonging to Stages IV or III (see Fig. 2.5). (☗) are glacigenically rounded hill cupolas up to which there lie remnants of ground moraine. (- - - -) shows the High to Late Glacial (Stage 0–III) glacier level of the Tibetan inland ice. Analog photo 30/08/1996 by M. Kuhle

of till there belongs a 9 km long, leaf-shaped remnant of till deposited in a polished depression ("Schliffwanne") (34° 03′ 33.83″N 84° 52′ 27.96″E, ca. 5000–5220 m asl; Kuhle 2013a: 153–154, Fig. 47: No. 8). Due to its light color on dark bedrock an erratic genesis is likely. The bedrock shows a necessarily older glacigenic ground polishing—a glacigenic abrasion of the layer edges ("Schichtkopfstreifenschliff") (Fig. 2.93, see also http://extras.springer.com: Character N).

SW of the Zangsar and Purog II, there is at 33° 58′ 04.22″N 84° 31′ 16.00″E a 5130 m high glacigenically polished rock ridge (Kuhle 2013a: Fig. 47: No. 4). It is surrounded on a large scale by till covers. They are decameters in thickness, so that several up to ca. 5500 m rising rockhills are "drowning" in them on their lower base (34° 00′ 05 16″N 84° 34′ 25.01″E, 5035 m; ibid.: Fig. 47: No. 5 and 34° 00′ 47.63″N 84° 25′ 30.85″E; ibid.: Fig. 47: No. 6). Their different conditions point to two ages, namely that of Stage IV and—lying next to it—Stage V. The younger till cover has been built up from a westward direction, namely the nearby Chagdo Kangri (6111 m). Even today, this mountain massif is still glaciated.

Somewhat further east (34° 09′ 35.14″N 85° 16′ 22.70″E, 5020 m asl) and south of the Laxiongcuo (lake), there is one of the rare "windows" (holes) in the moraine cover (ibid.: Fig. 47: No. 9). On its margins the margins of the till are diagonally exposed (Fig. 2.94, see also http://extras.springer.com: Character O). These have been reshaped solifluidally in the active layer of the permafrost. In the places where the

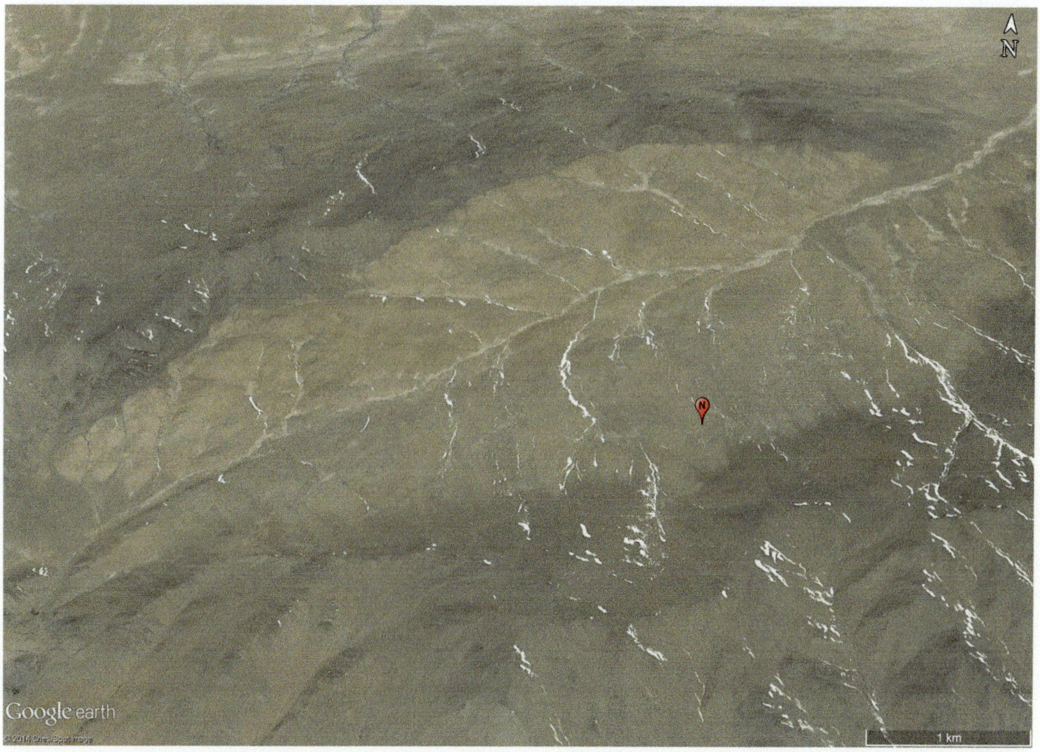

Fig. 2.93 Satellite image, Google Earth Pro: center 34° 3′ 55.00″N 84° 51′ 33.55″E; http://extras.springer.com: Character N indicates light ground moraine cover with irregular margins

till is absent, the older up to maximally High Glacial ground polishing is visible. It is a glacigenic abrasion of the layer edges ("Schichtkopfstreifenschliff") in a WNW to ESE direction. The marginal till cover has to be classified as belonging to Stage V or IV. If this were not a till cover but loose material weathered in situ, there would have to be visible a marked change in the regime from weathering in situ to fluvial (retrogressive) erosion. However, a change of the regime such as this cannot be proven because the Tibetan Plateau today as in the past has always shown only insignificant differences in altitude.

Again further east tills occur in the vicinity of the Garkung Caka (lake) south of the 6090 m high massif (34° 08′ 26.54″N 86° 36′ 58.36″E) with the typical state of preservation of Stage V. They have already been mentioned in Sect. 2.4.6. Also mentioned have been the till covers at 33° 54′ 00″N 86° 32′E, 33° 54′ 36″N 86° 34′ 13″E, and 33° 55′ 34″N 86° 32′ 54″, which reach up to the culmination of a hill up to 400 m above the lake level. They contain light-colored and meter-sized boulders, probably belonging to Stage IV. This assumption is true if the ice thickness during Stage V did not exceed those 400 m above the lake level.

After having discussed these till findings of Stage V, we now turn to till remnants that are also situated within the glaciation limits of this Stage (V) but are found in the eastern Zangsar piedmont area (Fig. 2.95, see also http://extras.springer.com: Character P). Since the deglaciation they have been fluvially dissected to depths of 80–100 m (34° 12′ 25″N 86° 10′ 17″E, 5220–5240 m asl). On the resulting plateau-shaped till remnant, one can find again the usual dead-ice ponds.

Also within the borders of the piedmont glaciation of Stage V, south of the Purog II, i.e., SW of the Zangsar (34° 11′ 58″N 85° 37′ 58″E

Fig. 2.94 Satellite image, Google Earth Pro: center 34° 9′ 32.40″N 85° 16′ 18.51″E; http://extras.springer.com: Character O shows one of several windows (holes) in this Stage V or IV ground moraine, situated at 34° 09′ 35.14″N 85° 16′ 22.70″E at 5020 m asl (see Fig. 2.5)

5180–5225 m asl), till material of Stage V is found on a mountain ridge of dark-reddish bedrock. It contains light-colored boulders with lengths of 1–3.5 m (Fig. 2.96, see also http://extras.springer.com: Character Q). They are located ca. 100 m above of two thalwegs.

Toward the south, at 33° 57′ 46.31″N 85° 17′ 12.21″E, there is an abraded hill that towers 600 m above the Mubucun lake. Up to its culmination at 5425 m asl, its surface consists of rust-colored rock. At its margins it consists of loose rock susceptible to solifluction. However, in its center there could also be a till containing components of iron. Due to the relative height of the hill a glacigenic transfluence for Stage V is not probable. This might have occurred for the last time during Stage IV.

Further distant in the southern foreland of the Zansgar Massif at 33° 52′ 44.32″N 85° 27′ 23.99″E at 5130 m asl is a till cover with dead-ice holes located alongside a preserved pond in a polished basin ("Schliffwanne"). In this till cover 1–3 m long, light-colored boulders are embedded in a fine matrix (Fig. 2.97, see also http://extras.springer.com: Character R). Where the till thins out or is lacking, the polished outcropping edges of the stratum are visible. Due to the presence of permafrost, tundra polygons (polygonal ice-wedge networks) occur in places. The till cover has to be classified as belonging to Stage IV.

A corresponding condition of till that must be classified as belonging to Stage IV also can be observed farther east at 33° 52′ 37.00″N 85° 49′ 28.19″E at 5070 m asl. Here, too, 1–3 m large, light-colored boulders are embedded in the matrix and have been exposed by solifluction.

Yet again farther to the south (33° 50′ 11″N 85° 32′ 55″E, 5050–5080 m asl) there is a field section covered by meter-thick till (Kuhle 2014a:

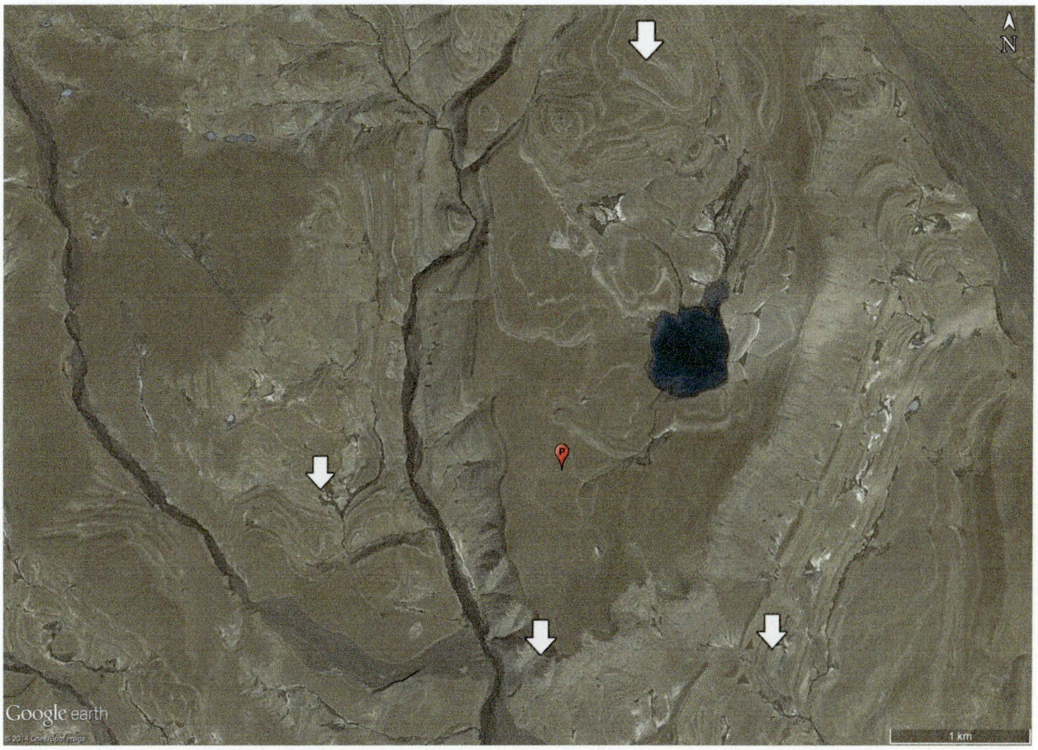

Fig. 2.95 Satellite image, Google Earth Pro: center 34° 12′ 25″N 86° 10′ 17″E; http://extras.springer.com: Character P indicates a ground moraine plateau near the Zangsar Kangri Massif

134, Fig. 6). The till contains meter-sized boulders that are lighter in color than their reddish matrix. In the oldest case, it must be classified as belonging to the late Late Glacial Stage IV (Sirkung stage). Approximately 1.1 km further SW (33° 49′ 49″N 85° 32′ 23″E, 5075 m asl) this till cover mantles a hill ridge which towers 250–400 m above its surroundings (Fig. 2.98, see also http://extras.springer.com: Character S). Its reddish color obviously results from the red color of the bedrock situated 11–40 km to the north. Thus the local direction of ice flow in Stage V or IV from N to S up to here is evidenced. The red coloration gets weaker in this direction.

An alternative genesis of this cover of loose stones in situ is made unlikely by the fact—as is also the case in many other places (see above and below)—that they show explicit signs of superimposed slope-downward reshaping by extant working edges. They prove a morphodynamics that cannot have developed in the past, i.e., synchronously with the development of a different weathering surface. Because the past periglacial weathering conditions—hypothetically without an ice cover—must have been the same as the extant ones, there would otherwise have occurred a syngenetic destruction. This, however, is not the case. Therefore, we must assume that we are dealing here with a transported till. This always also contains local material.

Also of reddish color is the till further west at 33° 45′ 22.33″N 85° 11′ 39.48″E at 4960 m asl. It shows dead-ice holes with a filling of water. In the fluvial thalwegs the fine-material matrix has been eroded so that lighter boulders have been concentrated there. This till has to be classified as belonging to Stage V or—in the oldest case—Stage IV.

The till preserved ca. 45 km farther west at 4990–5020 m asl at 33° 46′ 43.38″N 84° 42′ 29.95″E is of grey color. It covers and buries

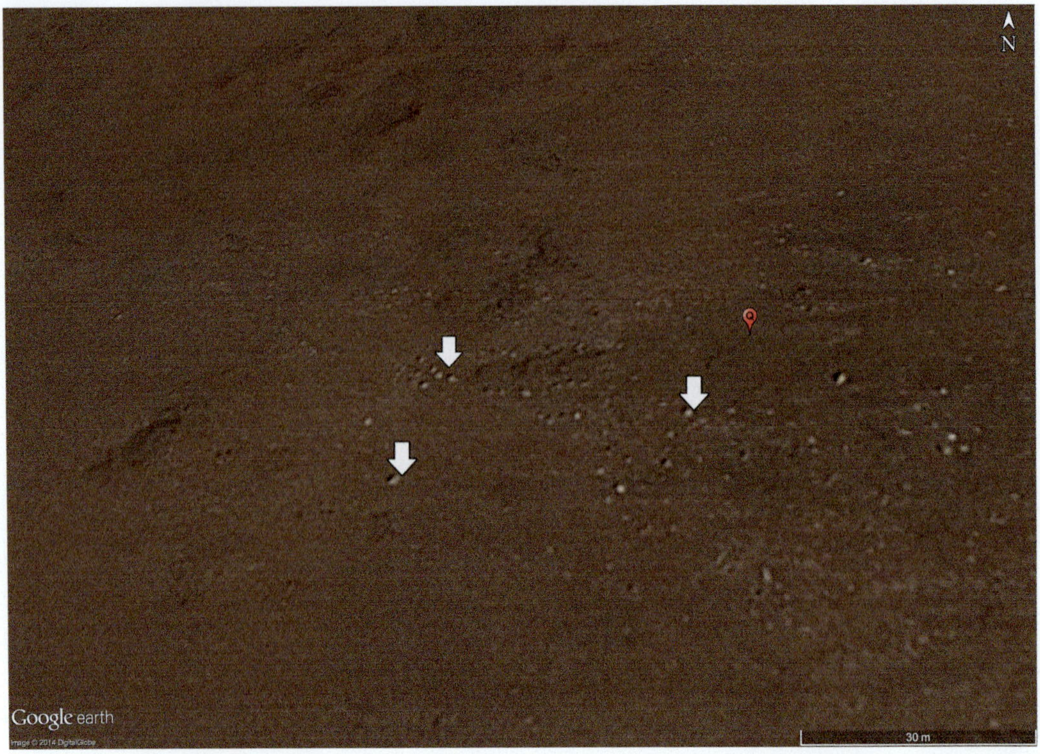

Fig. 2.96 Satellite image, Google Earth Pro: center 34° 11′ 58″N 85° 37′ 58″E; http://extras.springer.com: Character Q: Detail B of Fig. 2.133. Light blocks several meters in diameter and with lengths of 1–3.5 m in one ground moraine. This ground moraine cover has to be classified as belonging to the Holocene Stage V (Nauri stage, Fig. 2.5). It is situated on reddish bedrock in the western Zangsar Kangri Massif at 34° 11′ 58″N 85° 37′ 58″E at 5180–5225 m asl. Whether these are light boulders from quartz veins, are protruding quartzite boulders, or represent other material transported over long distances can only be clarified in situ

a relief of dark bedrock on a large scale (Kuhle 2013a: 154: Fig. 47: No. 3) for several 100 km^2. It belongs to Stage IV.

Some 10–20 km farther south (33° 37′ 28.89″ N 84° 48′ 45.84″E, 4850–5050 m asl) is a younger till cover (ibid.: 154: Fig. 47: No. 1). Its surface still shows dead-ice holes that have not filled with debris yet. It was accumulated from the opposite direction, that is, from south to north. Its catchment area is the northern side of the mountain threshold of the Pozarmalung and Dareyog massifs, still glaciated today. These reach 6260 m asl. Here, at a ΔELA of 300–500 m, large nourishing areas are accessible. Therefore, the till cover under discussion might be classified as belonging to Stage V. During Stage IV, however, it certainly was still covered by glacier ice. A nourishing area such as this, which due to its altitude at 5465–5710 m asl became accessible at an insignificant ΔELA, and remained so for some time, lies at 33° 31′ 12.47″N 84° 53′ 20.58″E. In this area there is a streaky till cover with several holes (“till windows”) in it (ibid.: 154: Fig. 47: No. 2). This till has to be classified as belonging to Stage V. This fjell-like, completely abraded 5710 m high mountain elevation probably carried its own small local ice cap during the Late Holocene, more precisely during Stage VI (see Fig. 2.5). Similar elevations tower to the west and east of the Tibetan basal area.

Fig. 2.97 Satellite image, Google Earth Pro: center 33° 52′ 44.32″N 85° 27′ 23.99″E; http://extras.springer.com: Character R shows ground moraine surface of Stage IV at 5130 m asl south of the Zangsar Massif with light-colored 1–3 m sized boulders

South of the still glaciated mountain threshold of the Pozarmalung and Dareyog massifs at 4900–5000 m asl lies a young till, interspersed with dead-ice holes (Fig. 2.99, see also http://extras.springer.com: Character T: No. 1 at 33° 06′ 51.58″N 84° 51′ 52.27″E, 4950 m asl, as well as Kuhle 2013a: Fig. 46: No. 3). Freshly preserved till can also be observed within the massifs up to ca. 5250 m asl. Thus it can be classified as belonging to Stage V.

This till cover basally mantles a 400 m high W-E oriented roche moutonnée, i.e., glacially streamlined hill, on its northern and eastern flanks (ibid.: 153: Fig. 46: No. 1). Up to its culmination (Fig. 2.99: Character T: 33° 04′ 08.32″N 84° 49′ 21.24″E, 5220 m asl) there is a differently colored till cover, penetrated by dark bedrock. Due to its different coloration, this till material must be erratic. It contains light-colored, meter-sized boulders that in part have developed periglacial patterns. According to its state of preservation as well as the genetically necessary minimum thickness of ice of more than 400 m, this till cover must be classified as belonging to the late Late Glacial Stage IV.

The glacially streamlined hill has a polyglacial (and most recently LGP to Late Glacial) genesis. Evidence for this includes the relatively large ice thickness that was necessary for its development, as well as the direction of ice discharge from east to west. Large exaration rills run in this direction.

The till cover at 33° 15′ 53.82″N 84° 18′ 30.18″E is assigned to Stage IV, or due to its state of preservation at most to Stage III. It is situated basally between the hill chains at about 4450–4830 m asl, and has in part buried

Fig. 2.98 Satellite image Google Earth Pro: Center 33° 49′ 49″N 85° 32′ 23″E; http://extras.springer.com: Character S; detail A of Fig. 2.133. Light-colored block material in reddish cover of fine material. The material is 50 km from the Zangsar Kangri Massif at an elevation that towers 250–400 m above its surroundings. If it is—as can be assumed—foreign material, its transport must have taken place due to super-regional glaciation. There is no other possibility

the relief of the hill. It stretches along the southern margin of the Chabo Cuo (lake) (ibid.: 153: Fig. 46 above No. 3).

South of the Mayer Kangri, Quomo Ri, and Peak 6045, there is at 32° 45′ 25.81″N 86° 46′ 58.65″E at about 5240 m asl erratic till from Stage IV. Its material is light-yellowish and has a thickness of ca. 2–5 m above the dark-greyish bedrock. It contains boulders with a size of ca. 1–3 m. Because of the considerable and increasing relief to the south, the extensive glaciation must have come from the north. Thereby the glacier cover in this area must have filled and buried differences in relief of more than 700 m. In some places this till cover is interrupted by rock surfaces without a cover of loose rock. In these windows (holes), mainly W-E-trending outcropping edges of the strata were dismantled by ground polishing. This applies for example south of the till margin at 32° 45′ 25.81″N 86° 46′ 58.65″E at about 5230 m asl and 32° 44′ 39.73″N 86° 45′ 53.42″E at 5140 m, as well as at 32° 43′ 51.05″N 86° 44′ 54.23″E at 5010 m and at 32° 41′ 52.94″N86° 45′ 56.31″E at 5100 m asl. All of these examples feature flat areas without fluvial (retrogressive) erosion.

Fig. 2.99 Satellite image, Google Earth Pro: center 33° 04′ 08.32″N 84° 49′ 21.24″E; http://extras.springer.com: Character T, 5220 m asl; No. 1: 33° 06′ 51.58″N 84° 51′ 52.27″E, 4940 m asl

The till observed on a large scale at 32° 34′ 25.62″N 84° 42′ 00.55″E and at 4950 m asl (ibid.: 152–153, Fig. 46: No. 2) contains boulders of a few meters in extent. It also has to be classified as belonging to the late Late Glacial Stage IV. A topographically accompanying till remnant is situated at 32° 25′ 20.84″N 84° 40′ 18.14″E up to the culmination of a hill at ca. 5400 m asl. The till, which has been transported a long distance, contains boulders of 1–2 m in size and is situated 400 m above the surrounding valley floors (Fig. 2.100, see also http://extras.springer.com: Character U). It is one of numerous remnants of till in the nearer and more distant environs. The highest rounded mountain summit in the vicinity reaches ca. 5850 m asl.

To summarize, it can be established that the Holocene tills of the southern foreland of the Zangsar Purog-II-Kangri, farther to the south up to the last-mentioned position, are increasingly replaced by late Late Glacial tills. Accordingly, the till covers of Stage V are gradually being supplanted by those of Stage IV.

The tills located farther to the south have already been introduced (see Figs. 2.91 and 2.92). The accompanying sediment analyses of the lowest occurrences of till at 4365 m asl (grain size spectrum and microscopic SiO_2 grain analysis of the medium sand fraction) confirm their specific till character (Kuhle 1998b: 8 and 9; Figs. 8 and 9: Nos. 59 and 61; 1999: 13 and 25: Figs. 5 and 21). Accordingly, the fine-material matrix shows a clay peak as well as the typical

Fig. 2.100 Satellite image, Google Earth Pro: center 32° 25′ 16.70″N 84° 40′ 15.74″E; http://extras.springer.com: Character U shows ground moraine, transported over long distances, containing 1–2 m sized erratic boulders and lying 400 m above the surrounding valleys at 32° 25′ 20.84″N 84° 40′ 18.14″E at an altitude of 5400 m asl. The light-colored, erratic boulders lie on darker bedrock

bimodal grain size cumulative distribution function. A composition of up to ca. 70% glacially crushed SiO_2 grains confirms the finding of a glacigenic development. Postglacial reshaping since deglaciation is also diagnosed sedimentologically by a glaciofluvial portion of 11–26% polished (lustrous/fluvially polished) SiO_2 grains and an eolian-frosted portion of the grains of ca. 10% (dull-eolian). The polymictic composition of the large components proves a moraine-typical mixture (Fig. 2.101). It confirms the large-scale, glacigenic catchment area of this region. By contrast, primarily fluvial accumulation typically only has a small-scale catchment area on this plateau landscape.

2.4.7.1.1 Hypothetical Reference to the Flow Direction of the Ice from the Southern Zangsar Purog-II-Kangri Foreland to the South During the Middle Holocene to Late Late Glacial Stages (Stages V and IV)

Thus far the concept of a former spreading and movement of a glacier ice cover (see Sect. 2.4.7.1) from north to south has been pursued. Supporting this assumption is the glacigenic long-distance transport of material to the south. An example of this is provided by the till in Fig. 2.98, which provides evidence of

Fig. 2.101 Typically representative ground moraine on a nearly horizontal plain in the southern, arid portion of Central Tibet at 31° 51′ 55.19″N 84° 58′ 29.87″E at 4850 (4910 aneroid measurement) m asl. (■) marks, in the ground moraine exposure, the clayey fine material matrix between the polymictic coarser, non-sorted components. (♦) signifies the edged, (⬟) round-edged, and (●) rounded rock components, which are here isolated from each other, within the matrix. The petrography of the rock components is also varied. A sorting of the grain sizes (classification) is completely lacking, so that a fluvial genesis and even fluvial shifting can be ruled out. Analog photo 30/08/1996 by M. Kuhle

transport from north to south. Its reddish color can be traced back to the bedrock situated decakilometers farther north (see above).

Such incongruities in coloration between till and bedrock also can be observed on a somewhat larger scale. It seems possible to exemplify this by a ca. 210 km long, slightly reddish to brownish, darkened till veil that stretches in a wide bend opening to the west, namely from 34° 5′ 58.84″N 86° 19′ 33.81″E up to 32° 41′ 17.66″N 87° 6′ 42.13″E in the east around the Qomo Ri massifs. This bend is formed of till from Stage IV. It also exhibits the necessary fading of color, based on the distance of transport. Accordingly, the flow of the glacier cover reached a distance of ca. 170–180 km from the Zangsar Kangri to the south.

However, the till surface of Stage IV in general is somewhat darker than that of Stage V. The cause of this lies in its more than twice as long exposure to weathering (see Fig. 2.5). Thus the described fading of the color toward the south is placed into perspective. In addition, at some locations along this N-S profile one finds further reddish rock outcrops, even further south. Therefore, too, the hypothesis of a continuous ice flow to the south remains somewhat uncertain, i.e., not completely warranted by the empirical data.

2.4.7.2 The Past, Historical to Late Glacial Glaciations of the Northern Gandise Shan (Nganclong Kangri) Along a S-N Profile to the Zangsar Purog-II-Kangri and Their Ice Discharge to the North

The ice also flowed down from the south, i.e., from the high areas of the 6815 (6742)–7093 (7070, Loinbo Kangri) m high mountain massifs of the Gandise Shan to the north. At the same time, it also flowed toward the northern glacier front of the ice cover reconstructed above (Sects. 2.4.7.1 and 2.4.7.1.1). Both glacier fronts therefore must have met during Stage IV or III. Precise information on where this might have occurred is difficult to provide. However, a fair approximation is possible from the current glaciation and the former glacier margins along a S-N profile in the Gandise Shan area under discussion. For this purpose, we now turn more thoroughly to the 6815 (6742) m high massif of the Shakham (Kuhle 1998b: Fig. 9: No. 57; 1999: 125–130, Photos 81: No. 1, 82: Nos. 1 and 2, cover photo).

2.4.7.2.1 The Historical to Late Glacial Glacier Stages of the Shakham Massif

In November 2012, the current glacier tongue of the southern Shakham-E-glacier (Kuhle 1999: Photo 81 ∇) terminated at ca. 5490 m asl. There are six other ice margin positions determinable by terminal moraines at ca. 5410 m, 5400 m, 5370 m, 5320 m, 5290 m and 5160 m asl, respectively. They represent the Historical Stages XII (ibid.: Photo 81 ∇), XI, X (ibid.: Photo 81 X), IX, VIII, and VII (see Fig. 2.5). The ΔELA of the "younger Dhaulagiri stage" (VII, where the ice margin was situated at 31° 41′ 22.42″N 85° 7′ 40.05″E) amounts to ca. 165 m ((5490 − 5160)/2 = 165) (ΔELA = (tim − tif)/2; see Sect. 2.1). With such a ΔELA it could also be the Neoglacial Stage 'VII ("middle Dhaulagiri stage"). In this case, the ice margin would possibly only be insignificantly older, namely ca. 300 years, than at Stage VII (Fig. 2.5). The glacier of the oldest Neoglacial Stage V had already built up a foreland glaciation down to ca. 4900 m asl (up to 31° 44′ 49.58″N 85° 9′ 31.21″E). The ΔELA amounted ca. 300 (295) m ((5490 − 4900)/2 = 295).

In the eastern piedmont area of the Shakham region under discussion, the lowest point of the terrain is located at ca. 4750 m asl. Down to this point, no further lowest ice margin position can be evidenced. A maximum ΔELA of at best ca. 370 m ((5490 − 4750)/2 = 370) would be needed to reach this position.

The next-older stage is Stage IV (Sirkung stage). With its ΔELA of ca. 700 m (see Fig. 2.5), it belongs already to the late Late Glacial. At that time, the entire plateau field in the nearer and more distant mountain foreland was lying underneath a glacier ice cover (ibid.: 22–27, 125–128, 129–144, Photo 81 IV, Photos 82–90; 1998b: Fig. 9: Nos. 57–61).This is evidenced by the covering tills—for example, in Fig. 2.102 or further north at 31° 58′ 4.07″N 84° 57′ 54.41″E at 4810 m asl, with an absolute till thickness of 6 m and boulders more than 4 m in size.

Also the basin of the Tung Hu (lake) located 70 km farther north and reaching down to 4400 m asl must still have been covered by inland ice during the same period. The ΔELA estimated for this lies at ca. 545 m ((5490 − 4400)/2 = 545). However, this estimate is too large, for due to the heightening of the inland ice, the accumulation of the ice was further reinforced. This feedback loop, however, cannot be quantified more precisely.

The ENE 6500 and 6400 m summit glacier in the southern part of the Shakham Massif (Fig. 2.102), which reaches down to 5340 m asl (11/15/12), shows five Historical to Late Neoglacial glacier stages. Its neighboring glacier to the north has six glacier stages. The last one terminated at 5570 m asl in 2012 and flowed together with the first one during Stage IX (Fig. 2.5). The oldest stage of this united glacier tongue reached down to ca. 5200 m. Due to their

Fig. 2.102 From 31° 49′ 28.22″N 85° 2′ 10.04″E at 4850 (4755 aneroid measurement) m asl facing south toward the Shakham (Shakangsham Kangri) Massif (6815 or 6742 m) at 31° 41′ 53.88″N 85° 3′ 41.91″E. (■) show ground moraines, which quite probably have to be classified as belonging to Stages V or IV (see Fig. 2.5). (▲) are glacigenic abrasion-forms of the inland ice cover that existed here as late as into the Late Glacial (Stages I–III); among them glacigenic triangular slopes on truncated spurs (▲ white). (◁) marks coalescing debris-flow fans of dislocated moraine material as well as the debris of backward (retrogressive) erosion in both of the small valleys above since deglaciation. (- - - -) denotes the High to Late Glacial (Stage 0–III) glacier level of the Tibetan inland ice between ca. 6110 and 6320 m asl. Analog photo 30/08/1996 by M. Kuhle

two respective ΔELAs of 70–173 m ((5340 − 5200)/2 = 70 and (5545 − 5200)/2 = 172.5), this stage has to be classified as belonging to the "middle Dhaulagiri stage" ('VII, Fig. 2.5). Above the very prominent lateral moraines of Stage X (Little Ice Age, LIA) lateral moraines of the older Historical to Neoglacial glacier stages are also preserved in many places (Kuhle 1999: Photo 81 V–X).

According to their age, these latter remains have been progressively solifluidally reshaped. The next-older terminal moraine remnant at 31° 38′ 57.70″N 85° 8′ 54.56″E provides evidence of an ice margin at 5120 m asl. Due to its accompanying ΔELA of ca. 213 m, it can best be classified as belonging to the "older Dhaulagiri stage" (VI). Because the lowest point of the mountain foreland is located at 5070 m asl, the next-older, medium Holocene Nauri stage (V) must have built up a still larger hammer-head-formed piedmont area. At a previous ΔELA of ca. 250 m, this must have united with the foreland glacier lobes of at least two neighboring glaciers to the south. Farther north, for the same period, a ΔELA of ca. 300 (295) m has already been reconstructed (see above).

The next-older Stage IV (Sirkung stage) shows a ΔELA of ca. 700 m in High Asia (Fig. 2.5). Even at a much smaller ΔELA, an inland ice glaciation still must have existed during the latest Late Glacial. Therefore, the till covers that completely cloak this entire section of the Tibetan Plateau and its hilly landscape (see above) also must be classified as belonging to Stage IV (ibid.: 22–27, 125–128, Photo 81 IV).

2.4.7.2.2 Where Might Have Been the Confluence of the Medium to Late Late Glacial Inland Ice Complexes of the Zangsar Purog-II-Kangri in the North and the Shakham Massif in the South on the N-S Profile Between the Two Mountain Groups?

A complete, largely overriding glacier cover between the two Central Tibetan mountain chains has been evidenced by an equally extensive till cover. This covering probably still occurred during Stage IV. None of the moraines of the area concerned has needed for its explanation an ΔELA of even only approximately 700 m, which represents the acknowledged ΔELA in High Asia for Stage IV. In order to explain the complete covering by ice, even a minor ΔELA is sufficient.

If—due to unknown factors such as, for instance, aridity—we first proceed from a minor ΔELA of less than 700 m, we still must assume a heightening effect for the inland ice. In other words, even a minor ΔELA would most likely still have been accompanied by a self-induced feedback loop that would reinforce the growth of the inland ice. However, in order to be conservative in our assessment, we re-date this Late Glacial glacier cover from Stage IV to Stage III. In High Asia, Stage III in many places is linked to a ΔELA of 800–900 m (see Fig. 2.5). This would thus constitute about twice the magnitude of the ΔELA necessary to sustain the stable nourishing of an inland ice sheet in the first place.

Apart from the existing, very high elevation forms of abrasion that cannot be classified chronologically and which could very well be older, the Late Glacial till covers prove ice thicknesses of 400–700 m (Sect. 2.4.7.1). They do so because it is up to these positions in height that they are located above the valley bottoms. These tills prove minimum thicknesses. Naturally they might have been and most likely were exceeded by the glacier cover many times over.

On this N-S profile, the Late Glacial inland ice cover had two centers: the massif of the Zangsar Purog-II-Kangri and the Skakham Massif. The latter is higher, but it is located 290 km farther south. Almost more important to glacier nourishing is the also described mountain threshold extending ca. 215 km from NW to SE between the Pozarmalung and Qomo Ri. It reaches an altitude of 5400–5700 m over a large area. It is situated clearly nearer to the Zangsar Purog-II-Kangri than to the Shakham Massif. Therefore, a junction of the two ice centres south of this intermediate mountain threshold is most likely to be correct. The seamline of the two ice complexes might have existed in the area of the W-E-extending depression north of Character U in Fig. 2.100 that reaches down to 4800 m asl. This would be the most northern occurrence. More likely, however, is a seamline in the 150 km long, W-E-extending depression north of Lumaringbo (Gertse). At the Tung Hu (lake) it reaches down to 4400 (4360 west of Lumaringbo) m asl. The explanation is this: in its western section, tills of Stages IV (at most Stage III) and III (at most II) shown in Fig. 2.91 have been identified; they here extend upslope toward the south (see also Kuhle 1998b: Fig. 9j and 9: No. 66) (Sect. 2.4.7.1).

2.4.7.3 The Level of the High Glacial and Old to Middle Late Glacial Inland Ice Cover from the Zangsar Purog-II-Kangri up to the Shakham Massif

North of the Zangsar Purog-II-Kangri, a High Glacial highest level of the Tibetan inland ice at about 7000 m is considered to be reconstructable (Sect. 2.4.6.1). Directly at the 6500 m high Zangsar Purog-II-Kangri Massif, a minimum level at 6420 m can be evidenced by the abrasion form of Purog II. Farther south there is an abraded high summit at 34° 08′ 26.54″N 86° 36′ 58.36″E that proves a past minimum of the inland ice at 6090 m asl. At 33° 29′ 37.21″N 85° 34′ 11.22″E there is another such summit providing evidence of a former ice level minimum at 6080 m asl. At 33° 27′ 46.68″N 86° 46′ 34.32″E the abraded cupolas of all summits of the Mayer Kangri are located. They tower up to 6250 m asl. At 33° 20′ 32.12″N 84° 40′ 23.19″E all summits of the 6050 m high Pozarmalung have been abraded. At 33° 20′ 33.37″N 85° 12′ 58.03″E, all summits of the Dareyog have been abraded up to the highest point at 6260 m. At 33° 11′ 03.44″N 86° 56′ 59.76″E, in turn, all summits of the Qomo Ri have been abraded up to the highest point at 6050 m. At 33° 28′ 36.40″N 87° 38′ 47.58″E, the 6100 m high summit cupola of the Janghgaida Rinag has been abraded as well. Consequently, in this enumeration the highest summits have been considered, and all have been abraded. The level lying above 6420 m has to be classified as belonging to Stage 0, perhaps also to the oldest Late Glacial Stage I.

Very probably the glacigenic abrasions reached below 6420 m not only during the High Glacial Stage 0 (LGM, LGP, ca. MIS 2), but also still during the Late Glacial Stages I–II and perhaps even still during Stage III (see Sect. 2.4.6.1).

Thus, High Glacial ice thicknesses of 1500 m are probable in this area.

In the 6815 (6742) m high Shakham Massif (Fig. 2.102), it seems, the lowest of the suspected high- abrasion mountain cupolas as well as rounded saddle profiles lay at ca. 6000 m asl (Kuhle 1999: 25, 125–128, Photo 81 - - - -0- - - -). On this basis, in 1996, the present author made a careful estimation of the minimum height of the inland ice surface. Due to the more exact topographic data are now available, that estimate has to be corrected upwards by a value of 100–200 m. Consequently, abraded saddle forms that can be interpreted as being transfluence passes are situated at between ca. 6110 and 6220 m asl. At ca. 6300 and 6500 m altitude (at 31° 38′ 26.04″N 85° 5′ 50.51″E) there are flattenings in the summit crest. They are reminiscent of the remnants of old piedmont areas that are of no use for the reconstruction of ice levels. The rounded summit cupolas of mountains connected to these areas immediately to the south reach 6320 m asl, and can be considered as the highest forms of abrasion (Fig. 2.102).

In the immediate vicinity, e.g., at 31° 44′ 02.18″N 84° 55′ 41.38″E, there are polished mountains 5845 m in height (ibid.: 125–132, Photos 81–83). In some places they are shaped similar to those in the past abraded fjell landscapes of Scandinavia (ibid.: 141–142, Photo 87a). On those mountain ridges (Fig. 2.92) one can find layers of till (Fig. 2.101) spread over large areas. Corresponding till covers are also found up to ca. 10 km farther WNW on the same, ca. 450 m high low mountain range. On its WNW-margin at 31° 53′ 27.32″N 84° 51′ 41.45″E, at about 5040 m asl, there is a 2–3 m thick till cover that is in the process of dislocation by solifluction and which shows terraced slide demarcations. Till slides such as these can be observed in High Asia in many places (Kuhle 1982a: 108, Abb. 38; 52, 55, 184; 2013a: 119–123, Photo 76, 77). At 31° 53′ 45.68″N 84° 51′ 21.48″E at about 5090 m asl, there are horizontal, glacigenic exaration rills in the till. Slides of till in the current active layer destroy these rills on the slopes in a downhill direction. Tills greater than 10 m in thickness, which partly contain large boulders, are situated at 31° 54′ 36.66″N 84° 50′ 28.14″E at 4940 m and at 31° 54′ 39.30″N 84° 50′ 45.59″E at 4930 m asl. They have been fluvially cut quite recently.

In small valleys, as for example at 31° 52′ 46.84″N 84° 53′ 40.78″E at 5090–4970 m asl, old basin-shaped fillings of loose rock material can be observed as valley bottoms. Today they are fluvially dissected. Dissection is the reverse of accumulation here. Dissection is recent, and is proved by fresh working edges on the upper margins of the cutting. The cause of this geomorphological change of regime can only have been a deglaciation. The loose material is thus identified as till.

Again 50 km farther to the west there is another group of mountains with a medium height of 800–1000 m above the basal plateau level of Tibet. It reaches an absolute height of ca. 5700 m, up to which one can observe preserved till covers and exaration rills. This applies, for example, to the locations at 31° 49′ 48.52″N 84° 17′ 18.17″E and 31° 50′ 13.87″N 84° 17′ 29.29″E. Between 5110 and 5175 m asl, meter-sized light-colored boulders lie ca. 100–300 m above the thalweg tills. They show evidence of slides as well as exaration rills and/or fresh fluvial traces of erosion, i.e., dissections.

Further till covers reach 5480–5580 m asl. They are situated at 31° 50′ 50.03″N 84° 16′ 53.99″E and 31° 51′ 10.44″N 84° 16′ 0.99″E and contain meter-sized boulders. They also have exaration rills and slides.

Further relevant glacio-geological and glacio-geomorphological findings come from the location at 31° 49′ 55.64″N 84° 15′ 45.27″E at about 5050 m asl. Here we find meter-sized boulders and postglacial furrows, as well as 1–2 m thick till with furrows at 31° 49′ 22.66″N 84° 17′ 14.79″E at 5000 m asl; further at 31° 52′ 13.70″N 84° 15′ 55.79″E at 5580–5590 m asl, at 31° 51′ 14.29″N 84° 13′ 16.32″E at 5600 m asl—here again with meter-sized boulders—and at 31° 52′ 53.32″N 84° 8′ 59.99″E at 5560 m asl, also with meter-sized boulders. And finally, farther north at 32° 0′ 1.57″N84° 15′ 38.06″E, at 4830–4940 m asl, light-colored till overlies dark bedrock.

The identification of till also requires evidence that no weathering in situ can be given as an alternative explanation. Such evidence can be provided by means of adjacently positioned polished rocks. For instance, there are glacigenic rock polishings at 32° 10′ 44.58″N 83° 52′ 15.57″E at 4930 m, and also at 31° 56′ 37.12″N 84° 13′ 29.15″E at 5020 m asl. They prove a former glacier overlay.

The two low mountain ranges set upon the Tibetan Plateau have valleys that are arranged concentrically. Under the inland ice cover they were filled with glacier ice. This was frozen to the permafrost ground surface below. Accordingly, the High Glacial movement of the inland ice took place mostly in the hanging glacier ice, but not in the basal ice lying below. The ice fillings of these small-scale valleys, which were frozen to the ground, were detached from the inland ice cover. They sealed the valleys without participation in the ice flow (see Sect. 2.3). Accordingly, their local stability was so great that it led to detachment from the overthrusting, massive inland ice cover. Because of this immobility of the ice in these small valleys, their V-shaped form can be explained, which has largely survived the Pleistocene glaciations. Only during the comparably short phases of the Early Glacial build-up and Late Glacial phases of decline could a short flow of the ice have taken place. Examples of the development of depression- or trough-like widenings of an otherwise V-shaped valley can be seen in the small valley at 31° 52′ 35.60″N 84° 12′ 56.47″E and 31° 52′ 22.07″N 84° 17′ 1.89″E. The rest of the small valleys of these low mountain ranges are mostly V-shaped.

This chapter can be summarized as follows: the author has classified the pertinent ice that last covered this terrain as belonging to the Late Glacial Stages IV–III or II. The level of this inland ice at that time must have reached at least ca. 5700–5800 m.

For the LGM (Stage 0), however, an ice level at at least 6000–6200 m asl must be assumed. This ice lay in the vicinity of the Shakham Massif ca. 1400–1700 m above the basal level of the Tibetan Plateau. Accordingly, the minimum ice thickness amounted to ca. 1400–1700 m. This is ca. 200 m more than the author formerly estimated based on the poor condition of the maps available at the time (Kuhle 1999: 25).

2.4.7.4 The South Section of the Profile: The Nganclong Kangri (Northern Gandise Shan) in the North up to the Tsangpo Valley in Southern Tibet with Regard to the Prehistoric to High Glacial Ice Cover

First we continue our westward movement in the Nganclong Kangri (northern Gandise Shan). Further till covers and ground polishings at latitudes of 32°–33° will thus be introduced here as further samples from this area.

Fig. 2.103 Satellite image, Google Earth Pro: center 31° 53′ 43.58″N 83° 57′ 55.53″E; http://extras.springer.com: Character V; Nos. 1–3. At 31° 53′ 43.58″N 83° 57′ 55.53″E there are erratic remnants of ground moraines up to 5800 m asl. Character V (32° 43′ 53.64″ N 81° 34′ 55.13″E), east of the 5921 m high Dorenza summit, is a light-colored ground moraine up to 5900 m asl and higher. At No. 1 (32° 29′ 21.28″N 82° 4′ 40.28″ E) and up to 32° 29′ 58.85″N 82° 05′ 01.83″E at 5500–5800 m asl there is light-grey ground moraine on rounded mountains consisting of dark bedrock. At No. 2 (32° 22′ 41.85″N 81° 45′ 2.23″E) there is ground moraine at ca. 5470 m asl. At No. 3 (32° 22′ 41.85″N 81° 45′ 2.23″E) there is a 5470–5740 m high plateau with a lighter ground moraine cover, some meters thick and full of holes on darker bedrock

In the eastern section of the currently non-glaciated 5921 m high Dorenza Massif, light-colored remnants of till covers are found at 32° 43′ 53.64″N 81.34′ 55.13″E at 5700 m asl. In places they even reach up to ca. 5900 m asl. They lie on rounded mountain ridges and cupolas of dark, abraded rock. These moraines obviously are erratic. Accordingly, they were transported up to here at an ice thickness greater than 1400 m. This ice thickness is needed in order to reach these places of deposition. Accordingly, we must be dealing here with Late Glacial (Stage I or II; Fig. 2.5) till. Later, i.e., during the Holocene to Historical local glaciations of the massif, it was removed in part. Therefore, at present, it is punctured in places by outcropping rock faces (Fig. 2.103 Character V).

A further till cover, also preserved in remnants, is situated at 32° 34′ 2.19″N 83° 7′ 1.17″ E at about 4850 m asl (Fig. 2.104 Character W: No. 1). Their light color supports the idea of an erratic substrate, i.e., a larger distance of transport. It lies on dark loose material. Possibly, this is older detritus of till. In places it is also interrupted by bedrock and contains 1–2 m sized boulders. Adjacent, at 32° 33° 27.96″N 83° 4′ 57.09″E, there is at 4900 m asl also a remnant of till cover with meter-sized boulders (Fig. 2.104 Character W).

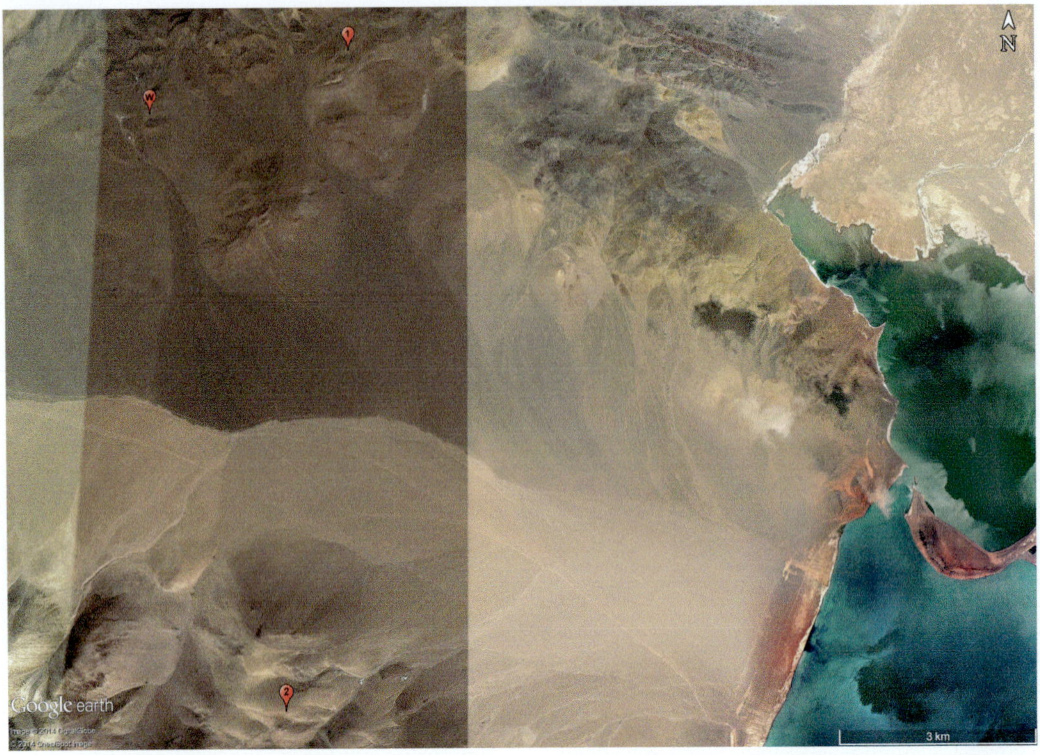

Fig. 2.104 Satellite image, Google Earth Pro: center 32° 31′ 12.56″N 83° 8′ 46.91″E; http://extras.springer.com: Character W; Nos. 1–2. No. 2 marks at 32° 28′ 34.09″N 83° 6′ 32.94″E at ca. 4700–5010 m asl further meter-thick ground moraine. This lies up to ca. 600 m above the neighboring, broad, basin-formed valley bottom and contains meter-sized boulders

These deposits require a thickness of the ice cover of only 500 m. Therefore they can be classified as belonging to one of the late Late Glacial stages, Stage IV or III.

At 32° 30′ 43.78″N 80° 9′ 12.11″E at about 4360 m asl there is dislocated till on a dark rock pedestal east of the settlement of Ali (Shiquanha) (Fig. 2.105) (Kuhle 1999: 6–7, Fig. 2: No. 101).

At 32° 31′ 46.58″N 81° 33′ 41.33″E at about 4920 m asl, lighter till lies on the dark bedrock of a rounded mountain ridge ca. 120 m above the valley bottom. This loose rock, reshaped by solifluction since the deglaciation, must not necessarily have been transported here from far away.

A roche moutonnée located at 32° 30′ 43.19″ N 82° 26′ 30.42″E, at 4425 (aneroid measurement 4260) m asl (see Kuhle 1999: Photos 106 and 107), breaks through an extensive till cover in a 14 km wide valley bottom (ibid.: Photo 103). The till cover and the roche moutonnée were modified by limnic activity of the Holocene lake phase up to a level of 4460 m. Sediment analysis of the cover of loose stones outside of the former extension of the lake provides evidence that it is till (ibid.: 28–30, Fig. 24). Its granulometric characteristics—such as the trimodal progression of the cumulative distribution function, the clay content, etc.—all correspond with those of the Western Canadian till of the Laurentide Ice Sheet. For comparison samples were taken from here as well (ibid.: Fig. 25). The microscopic form analysis of the SiO_2 grains also confirms the loose material as being till. It contains ca. 91% glacially crushed grains (ibid.: 13, 28, Fig. 5: No. 31.08.1996/1). The remaining 9.1% of the sample consists of dull (eolian) grains.

Fig. 2.105 At 32° 30′ 43.78″N 80° 9′ 12.11″E, ca. 4360 (4130 aneroid measurement at high pressure) m asl facing SW looking into the orographic left valley flank of the Tsenko Zangbo (or Gar Zangbo), which leads down to the west into the basin of Shiquanha (Ali). (■) marks several-meter-thick, lighter Late Glacial ground moraine on dark bedrock (◆) that since the deglaciation has been reshaped and displaced. (◆) at the same time signifies an outer bank in the rock, steeply undercutting the valley flank. This undercutting has taken place in a subglacial meltwater tunnel that followed the valley course during the Late Glacial inland glaciation in Stages II–III (Kuhle 1999: 199, 200). (□) marks the present-day gravel floor of the valley that developed from a glacio-fluvial gravel field (sandr) of the Late Glacial Stage IV. Analog photo 01/09/1996 by M. Kuhle

From this it is clearly recognizable that even immediately after the deglaciation in this broad valley bottom, the influence of meltwater must have been insignificant.

Above the Holocene lake level mentioned above, there is at 32° 30′ 23.03″N 82° 28′ 45.43″E also a roche moutonnée that consists of dark, glacigenically polished bedrock. Between ca. 4700 and 4820 m asl it is covered in part by light-colored till. Basal abrasion forms such as these can also be identified in the adjacent region to the south. The necessary ice thickness of at least 400 m allows a classification into Stages IV or III.

A ca. 0.5 km long, ca. 80 m high roche moutonnée that extends from NE to SW is located at 32° 30′ 8.50″N 82° 38′ 33.66″E and reaches ca. 4480 m asl. Its streamlined form was shaped into nearly vertical-standing metamorphites.

North of the presently non-glaciated 6192 m high Yaria an extended light-grey colored cover of till can be found at 32° 29′ 58.42″N 81° 31′ 13.59″E and at 5200–5900 m asl. It is interrupted in places by dark bedrock.

At 32° 29′ 21.28″N 82° 4′ 40.28″E and farther away, e.g., also up to 32° 29′ 58.85″N 82° 05′ 01.83″E in an area at between 5500 and 5800 m asl, there are large, light-grey till covers spanning several square kilometers. They lie on rounded mountains of dark bedrock. This bedrock breaks through the covers at regular intervals and with similar extents. Thereby the surface forms an irregular pattern of loose rock deposits

Fig. 2.106 Ca. 125° panoramic photograph from the bottom of the Ngari Chu (the valley that leads down to the west to the Ngari settlement) from ca. 4380 m asl at 32° 31′ 39.88″N 80° 34′ 46.04″E (see Fig. 2.107: http://extras.springer.com: Character X) looking SSW diagonally downvalley (to the right). (▲) indicate glacigenic abrasion-forms such as glacially streamlined hills and roches moutonnées as well as glacigenic flank-polishings that have produced truncated spurs. (■) mark meter- to decameter-sized ground moraine covers, e.g., at 4540 m asl at 32° 31′ 8.23″N 80° 34′ 14.36″E (■ black on the left). (◁, ▷ and Δ) show debris-flow fans of ground moraine material dislocated since the deglaciation. (□) indicates the valley bottom of glacio-fluvial gravels (sandrs) that still are currently still being displaced. (- - - -) denotes the approximate High Glacial glacier level at about 5900 m asl. Analog photo 01/09/1996 by M. Kuhle

and rock (Fig. 2.103: Character V: No. 1). Here, the roundings of abrasion and the tills give evidence of ice thicknesses greater than 1400 m. Because the deposition of till must have taken place after the ground polishing, it is classified as belonging to the Late Glacial (Stages IV–I). Only a little to the south but far to the east, we find at 32° 29′ 9.19″N 83° 37′ 41.53″E and at 4635 m asl light-colored, loose material of minor (although still meter-thick) thickness, lying on darker, anthracite-colored bedrock. Here, too, we are most likely dealing with a till cover. At 32° 28′ 34.09″N 83° 6′ 32.94″E, at 4700–5010 m asl, we again find meter-thick till cover. It lies up to ca. 600 m above the adjacent, broad, basin-shaped valley bottom and contains large meter-sized boulders (Fig. 2.104, see also http://extras.springer.com: Character W: No. 2).

At 32° 27′ 38″N 80° 34′ 00″E and vicinity, there are very similar developed and preserved till sheets containing erratic material (Figs. 2.106 and 2.107 Character X). This till cover lies east of a 5930 m high peak. On its massif up to ca. 5820 m asl, further light-colored remnants of till are preserved (32° 28′ 45.72″N 80° 30′ 18.51″E). They lie on abraded and therefore rounded rock ridges (Fig. 2.107 Character X: No. 1).

At 32° 26′ 19.80″N 82° 08′ 28.67″E up to 32° 27′ 09.64″N 82° 08′ 42.46″E there are light-grey erratic tills, likewise lying on rounded mountain surfaces. They are located at between 5400 and 5650 m asl on very dark rock faces. Also on this N-S profile with its W-E oscillations, one finds light-yellow till lying at 32° 26′ 01.93″N 82° 14′ 23.76″E. It contains up to 5.5 m long light-colored boulders. It lies on darker rocks of rounded hills up to 5650 m and farther south, in the Gangmo Shan, in places even up to approximately 6040 m asl. At 32° 25′ 44.43″N 83° 18′ 54.79″E, at ca. 4670 m asl, there is local till material of minor thickness interspersed with holes. At 32° 23′ 10.54″N 83° 28′ 15.66″E, at about 5000 m asl, a light-grey cover of loose material has been observed that also looks very much like till. It lies on dark-grey bedrock. At 32° 22′ 41.85″N 81° 45′ 2.23″E a 5470–5700 m high plateau is located, with light-colored remnants of a till cover. These are only a few meters thick and, again, interspersed with numerous holes. As a result, in many places the darker bedrock of the underground becomes visible. Nearby and surrounding 32° 22′ 40.24″N 81° 27′ 7.74″E, at ca. 5240 m asl, a

Fig. 2.107 Satellite image, Google Earth Pro: center 32° 36′ 49.44″N 80° 27′ 37.88″E; http://extras.springer.com: Character X; Nos. 1–3. No. 1 marks at 32° 28′ 45.72″N 80° 30′ 18.51″E and 5820 m asl ground moraine on dark rock on the 5930 m massif, adjacent to the terrestrial photo of Fig. 2.106. No. 2 marks at 32° 46′ 28.33″N 80° 23′ 59.43″E and 5640 m asl further light remnants of ground moraine south of round, ca. 6000 (6020) m high, dark bedrock cupolas. No. 3 marks at 32° 49′ 53.23″N 80° 25′ 23.06″E a light remnant of ground moraine (Stage III or IV) on the culmination of an abraded dark rock hill at 5750 m asl

light-colored, in places transparent till veil covers the abraded rock extensively. In some places it is pierced by the dark rock from below in the shape of low tors.

On the bottom of the broad depression running from W to E, ca. 38 km WNW of the Gertse settlement at 32° 22′ 33.25″N 83° 40′ 7.97″E at about 4410 m asl, there is a km^2 wide cover of till (Kuhle 1999: 147, 148, Photo 94; 1998b: Fig. 9: No. 67). On its surface are remnants of dead-ice holes. The till cover reaches more than 20 m in thickness, and has to be classified as belonging to Stage IV. At 32° 21′ 42.78″N 81° 38′ 32.64″E at about 5120 m asl, one detects light-colored till interspersed with holes on dark bedrock of the local high plateau.

At 32° 21′ 8.53″N 83° 40′ 14.26″E at about 4540 m asl, ca. 150 m above the valley bottom, light-colored till can be observed. It lies on a rounded mountain ridge of dark outcropping metamorphites (Kuhle 1999: 6, 7, 149–158, Fig. 2: Nos. 68–74).

At 32° 21′ 6.67″N 83° 7′ 8.26″E at ca. 5600 m asl, grey-colored till has been found. It contains boulders more than two meters in diameter and is deposited on dark bedrock. The up to 5960 m high mountains on the slopes of which this till lies are rounded (ibid.: 6, 7, 158–162, Fig. 2: Nos. 75–76). A little farther to the south, at 32° 20′ 51.33″N 81° 33′ 26.00″E, at ca. 5150 m asl, a lighter-colored cover of loose rock material on a dark underground can be

Fig. 2.108 Looking NE from a ground moraine plain at 32° 4′ 24.56″N 81° 49′ 41.57″E and 4600 m asl (locality: Fig. 2.109: Character Y). (■) mark up to decameter-thick ground moraine covers that are light-colored in the middle- and background and in some places are protruded or pierced by abraded mountain ridges, rock ribs, and roches moutonnées (▲) of dark bedrock. (- - - -) indicates the Glacial inland ice surface at ca. 6000 m asl. Analog photo 31/08/1996 by M. Kuhle

recognized. In places it is interrupted by this dark bedrock face. Thus, the till probably is erratic, resembling the adjacent till covers (Figs. 2.108 and 2.109, see also http://extras.springer.com: Character Y) (see also ibid.: 6, 7, 180, Fig. 2: No. 85, Photo 125).

At 32° 18′ 12.82″N 82° 31′ 6.83″E at between ca. 5350 and 5470 m asl, a light-colored till sheet again covers the dark outcropping rock with gaps. Here, the dark rock forms an abraded mountain ridge, i.e., a remnant of an old plateau surface. Here one can differentiate a lighter-colored and a somewhat less light-colored cover. 11.5 km to the E, on a mountain cupola at 32° 17′ 2.91″N 82° 38′ 13.55″E, in the 5966 m high massif up to an altitude of ca. 5530 m, there lies a light-grey till veil interspersed with numerous holes. It is situated on dark bedrock that is visible through holes in the till. In the same massif, but 3.5 km farther south (32° 15′ 7.32″N 82° 38′ 34.80″E), a light-grey till veil exists up to ca. 5700 m asl. It also lies on dark rock. Due to the glacio-geomorphological arrangement of their positions, tills such as these have to be classified as belonging to Stages 0–III (Fig. 2.5). Here we are 23–34 km east of a main valley bottom whose width ranges from 2 to more than 19 km. It lies at ca. 4440–4530 m asl (aneroid measurements). Figures 2.110 and 2.109 (see also http://extras.springer.com: Character Y: No. 1) show a similarly extensive main valley bottom located 26 km farther WSW. The till there, which for topographical reasons belongs to the very glaciation under discussion here, lies 1000 m further below (see also ibid.: 6, 7, 173–175, Fig. 2: No. 84, Photo 118).

Fig. 2.109 Satellite image, Google Earth Pro: center 32° 4′ 48.00″N 81° 58′ 16.31″E; http://extras.springer.com: Character Y; Nos. 1–4. Y is situated at 32° 4′ 24.56″N 81° 49′ 41.57″E at 4600 m asl (see Fig. 2.108). No. 1 marks at 32° 7′ 43.17″N 82° 5′ 45.54″E ground moraine, shown at 4710 m asl facing west on the terrestrial photo of Fig. 2.110. No. 2 marks at 32° 3′ 50.83″N 81° 57′ 58.49″E the position of the terrestrial photo of Fig. 2.111 that shows a genetically corresponding ground moraine at 4600 m asl. No. 3 is a locality at 32° 9′ 5.21″N 82° 7′ 6.91″E at 4790 (4810 aneroid) m asl with glacigenic ground-polishing and a roche moutonnée towering above a ground moraine (see terrestrial photo of Fig. 2.112). No. 4 marks at 32° 4′ 42.49″N 82° 2′ 50.48″E at 4620 m asl the locality of the terrestrial panoramic photograph of Fig. 2.114

Samples were analyzed from the corresponding adjacent tills (ibid.: 13, 32, Fig. 5 diagram 31.08.96/2, Fig. 26). The till character of the material is proven by sedimentological analysis. Based on previous work, the analysis reads here as follows: "Fig. 26 shows the bimodal course of the grain size columns in the ground moraine matrix. The peak in the clay is comparatively inconspicuous. The high limestone portion of 32.4% is specific to local moraine, because limestone outcrops in the underground rock (cf. Photo 115). Figure 2 diagram 31.08.96/2 shows the high portion of 83.5% glacially crushed quartz grains, triturated under a heavy ice pressure. The examination under the microscope has made clear that freshly crushed and sharp-edged material is concerned. This points to a Late Glacial (LGM = 0) ground moraine and not to an older one. The sampling took place on a flat locality with nearly no incline. 200 m away from the bedrock (cf. Photo 115). Thus, the alternative possibility of freshly weathered bedrock is inapplicable." (Kuhle 1999: 31).

Ca. 53 km farther west, at 32° 14′ 55.48″N 81° 42′ 40.01″E, there is light-colored till with boulders of over 4 m in size at 5340 m asl. It consists of local moraine. Similar colored bedrock is in the immediate vicinity. On its margin the till is punctuated so that the much darker

Fig. 2.110 At 32° 7′ 43.17″N 82° 5′ 45.54″E from ca. 4710 (4530 aneroid measurement) m asl looking across a ground moraine plain (locality: Fig. 2.109: Character Y: No. 1) facing west. (■) mark up to decameter-thick, in part probably over 100 m thick, erratic, light-colored ground moraine covers that have filled a dark-colored (in places outcropping) rock landscape (background). (▲) are abraded mountain ridges and roches moutonnées consisting of that same dark bedrock. In places where rock edges have developed, edges formed by ongoing erosion ("Arbeitskanten") are identified, formed after the deglaciation by frost weathering. (- - - -) indicates the Glacial inland ice surface at ca. 6000 m asl. Analog photo 31/08/1996 by M. Kuhle

underground is visible. Ca. 6–10 km farther east, at 32° 13′ 7.95″N 81° 48′ 3.64″E at 4910 m asl, there is similarly light-colored till. Here, too, it obviously contains material that must have been picked up from the adjacent, equally light-colored rock faces, for in its current location, it only lies on dark bedrock.

At a horizontal distance of ca. 23 km farther SE at 32° 3′ 50.83″N 81° 57′ 58.49″E, there lies the same till material at 4600 m asl in a locally ca. 10 km wide main valley bottom. It is widely covered by postglacial, glacio-fluvial gravels and limnic sediments, and therefore is only visible on the framing slopes or at the foot of the mountains (Fig. 2.111; Fig. 2.109, see also http://extras.springer.com: Character Y: No. 2).

The surfaces of the narrower valley bottoms necessarily consist of gravel-floor sheets. Here the glacio-fluvial substrate consists of sandrs (outwash plains) that were deposited during the late Late Glacial to Holocene process of meltdown. The meltwater washed through the tills during the decline of the inland ice, and consequently the surfaces of the tills in the thalwegs in some places are now covered by gravel floors.

Therefore, in places, glacigenic forms of ground polishing such as roches moutonnées protrude from the tills (Fig. 2.112; Fig. 2.109, see also http://extras.springer.com: Character Y: No. 3) but, in others, also from such gravel floors (Fig. 2.113). Traces of subglacial meltwater, which naturally flowed down under great pressure and therefore at high velocity, are still reconstructable in the bedrock. At their margins, it undercut several roches moutonnées by cavitation corrasion (Figs. 2.112 and 2.113).

Fig. 2.111 From 32° 3′ 50.83″N 81° 57′ 58.49″E, ca. 4600 m asl looking SSW across a ca. 10 km long ground moraine plain (locality: Fig. 2.109: Character Y: No. 2). (□) mark a ground moraine covered on a large scale by glacio-fluvial, post-glacial gravels and lake sediments. (■) show at least decameter-thick, light-colored, and in many places erratic ground moraine-covers, covering on a large-scale a dark-colored, outcropping rock landscape. (☗) roches moutonnées and abraded mountain ridges such as glacially streamlined hills of the same dark bedrock. (- - - -) points to the Glacial inland ice surface in the middle foreground at ca. 6000 m asl. Analog photo 31/08/1996 by M. Kuhle

In an abraded mountain massif consisting of high-plateau remnants and rounded summits, and situated ca. 40 km further east, there lies a protruding roche moutonnée at 32° 3′ 13.96″N 82° 25′ 19.71″E. At ca. 5640 m asl, its rusty-brown rock breaks through a light-colored till. In the immediate neighborhood, there also lies a somewhat darker till sheet containing meter-sized boulders. Due to the flat incline of the surface of the area, transport of the material by solifluction can be ruled out.

In the same area there is at 32° 2′ 46.96″N 82° 26′ 43.02″E and at 5700 m asl a light-colored erratic till veil. It lies on an abraded surface of dark bedrock. Accordingly, here too, solifluction can be ruled out for transport of the loose material. In addition, an alternative genesis by in situ weathering can be ruled out, because its distribution shows many punctuations (windows).

At 31° 58′ 50.83″N 83° 55′ 33.58″E, light-colored till is found at 4980–5000 m asl. It contains light-colored boulders up to more than 2 m in size and lies on hilly terrain consisting of dark bedrock. 1.8–1.9 km further south, till lies exposed at 31° 57′ 51.73″N 83° 55′ 40.43″E at 5140 m asl as well as at 31° 57′ 52.76″N 83° 55′ 10.76″E at 5090 m asl. At the first-mentioned location, this coarse, blocky cover of loose stones is 1–2 m thick and recognizable as being erratic, i.e., not weathered in situ. At the second location, this light-colored sheet, too, covers dark rock and is 6–7 m thick.

At 31° 53′ 43.58″N 83° 57′ 55.53″E there are remnants of loose rock covers up to a height of ca. 5800 m asl. They are lighter in color than the

Fig. 2.112 Looking from 32° 9′ 5.21″N 82° 7′ 6.91″E at 4790 (4625 aneroid measurement) m asl onto a roche moutonnée that pierces through the ground moraine cover (locality: Fig. 2.109: Character Y: No. 3) facing NW. (■) marks the ground moraine surface that—by rising—clings to the roche moutonnée. (▲) is a rock surface polished by the inland ice. The edge formed by ongoing erosion ("Arbeitskante") and steepening below (▲) must be attributed to subglacial undercutting by meltwater (see also Fig. 2.113). Analog photo 31/08/1996 by M. Kuhle

underlying solid bedrock and contain light-colored boulders 1 m to more than 2 m in size. They are situated on rounded mountain ridges and so are exposed to fluvial (retrogressive) erosion.

At 31° 52′ 16.71″N 82° 8′ 31.05″E (Fig. 2.115, see also http://extras.springer.com: Character Z: No. 6) there is a ca. 4895 m high hill that is covered by till with 1- to over 2 m sized boulders. It towers above the main valley bottoms in the north by ca. 300 m (Fig. 2.114). Not far away to the west there lies at 31° 50′ 40.64″N 82° 3′ 21.89″E (Fig. 2.115: No 1) from 4620 to 5120 m asl conspicuously light-colored till (Fig. 2.114). It is several meters thick and contains 1–2.4 m sized blocks. In the vicinity of the location at 31° 47′ 37.08″N 81° 55′ 55.45″E (Fig. 2.115: No. 2), at 4850–5130 m, there are several meter-thick, lighter-colored till covers with boulders up to more than 2 m in size.

At 31° 45′ 25.42″N 82° 0′ 20.95″E (Fig. 2.115: No. 3), at 4790 m asl, ca. 40 m above the thalweg, there lies a light-colored, probably erratic till cover with meter-sized boulders. Somewhat WSW at 31° 44′ 43.81″N 81° 58′ 6.84″E (Fig. 2.115: No. 4), at 4925 m asl, lies streaky and at least 4 m thick, also light-colored till. It contains 1–2.5 m sized light-colored boulders. Here, too, moraine windows opening toward the darker debris below are preserved. Possibly this is an older till cover originating from a different flow direction of the inland ice. This might explain the differences in color of the loose rock material.

Again WSW, at 31° 44′ 3.82″N 81° 53′ 15.93″E (Fig. 2.115: No. 5) at 5010 m asl, there lies a light-colored till veil. It contains

Fig. 2.113 Facing west from ca. 7.7 km NNE of the photo locality of Fig. 2.112 from 4630 (4430 aneroid) m asl. (■) mark light-colored and in many places erratic ground moraine covers that on a large scale cover a dark-colored (and outcropping) rock landscape. On the valley bottoms (■ in the foreground) the ground moraine surface has been glaciofluvially flushed during and after the deglaciation. (☗) mark roche moutonnée-like rock hills; some of them (e.g., ☗ on the left) later were undercut by subglacial meltwater. (- - - -) indicates the Glacial inland ice surface at ca. 6000 m asl. Analogue photo 31/08/1996 by M. Kuhle

Fig. 2.114 Ca. 140° panoramic photograph from ca. 4620 (4440 aneroid measurement) m asl at 32° 4′ 42.49″ N 82° 2′ 50.48″E (locality: Fig. 2.109; http://extras.springer.com: Character Y: No. 4). This is a glaciofluvial plain (□) above which tower ground moraine ridges (■ to the very left) and high plains with up to at least decameter-thick ground moraine covers (the 5 ■ symbols on the right). For instance, south of the point where the photo was taken, at a distance of ca. 24 km, there is a ridge culmination with a ground moraine cover at 31° 52′ 16.71″N 82° 8′ 31.05″E and 4895 m, which towers up to 300 m above the plain (Fig. 2.115 on http://extras.springer.com: Character Z: No. 6). To the west at 31° 50′ 40.64″N 82° 3′ 21.89″E (Fig. 2.115: No. 1) there is at 4620–5120 m asl light-colored, erratic ground moraine (second to fourth ■ from the left). It is pierced by darker outcropping rocks and marginally demarcated (see Fig. 2.115: Character Z). (☗) signify glacigenically abraded rock hills, in places with a cover of ground moraine. (- - - -) indicates the Glacial inland ice surface at ca. 6000 m asl. Analog photo 31/08/1996 by M. Kuhle

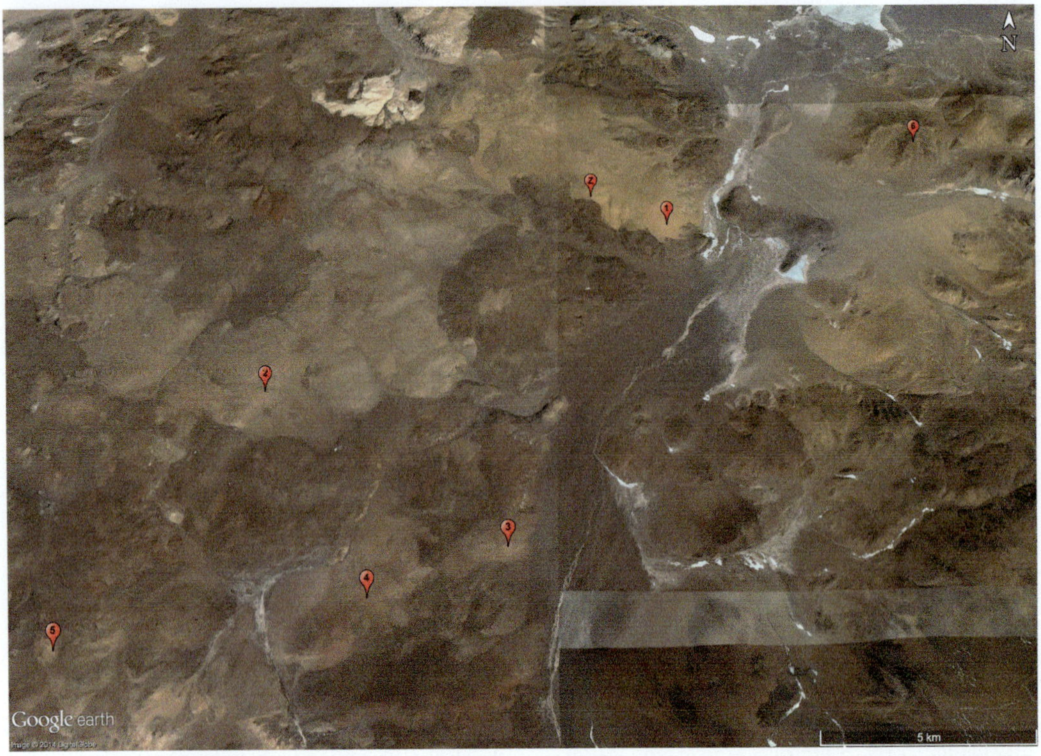

Fig. 2.115 Satellite image, Google Earth Pro: center 31° 47′ 58.18″N 82° 0′ 33.29″E; http://extras.springer.com: Character Z; Nos. 1–6. Character Z marks at 31° 51′ 8.11″N 82° 1′ 54.30″E and 4930 m asl the margin of a ground moraine cover, and No. 1 at 31° 50′ 40.64″N 82° 3′ 21.89″E at about 4620–5120 m asl shows a strikingly light-colored ground moraine (see Fig. 2.114). Further meter-thick ground moraine covers can be seen at No. 2: 31° 47′ 37.08″N 81° 55′ 55.45″E at 4850–5130 m asl; No. 3: 31° 45′ 25.42″N 82° 0′ 20.95″E at 4790 m asl; No. 4: 31° 44′ 43.81″N 81° 58′ 6.84″E at 4925 m asl; No. 5: 31° 44′ 3.82″N 81° 53′ 15.93″E at 5010 m asl; and No. 6: 31° 52′ 16.71″N 82° 8′ 31.05″E at 4895 m asl. They are pierced in places by up to ca. 4910 m asl towering, darker, heavily weathered heads of rock

boulders larger than 2 m in size, and is frayed on its margins. Its margin is surrounded by darker material of loose stones. Here, too, meter-sized boulders can be found. The following explanation can be given for this arrangement: due to its intermittent and inhomogenous flow dynamics, the overlying ice brought together local till covers and in places mixed them together or covered them up. This view is supported by the colored tracks of loose stones that are observed in many places and lead to kilometer-distant outcrops of bedrock from which the material of the particular local moraine originates. In many places, transport via solifluction can be ruled out due to the large distances involved. Furthermore, an equidirectional or even just sufficient incline is mostly absent. Both of these factors provide the crucial preconditions for solifluidal transport.

In the western mountain region of the 5910 m high massif at 31° 29′ 30.73″N 82° 43′ 21.83″E at 5144 m asl a light-colored till lies on the culmination of a mountain ridge. The locality lacks a higher catchment area with correspondingly light rocks, so that this must be erratic till. It contains boulders exceeding 1 or 2 m in size and is situated 420 m above the Xianacun (lake). The former shoreline of this lake at 4900 m asl has caused the formation of a cliff at 31° 15′ 42.43″N 82° 41′ 45.07″E. The ca. 7200 years-old shoreline (see Li and Zhu 2001) (see also Sects. 2.2.2.2 and 2.4.5.1) has here been built on stratified bedrock. Till lies on it. The reworking of the cliff that in the meantime (i.e., since ca. 7200 YBP) has taken place due to further weathering and solifluction.

At 31° 05′ 42.24″N 83° 2′ 39.88″E at about 5568 m asl, a light-colored till covers a darker rock cupola of bedrock. The latter surfaces in many places through the solifluidally structured cover of loose stones.

Immediately to the south at 31° 4′ 46.28″N 83° 2′ 33.51″E up to 31° 04′ 36.25″N 83° 3′ 1.80″E, there is also light-colored till on a dark underground. It lies on flat areas up to the summits of abraded mountain ridges at 5660–5712 m asl. This is 600 m above the nearest thalweg. The till contains 1–2 m sized, partly light-colored boulders.

On the southern valley side 5.2 km to the SSW there is at 31° 2′ 8.22″N 83° 1′ 40.29″E at 5722 m asl a further till sheet. It lies at the top of a mountain ridge and covers darker solid rock, 400 m above the thalweg.

The different colorations of the introduced till covers can be traced back to the varying mixtures of unconsolidated material they contain, representing a mixture of moraines transported long distances and sourced locally. Thus, the solid and (due to detersion) abraded rock underlying these moraines also determines the relevant color.

2.4.7.4.1 A Short Digression on the Holocene Glaciation on the Profile as Far as the Tsangpo Valley in Southern Tibet in Its Relation to the High Glacial Glaciation

In the previous main section (Sect. 2.4.7.4), the focus was on Late Glacial tills and High Glacial or polyglacially rounded mountain forms. This is due to the fact that on the respective section of the N-S profile, no considerably glaciated mountain massifs exist at present. However, south of the locality treated last, this changes. From here onward, Historical to Neoglacial glacier stages together with their ELA depressions are in fact available. Therefore, one can here reconstruct how far the past ELA must have been lowered into the terrain so that a glaciation took place.

On the eastern slope of the 6170 m high massif, at 30° 46′ 46.01″N 82° 55′ 46.59″E (Fig. 2.116 Character AA) at 5230 m asl, lies the bend of a former end moraine. The accompanying current glacier tongue termini lie at 5670–5600 m asl, i.e., on average at 5635 m asl. Accordingly, the mathematical ΔELA amounted to 203 m ((5635 − 5230)/2 = 202.5). This ΔELA classifies the ice margin as belonging to Stage VI–V (see Fig. 2.5). The lowest local former ice margin position is located at 5160 m asl. Here it reached the largest main valley at 30° 43′ 45.08″N 83° 0′ 31.61″E. For this an ΔELA of 238 m ((5635 − 5160)/2 = 237.5) is calculated. That is the ΔELA of Stage V. The lowest area of the main valley bottom is located at ca. 5070 m asl and does not show a further ice margin position. Accordingly, the terrain was filled by glacier ice at an ΔELA exceeding ca. 283 m ((5635 − 5070)/2 = 282.5). During the Holocene Stage V (Neoglacial), an ELA such as this was still quite possible. However, it is absolutely certain for the late Late Glacial. During the late Late Glacial Stage IV (Sirkung stage), the ΔELA even amounted up to 700 m (Fig. 2.5). At this time, the extension of the glaciers might still have had the nearly covering character of an ice cap. Due to its flattened form, the 6170 m high main summit indicates a former abrasion. This certainly must have taken place during the LGP (Stage 0, Fig. 2.5), and presumably also still during the Late Glacial.

Roughly 140 km further west is the adjacent massif of the Kailas (6660 m). The easternmost northern glacier of the Kailas terminates at 31° 4′ 55.58″N 81° 19′ 49′ 40″E at 5280 m asl (Fig. 2.117). The next-lower ice margin, marked by an end moraine, is situated at 31° 5′ 25.06″N 81° 19′ 53.47″E at 5220 m asl. This corresponds to a ΔELA of ca. 30 m ((5280 − 5220)/2 = 30),

Fig. 2.116 Satellite image, Google Earth Pro: center 30° 43′ 7.12″N 82° 47′ 56.04″E; http://extras.springer.com: Character AA; Nos. 1–4. Localities: Character AA: 30° 46′ 46.01″N 82° 55′ 46.59″E at 5230 m asl; No. 1: 30° 37′ 17.73″N 82° 50′ 12.67″E at 5300–5850 m asl; No. 2: 30° 39′ 1.42″N 82° 49′ 14.62″E at 5860 m asl; No. 3: 30° 38′ 28.13″N 82° 48′ 23.94″E; No. 4: 30° 38′ 38.69″N 82° 48′ 24.42″E at 5900 m asl

which has to be classified as belonging to the Historical Stage X (cf. Fig. 2.5).

A lowest further Historical ice margin of this glacier is found at 31° 5′ 52.25″N 81° 19′ 48.67″E at 5160 m asl. With a ΔELA of 60 m ((5280 − 5160)/2 = 60), it belongs to the Historical Stage VII.

In the valley adjacent to the west, the western north glacier of the Kailas shows approximately the same advances. Its Stage VII lies at 31° 5′ 40.99″N 81° 19′ 6.24″E at 5240 m asl (Fig. 2.118). Its tongue terminated during Stage X or XI at 5350 m asl at 31° 5′ 10.89″N 81° 19′ 0.04″E, so that here the ELA difference up to this stage amounted to 55 m instead of 60 m ((5350 − 5240)/2 = 65). The cause of this lies in the steeper catchment area of the direct north face of the Kailas, which has a smaller horizontal basal face (Kuhle 2014c). The respective end moraines of Stage VII are closely synchronized with the corresponding alpine meadow vegetation.

Without discussing in detail the former ice margin positions situated in the area between and in the vicinity (see Kuhle 1999: 6–7, 42–43; 235–247; Fig. 2: No. 120), we now turn to the almost perfectly preserved ice margin at the exit of the NW Kailas valley at 30° 59′ 8.64″ N81° 14′ 41.07″E at 4680 (aneroid 4575) m asl. It was also studied by the author in the field, and there classified as belonging to Stage IV (ibid.: 6–7, 42–43, 234–236; Fig. 2: Nos. 121–123;

Fig. 2.117 Looking south from ca. 5230 m asl toward the easternmost northern glacier of the Kailas (NE Kailas glacier). Its tongue ends at 31° 4′ 55.58″N 81° 19′ 49.40″E at 5280 m asl. Its highest catchment area is the Kailas summit and reaches 6660 m altitude. (■) signifies the ground moraine surface belonging to the Historical Stage VII (Fig. 2.5). Like all historical ground moraines of small mountain glaciers, it is comparably coarse grained and marked by edged boulders and a slightly water-scoured surface. (▲) shows rocks polished by the Neoglacial (Holocene) flank-polishing of this side-valley glacier. The edge formed by ongoing erosion ("Arbeitskante") and steepening below (▲) represents historical undercutting (Stage VII) caused by the glacier tongue at the "black-white boundary." Due to the ever-present humidity, frost weathering is especially intensive at the black-white boundary. Analog photo 05/09/1996 by M. Kuhle

Photo 170). It reached the southern foreland of the Kailas Massif and represents a mathematical ΔELA of ca. 300 m ((5280 − 4680)/2 = 300). Although this ΔELA would even qualify this ice margin for the Holocene Stage V (Fig. 2.5), it has been classified as belonging to the last pre-Holocene stage (IV). Here the ΔELA reconstruction due to the topography and its diverse exposures is rather complicated (Kuhle 1988a). The calculation was based on the N-facing, current glacier terminus at only 5280 m asl, that is, the very low glacier terminus of the NE Kailas glacier (see above). It was then set in relation to the SSW-facing former ice margin at a position 600 m further below. The resulting ΔELA estimation is therefore somewhat too small and probably exceeded 300 m. At the same time, the exponential filling-up of ice of the there very gently inclined trough valley network probably had the effect of autonomous amplification. This was caused by the feedback from the enlarged nourishing area. Both of these opposing factors

Fig. 2.118 Looking south from ca. 5160 m asl toward the north face of Kailas (6660 m). The summit towered ca. 650–850 m above the inland ice surface. Its form, with steep flanks, was shaped by its polyglacial undercutting by repeated inland glaciations in the course of the Quaternary (see also Fig. 2.119). In the side valley below is the westernmost northern glacier of the Kailas (NW Kailas glacier), the two uppermost ■ symbols signifying its Stage VII end moraines at 5240 m asl on both sides at ca. 31° 5′ 40.99″N 81° 19′ 6.24″E. The two lowermost ■ symbols mark fluvially flushed (i.e., cut) Holocene remnants of ground moraine. (□) is the corresponding present-day streambed of glacial meltwater. (⊠) are roughenings due to frost weathering that have taken place since the Holocene deglaciation. (◁) is a detritus cone, also formed by weathering, that also contains a moraine core. Analog photo 05/09/1996 by M. Kuhle

have to be weighed against each other. With a dating into the Late Glacial Stage IV (Sirkung stage), a ΔELA of 700 m would be possible. Obviously this is too large. However, to be safe, it is nevertheless posited here. During the preceding Late Glacial Stage III (Dhampu stage with a ΔELA = 800–900 m; Fig. 2.5), the entire mountain foreland (Tibetan Plateau) between the Kailas and the Gurla Mandhata Massif (7690 m) 54 km farther south was already covered by the inland ice (see Kuhle 1999: 225–234). This is proven by the till sheet between the two massifs that in places is covered by lakes (Fig. 2.119).

2.4.7.4.2 Continuation Towards the Southern Section of the Profile: From Nganclong Kangri (Northern Gandise Shan) in the North up to the Tsangpo Valley in Southern Tibet with Regard to the Former up to High Glacial Ice Cover

During the High Glacial (LGP; Stage 0), the inland ice level reached a height of at least 5800 m asl in the Kailas area. This has been detected on the southern slope of the Kailas as well as on the northern flank of the Gurla Mandhata (ibid.: 42, 229–230, 233–234; Photos 163, 165, 168, 169).

Fig. 2.119 From ca. 4800 (4620 aneroid measurement) m asl at ca. 30° 48′N 81° 22′E looking NNW toward the 6660 m high Kailas. Visible are ground moraine covers that probably far exceed 100 m in thickness. (■ black) marks ground moraine of the Late Glacial Stage III (Dhampu stage; see Fig. 2.5), which continue between Langa Co and Mapam Yumco. (■ white) mark further remnants of ground moraine up to ca. 700 m above those two large lakes. (☗) are glacigenic abrasion-forms up to 5740–5820 m asl. (- - - -) is the High Glacial glacier level (Stage 0, LGP, LGM) of the Tibetan inland ice at ca. 5810–6010 m asl. Analog photo 03/09/1996 by M. Kuhle

Further north, i.e., north of the Kailas, there is e.g. at 31° 16′ 35.79″N 81° 20′ 31.17″E (Fig. 2.120 Character AB) a light-colored till sheet covering the area from 5530 m asl at the lower end of the slope as far as up to 5660 m asl. This material is obviously erratic matrix on dark rock.

At, for example, 31° 15′ 56.88″N 81° 21′ 42.40″E, there is a 5990 m high abraded mountain (Fig. 2.120: No. 1) that shows a light-colored cover of loose stones that looks like till. A corresponding overlay of loose stones is located at 31° 13′ 35.98″N 81° 21′ 15.65″E on a roundly abraded mountain at 5870 m asl (Fig. 2.120: No. 2). At 31° 12′ 6.32″N 81° 17′ 37.50″E (Fig. 2.120: No. 3) and 31° 12′ 21.73″N 81° 19′ 44.45″E, and likewise at 5870 m asl (Fig. 2.120: No. 4), there are remnants of till on the summits of abraded mountain ridges and cupolas. Accordingly, for the High Glacial Stage 0 (LGP), ice levels of at least 6100 m asl are evident there.

The form of the solitary Kailas summit is therefore remarkable in a geomorphological sense. With its steep flanks it must have towered above the inland ice surface by ca. 650–850 m. At elevations 650–850 m below the highest point, the highest flattening cracks and crest ridges begin to show (Figs. 2.118 and 2.119). Necessarily these are tied to the nearly horizontal layers of conglomerate rock. However, the steep summit form cannot be explained in this way. Instead, it must be traced back to the polyglacial undercutting of the mountain flanks due to the

Fig. 2.120 Satellite image, Google Earth Pro: center 31° 15′ 21.38″N 81° 18′ 20.33″E; http://extras.springer.com: Character AB; Nos. 1–4. Character AB: At 31° 16′ 35.79″N 81° 20′ 31.17″E from 5530 m at the footslope up to 5660 m, ground moraine with a light, obviously erratic matrix on dark rock. No. 1: 31° 15′ 56.88″N 81° 21′ 42.40″E, a 5990 m high, abraded mountain. No. 2: lighter overlays of loose stones also at 31° 13′ 35.98″N 81° 21′ 15.65″E up to 5870 m asl. No. 3: at 31° 12′ 6.32″N 81° 17′ 37.50″E and No. 4: at 31° 12′ 21.73″N 81° 19′ 44.45″E up to 5870 m asl, ground moraine remnants on abraded mountain cupolas

inland glaciations in the course of the Quaternary. The resulting, by ca. 200 m lower, not very distinct demarcation line must be due to the repeated rises and falls of the ice level during the period from the Early to the High to the Late Glacials.

We now turn to the original N-S profile ca. 150 km farther ESE from the Kailas Massif. The localities already belong to the immediate northern catchment area of the upper Tsangpo. On the profile in question there are further paradigmatic localities with Late Glacial till (Stage IV or III) which shall be introduced here:

On a mountain slope at 30° 37′ 17.73″N 82° 50′ 12.67″E (Fig. 2.116, see also http://extras.springer.com: Character AA: No. 1), from 5300 m to ca. 5850 m asl stretching towards the NNW, lies a nearly continuous till cover. It contains 1–2 m sized boulders and in places is a decameter in thickness. It is classified as belonging to Stage III. At the upper end of the slope at 5860 m, at 30° 39′ 1.42″N 82° 49′ 14.62″E (Fig. 2.116, see also http://extras.springer.com: Character AA: No. 2), there lies a till-like cover of loose stones. It contains boulders up to 3 m in size embedded in a fine matrix, which might have been plucked off in situ by the ice.

In the same currently non-glaciated mountain area, somewhat further west at between 30° 38′ 28.13″N 82° 48′ 23.94″E (Fig. 2.116, see also http://extras.springer.com: Character AA: No. 3) and 30° 38′ 31.52″N 82° 48′ 36.57″E, at 5760 m asl, we find a lateral moraine or

Fig. 2.121 Satellite image, Google Earth Pro: center 30° 16′ 5.84″N 83° 8′ 53.53″E; http://extras.springer.com: Character AC; Nos. 1–2. Locality of Character A: 30° 17′ 13.89″N 83° 8′ 25.75″E. No. 1: 30° 18′ 17.36″N 83° 8′ 22.39″E, light-colored ground moraine up to a height of ca. 5420 m asl. No. 2: 30° 14′ 37.58″N 83° 8′ 53.24″E, light sections of ground moraine between 5800 and ca. 5090 m asl

kame margin. It runs together in the confluence area of two valleys ca. 300 m above the joint thalweg and has to be classified as belonging to the period of melting of the late Late Glacial Stage IV.

Above it stretches a till cover up to 30° 38′ 38.69″N 82° 48′ 24.42″E and 5900 m asl (Fig. 2.116, see also http://extras.springer.com: Character AA: No. 4). Accordingly, stratigraphically this belongs to the completely developed Stage IV or the next-older Stage III. It contains boulders from 1 m up to more than 2 m in diameter. Due to the melting of snow, in places the fine-material matrix has been washed away from it unevenly, leaving behind a patchy distribution. The mountain cupola, abraded by the inland ice and therefore rounded, has been slightly sharpened since its deglaciation. This occurred through the development of a northwardly exposed, flatly scoured-out cirque.

At 30° 35′ 27.07″N 82° 42′ 18.38″E lies a decameter-thick remnant of till at an altitude of 5070–5200 m. It contains 2–3 m large boulders and is classified as belonging to the Late Glacial Stage IV or III (see Fig. 2.5).

At 30° 18′ 17.36″N 83° 8′ 22.39″E (Fig. 2.121, see also http://extras.springer.com: Character AC: No. 1), up to a height of ca. 5420 m, there lies a light-colored and at most a few meters thick till on extensive mountain ridge summits. It contains light-colored, 1 m to over 3 m sized boulders and lies on dark bedrock. Equally light-colored and towering rock tors are lacking. Thus, this overlay of loose rock must have been horizontally dislocated by at least several hundred meters. Somewhat to the south, at 30° 17′ 13.89″N 83° 8′ 25.75″E (Fig. 2.121, see also http://extras.springer.com: Character AC: A) up to ca. 30° 14′ 37.58″N 83° 8′ 53.24″ E (Fig. 2.121: No. 2), further light-colored sections of till cover have been found in positions at between 5800 down to 5090 m asl. They are also located here on dark bedrock and contain up to 2.7 m sized boulders.

Ca. 77 km further SE down the Tsangpo Valley at 29° 50′ 55.45″N 83° 47′ 4.64″E and 5017 m asl, a light-colored remnant of till is found on a hill cupola, ca. 450 m above the Tsangpo River. It overlies a dark surface and contains boulders up to 3.5 m in size. Somewhat further north, till like this also persists at a large scale on the gently inclined slope; in steeper sections it has been removed since the deglaciation. Since that time fine slope rills have been cut into the till surface.

On the opposite, orographic left side of the valley, there is at 29° 46′ 17.87″N 83° 46′ 46.75″E a corresponding light-colored local till. It is situated at about 5310 m asl and contains up to 2 m sized boulders. Ca. 290 m farther west, at 29° 46′ 17.52″N 83° 46′ 36.88″ E, and 10 m higher (5320 m asl) this till has been exposed by fresh rills. It lies ca. 750 m above the thalweg of the main valley of the Tsangpo River.

At 29° 49′ 33.23″N 83° 8′ 45.63″E, at about 5040 m asl, lies a cover of loose rocks containing a large portion of fine matrix. In many places where it has been cut by retrogressive erosion, small valleys have developed. They prove the insignificant resistance offered by the material that is characteristic of tills. Their spotted patterning corresponds with that of typically arctic tills. The underlying surface mainly consists of dark bedrock. In places the bedrock also penetrates the large-scale cover of loose stones and shows the same color. Thus, at least in part, the till is local here.

At 29° 38′ 53.12″N 84° 25′ 18.34″E at about 5230 m asl, there is a till cover built up of mixed detritus of sedimentary rock. It is located ca. 520–600 m above of two adjacent thalwegs and might even have been located at the level of the snowline still during the late Late Glacial Stage IV. In any case, for the last time it was covered during this stage by allochtonous glacier ice (see below). Immediately, only 20–24 km, NE of one of the two thalwegs is the still currently glaciated Lunpo Kangri Massif. This was therefore the last glacier catchment area of the locality just mentioned, which still towered above the ELA (snowline).

2.4.7.4.3 A Second Digression on the Holocene Glaciations on the Profile Towards the Tsangpo Valley in Southern Tibet in Their Relation to the High Glacial Glaciations

The 7172 (7327) m high Lunpo Gangri (Loinbo Kangri) Massif (Kuhle 2011b: 196: Fig. 9; 2013b: 20: Photos 1 and 2: No. 1) has a SW glacier that flows down to 5410 m asl. Downvalley from this glacier terminus at 29° 47′ 6.32″N 84° 35′ 13.58″E at 5280 m asl and at 29° 47′ 4.36″N 84° 34′ 13.24″E at 5220 m asl, there are two complete end moraines. They prove an ΔELA of 65 m ((5410 − 5280)/2 = 65) and one of 95 m ((5410 − 5220)/2 = 95). The ice margins have to be classified as belonging to the Historical Stage VII (younger Dhaulagiri stage, Fig. 2.5) and Late Holocene Stage 'VII (middle Dhaulagiri stage).

A further glacier stage that due to end moraines is unmistakable can be diagnosed on the 8 km wide valley bottom of the orographic left side of the Tsangpo Valley at 29° 44′ 30.84″N 84° 28′ 12.00″E at 4880 m asl. Its end moraine confirms a ΔELA of 265 m ((5410 − 4880)/2 = 265). In the immediate vicinity both to the north and south, one can observe ice margin positions of parallel-running glaciers that show the same ΔELA. They, too, exhibit the geomorphologically same fresh state of conservation and belong to the middle Holocene (Neoglacial) Stage V (Nauri stage) (Fig. 2.5). The other three expositions of the massif also exhibit corresponding tongue basins with end moraines, i.e., the same glacier stages.

Here, the lowest possible point of the relief that can have been reached by the former glacier cover lies at ca. 4550 m asl, well over 40 kmfrom the Lunpo Gangrii. When this point was reached by the glacier ice, the filling with ice of the local basins must inevitably have taken place. The relief then became completely glaciated. This stage was reached at a ΔELA of ca. 430 m ((5410 − 4550)/2 = 430). The next-older stage (after Stage V) was the late Late Glacial Stage IV (Sirkung stage). It had a ΔELA of ca. 700 m. Accordingly, at this time the terrain still was filled and overridden by an ice cover, so that only the Lunpo Gangri towered above it. The mere surplus of the ΔELA relative to the ΔELA that only would have been necessary for complete filling with ice already amounted to 270 (700 − 430 = 270) m. That is the same amount of ΔELA as is attributed to the strongest Holocene advance. To be precise, the ΔELA of this Holocene Stage V all in all amounted to 265 m (see above). For Stage III, the reserve buffer for a past inland ice cover would even amount to 370–470 m, for Stage II to 570 m, for Stage I 670 m, and for the High Glacial Stage 0 it would be 870 (1300 − 430 = 870) m (see Fig. 2.5). These are indeed very large reserves that provide great support for the geomorphologically and sedimentologically reconstructed complete inland glaciation.

2.4.7.4.4 Further Continuation of the Southern Section of the Profile from the Nganclong Kangri (Northern Gandise Shan) in the North up to the Tsangpo Valley in Southern Tibet with Regard to the Late to High Glacial Ice Cover

The adjacent locality at 29° 38′ 28.53″N 84° 2′ 21.36″E at 4565 m asl is situated 15 m higher than the lowest relief point indicated above. Here lies till with boulders of 1–3 m in size a few meters above several pools filled with water. It has to be classified as belonging to Stage IV or III. A little to the east at 29° 37′ 50.61″N 84° 8′ 27.78″E, at between 4560 and 4590 m asl and also on the floor of the Tsangpo Valley, are found further dead-ice holes with water pools. They probably also originated during Stages IV or III. Here we are only 15–45 m below the Kore La (pass) (Fig. 2.122). This is a glacial transfluence pass over which a 26–31 km wide and at least 1400 m thick glacier arm from Tibet flowed southward to the Himalaya breakthrough (Kuhle 2007b: 123; 2011b; 2013b: 90, Photo 60). This was the at most 120 km long Mustang Thak Khola outlet glacier (Kuhle 2004; 2011b: 178, Fig. 3), which terminated on the southern side of the Himalaya at ca. 1000 m asl, south of the Dhaulagiri and Annapurna (Kuhle 1980, 1982a).

The Kore La (pass) is situated at 29° 19′ 19.40″N 83° 58′ 47.19″E at 4610 (4603) m asl and has been formed in till, containing meter-sized boulders. It was overridden still during the late Late Glacial Stage IV and most certainly during the older Stage III. Here, at the southern margin of Tibet, the retrogressive erosion of the Ganges River system reaches farthest north into the high plateau (Fig. 2.122).

It is the Ganges tributary called Kali Gandaki that by now has already eroded back nearly as far as the Kore La (pass) (Kuhle 2011b: 177, Fig. 2, 182–185). Therefore, the loose rock of the only slightly resistant till south of Kore La at 29° 16′ 17.67″N 83° 59′ 24.74″E at 4500 m asl is already exposed (cut into) to depths of several decameters (Kuhle 2007b: 120, Photo 1, 123; 2013b: 10, 11, 19–38, 44, 51; Fig. 5; Photos 1, 2, 5–15, 24, 36). At 29° 15′ 2.25″N 84° 8′ 8.88″E, at 4960 m asl, lies a streaky, light-colored, local till. It lies slightly north of the watershed and ca. 30 m above the thalweg of a small side valley.

In the same area at 29° 33′ 47.40″N 84° 20′ 47.83″E on the culmination of a hill at 4890 m asl and thus 340 m above the Tsangpo thalweg, a light-colored remnant of till has been found. Correspondingly light remnants of a till

Fig. 2.122 From 4835 m asl (aneroid measurement) (29° 10′ 45″N 83° 52′ 50″E) facing NNW from the Nhubine Himal-NE summit (6386 m) up to facing east (right margin). The 7327 (7170) m high peak and the 6821 (6568) m peak belong to the Lunpo Gangri (Loinbo Kangri) massif; No. 3 is the 4610 m high Kore La (pass). (■) are meter- to decameter-thick ground moraine covers and -pedestals. (□) is a glaciofluvially modified ground moraine surface. (O) are coarse granite blocks on a surface of ground moraine. (↓) are small source valleys and -channels of the Kali Gandaki (Mustang-Thak Khola) developed by regressive erosion. They are cut into and through the ground moraines. (▲) mark glacigenic abrasions and roundings on outcropping metamorphic sedimentary rocks (phyllites). (- - - -) shows the Ice Age surface of the inland ice at about 6300–6000 m (from the middle up to the right). (⇨) is the transfluence area of the inland ice with its Mustang-Thak Khola outlet glacier across the Kore La. (▽ ▼) are active nivation niches and covers of solifluction debris above the alpine meadow vegetation. (① and ②) are the sampling localities of 29.03.05/2 and 27.03.05/1 (Kuhle 2011b: 179, 180, 192, Fig. 4a, samples No. 25 and 29, Fig. 5/1; 2013b: 24, 26, 61, Fig. 34, Photos 3 and 4). *Across the transfluence pass of Kore La flowed down a 26–31 km broad and at least 1400 m thick glacier arm from Tibet southward to the Himalaya break-through* (Kuhle 2013b). Analog photo 29/03/2005 by M. Kuhle

cover also are preserved in the surrounding area. They are erratic, because they lie on dark, metamorphic bedrock. In many places the light till has been removed from the bedrock surfaces and the bedrock abraded. Somewhat to the south, at 29° 32′ 20.38″N 84° 23′ 47.07″E, lies a 4640 m high glacigenic, rounded knob. It consists of horizontally arranged metamorphites with layers of quartzite. Erratic till overlies it up to nearly 100 m above the Tsangpo River. It contains boulders of 1–3 m in size.

On the mountain area rising 800 m SW above the Tsangpo and reaching 5450 m asl, in the vicinity of 29° 25′ 24.86″N 84° 17′ 41.24″E, light-colored remnants of till are widespread. Upslope there are equally light colored outcrops of bedrock. Yet, the covers of loose rock with 1–2 m sized boulders are also found lying on dark rock at, for example, 5080–5190 m asl.

Roughly 50–60 km further south, the till cover at 28° 58′ 54.76″N 84° 11′ 06.57″E is worth mentioning due to its immense thickness. Here, at 5410 m asl, the glacigenic unconsolidated material reaches a thickness of ca. 120 m (Fig. 2.123, see also http://extras.springer.com: Character AD). Sporadically it contains boulders up to ca. 3 m in size. The strikingly light color of the till proves its origin from the north. Only there does solid rock crop out with mainly this light color. In the south, however, these rocks are lacking. Accordingly, the Tibetan inland ice must have transported the glacially abraded material here, from the north to the south.

This locality is situated on the northern margin of the Damodar Himalaya, a mountain group of the Inner Himalaya (also called the Tibetan Himalaya). The W-E oriented marginal chain of the Damodar Himalaya in many places reaches 6500 m asl. It is at most 6699 or 6740 m (Khumjung or Khumjungar), 6899 m (Lugula Himal: see Kuhle 2013b: 123, 124, Photos 97 and 98: No. 45), and 6995 m (Ratna Chuli) high.

The author has investigated in detail the past glaciation of the Damodar Himalaya, and reconstructed an ice stream network there (Kuhle 2012c). It was connected to the Marsyandi Nadi ice stream network. During Stage 0 (LGP) the terminus of its parent glacier extended into the

Fig. 2.123 Satellite image, Google Earth Pro: center 28° 56′ 52.22″N 84° 14′ 15.16″E; http://extras.springer.com: Character AD; Nos. 1–10. Character AD: At 28° 58′ 54.76″N 84° 11′ 06.57″E in 5410 m asl, 120 m thick, glacigenic loose material. No. 1: 28° 56′ 30.33″N 84° 12′ 19.73″E, 5280 m asl, lower end of the Khumjung (Khumjungar)-North glacier tongue in the year 2012. No. 2: 28° 57′ 30.36″N 84° 11′ 46.22″E, 5110 m asl, the lowest oldest historical ice margin of this glacier. No. 3: 28° 56′ 55.91″N 84° 14′ 6.90″E, 5510 m asl, lower end of the current NW Lugula glacier tongue in the year 2012. No. 4: 28° 58′ 23.16″N 84° 14′ 52.52″E, 5390 m asl, the lowest oldest historical ice margin of this glacier. No. 5: 28° 59′ 15.90″N 84° 18′ 48.04″E, ca. 5140 m asl, lowest ice margin of the youngest piedmont glacier on this section of the Tibetan Plateau. No. 6: 29° 4′ 21.94″N 84° 21′ 42.22″E, 4870 m asl, next-deeper ice margin of the Damodar-North glacier. No. 7: 29° 6′ 14.87″N 84° 17′ 50.40″E, downslope-displaced exaration rills. No. 8: upper margin of ground moraine between 29° 1′ 43.43″N 84° 11′ 35.51″E at 5470 m and No. 9: 29° 0′ 15.82″N 84° 14′ 7.10″E, 5390 m asl. No. 10: 29° 0′ 17.00″N 84° 15′ 34.41″E, 5330 m asl, end moraine of late Late Glacial ice cap in Tibet, thrust from the north

southern Himalayan foreland as far as down as 400 m asl (Kuhle 1996b, 1998a: 87; 2014b).

During the LGP (Stage 0), the Damodar ice stream network reached glacier thicknesses of ca. 2000 m and ice levels of 6600 m asl (Kuhle 2012c: 9, 28–32, 34–36, Photos 6–10). The Damodar ice stream network received supply from the southern margin of the Tibetan ice across four transfluence passes (Kuhle 2013b: 124, 179, 208, 209, Fig. 167a: No. 79 and on both sides below No. 45; Photo 97 on both sides below No. 45). First, the former transfluence pass at 28° 53′ 30.65″N 84° 10′ 30.14″E between the Khumjungar and Lugula Himal needs to be pointed out. It is situated at 6060 (6057) m asl (Fig. 2.123, see also http://extras.springer.com: Character AD). Currently it is glaciated on both exposures—north and south (here SE). Also evident is the former ice spillway east of Lugula at 28° 53′ 10.91″N 84° 16′ 33.75″E at 6190 m asl. It extended through a trough valley leading gently upward from southern Tibet from ca. 5800–5850 m asl (Fig. 2.123). Still more clearly discernible and 240 m lower is the 5850 m high glacial transfluence pass coming from Tibet, situated at 28° 51′ 55.84″N 84° 19′ 44.47″E (Fig. 2.123). The highest transfluence pass at 6440 m asl can be found at 28° 51′ 45.18″N 84° 21′ 58.93″E (Fig. 2.123). It also has a trough-like shape and is located west of the Ratna Chuli. Accordingly, all four transfluence passes are gently rising trough valleys, but do not owe their glacigenic forms to the current glaciation, which is by far too shallow to form a trough. Rather an inland ice sheet was needed for this, rising and overflowing from the north. Farther SE between the Ratna Chuli and Panbari Himal (ca. 6900 m), there are five further transfluence passes, reaching ca. 6090–6250 m asl (Fig. 2.123).

The above-mentioned thick-layered and light-colored till was shifted upward from the north against the Damodar by several hundred meters (see above). On the northern slope of the Damodar it reached a height of at least 5800–6000 m asl, its highest preserved occurrences. In the forefields of the current glaciers this long-distance transported till has been taken up again. It has been shifted back a little to the north by the local glacier tongues. Thereby its material inevitably has been integrated into Historical to Holocene (Neoglacial) end moraines (Fig. 2.123, see also http://extras.springer.com: Character AD: e.g., No. 3).

Accordingly, TCN datings of boulders on these end moraines must indicate a clearly older than Historical or Neoglacial (Holocene) age. This mixing of polyglacial material is a general problem for the dating of tills in High Asia.

The former ice sheet glaciation, which increasingly transported till from the north to its current locations, probably still existed during the late Late Glacial Stage IV (Sirkung stage; see Fig. 2.5) (see Sect. 2.4.7.4.5) and definitely during the Late Glacial Stage III (Dhampu stage) or II (Taglung stage). During the last phase of its ice flow towards the south, the deposition of till was most intensive. The cause of this was the already strongly uplifted ELA at that time.

2.4.7.4.5 A Third Digression on the Historical and Holocene Glaciations on the Profile to the Tsangpo Valley and the Tibetan Himalaya in Southern Tibet in Their Relation to the High Glacial Glaciation

This digression concerns the geomorphological-chronological classification—that is, dating—of the Late Glacial, i.e., last superior inland glaciation, as well as the youngest glacier stages. In the glacio-chronology cupolas are formed by means of snowline depressions (ΔELA).

First example: In 2012, the current Khumjung–(Khumjungar) North glacier tongue reached up to 28° 56′ 30.33″N 84° 12′ 19.73″E, and down to 5280 m asl (Fig. 2.123, see also http://extras.springer.com: Character AD: No. 1).

Its oldest Historical ice margin was located at 28° 57′ 30.36″N 84° 11′ 46.22″E at 5110 m asl (Fig. 2.123: Character AD: No. 2). Its ΔELA amounted to ca. 85 m ((5280 − 5110)/2 = 85). Accordingly, it has to be classified as belonging to the Historical Stage VII (“younger Dhaulagiri stage,” Fig. 2.5) (see Kuhle 1982a: 163–165). This chronological classification due to the ELA depression is confirmed by the still very sparse plant life (pioneer vegetation). The lowest former ice margin of the Khumjung-North glacier reconstructable by end moraines is situated at 4910 m asl at 28° 58′ 17.30″N 84° 10′ 7.53″E. The ΔELA amounts to 185 m ((5280 − 4910)/2 = 185) so that the ice margin has to be classified as belonging to Stage VI or even ‘VII in the late Holocene (4000–1700 YBP) (ibid.: 161–163), or—in the oldest case—to Stage V in the middle Holocene (Fig. 2.5). During the next-older stage, the late Late Glacial Stage IV, an inland glacial ice cover still existed in the more distant northern mountain foreland. At the same time, immediately to the west was the decakilometer-wide Mustang Thak Khola outlet glacier (Kuhle 2011b: 178, Fig. 3). It completely covered the area of the respective lower Khumjung-North glacier valley and thereby overflowed it diagonally from N to S (Kuhle 2013b: 7, 10, 11, 19–51, Fig. 3; Photo 5; Photo 6: No. 6; Photo 7 on the left below No. 12; Photo 10 on the left of No. 13; Photos 23, 24, 28, and 36: on the left below No. 12).

Second example: The extant NW Lugula glacier terminated in 2012 at 5510 m asl at 28° 56′ 55.91″N 84° 14′ 6.90″E (Fig. 2.123: Character AD: No. 3). The oldest Historical position of its tongue end (Stage VII) reached down to 5390 m asl at 28° 58′ 23.16″N 84° 14′ 52.52″E (Fig. 2.123: Character AD: No. 4). This dating is confirmed by a ΔELA of 60 m ((5510 − 5390)/2 = 60) (see Kuhle 1982a: 163–165).

The next-older stage—together with the four Lugula-North glaciers adjacent to the east—still formed a 9.4 km wide, interconnected piedmont glacier on the Tibetan Plateau. Its lowest ice margin lay at ca. 5140 m asl at 28° 59′ 15.90″N 84° 18′ 48.04″E (Fig. 2.123: Character AD: No. 5). Its dating to the Late Holocene (Neoglacial) Stage 'VII (ibid.: 161–163) or VI (middle or older Dhaulagiri stage, Fig. 2.5) is evidenced by a ΔELA of ca. 185 m ((5510 − 5140)/2 = 185). The Late Holocene age of the piedmont glacier can be diagnosed by the condition of the till here. The till surface shows freshly preserved forms of an ablation landscape ("Niedertaulandschaft"), such as, for example, depressions and run-off rills as they develop due to the melting. In part they have been destroyed locally by erosion. In addition they also provide—due to the long-term denudation—an already different picture of the surface.

At 29° 4′ 21.94″N 84° 21′ 42.22″E at 4870 m asl, the next-lower ice margin of this Damodar-North glacier appears (Fig. 2.123: Character AD: No. 6). Here, the tongue terminus extended down into a relatively narrow valley. Accordingly, it flowed down as a narrow valley glacier from the piedmont glacier in a northward direction. Its ΔELA amounted to ca. 320 m ((5510 − 4870)/2 = 320) (Fig. 2.5), which properly corresponds to Stage V (Holocene Nauri stage). 30–65 km farther SSW to SW, the ΔELA of this Stage V was surpassed by ca. 250 m. Namely, here in the precipitation shadow of the Dhaulagiri and Annapurna, a ΔELA of ca. 570 (567) m was assessed based on the study of 17 north-facing glaciers of the Tibetan Himalaya (ibid.: 159, 160).

Somewhat further north, at 29° 6′ 14.87″N 84° 17′ 50.40″E in an orographic left side valley, there are downslope-shifted exaration rills (Fig. 2.123: Character AD: No. 7) created under the influence of permafrost. The valley has no connection to a high, i.e., still today glaciated catchment area. The exaration rills occur on the valley slope in a till cover at between ca. 4950 and 5090 m asl. Remnants of till here reach up to 5400 m asl. Obviously these remnants must have been deposited by means of a local ice cap in the area of Tibet. An ice cap such as this is confirmed by an ice level sloping downward to the south. This is evidenced by a light-colored—very likely synchronous—upper margin of till at between 29° 1′ 43.43″N84° 11′ 35.51″E at 5470 m (Fig. 2.123: Character AD: No. 8) and 29° 0′ 15.82″N 84° 14′ 7.10″E at 5390 m asl (Fig. 2.123 Character AD: No. 9) a little to the south. Thus, here a valley glacier tongue grew thinner from north to south, i.e., flowed to the south. The till cover continues to the north upvalley. Whether it still existed during Stage V or belonged to the glaciation of the late Late Glacial Stage IV cannot be determined with certainty.

The normal difference of the ΔELA of Stage V to Stage IV is from ca. 300 to 700 m, that is, 400 m (Fig. 2.5). This is too great a difference given the only very insignificant geomorphological discrepancy. Thus, the older till at 29° 2′ 6.58″N 84° 10′ 11.77″E at 5300–5500 m asl also shows very fresh traces of preservation in the form of dead-ice and ablation structures. Therefore, the glacier ice cover of this till has to be classified as being older than Stage V, but younger than Stage IV. Accordingly, it belongs most likely to the ablation (meltdown) phase of IV, i.e., to the transition from Stage IV to V.

This till in places is at least 2 m thick and contains 1–3 m sized boulders. The ramp-like slopes of till in the valley under discussion are similar to those in the valley of Chamonix in the European Alps. However, these Tibetan tills appear to be younger. This till corresponds in terms of its level with the till in the adjacent valley to the east. There, the till even reaches at least 10 m in thickness.

A further informative remnant of till is situated at 29° 2′ 1.31″N 84° 17′ 58.29″E at 5460 m asl. This is local moraine with light-colored portions, i.e., quartzite debris carried over small distances. It forms an orographic left till ledge that was transported up to here from the west. This till ledge falls away toward the NE downvalley and reaches down to ca. 5150 m asl at 29° 2′ 49.11″N 84° 19′ 15.58″E.

At 29° 0′ 17.00″N 84° 15′ 34.41 there lies an end moraine at 5330 m asl. It was overridden from the north, from Tibet, by a valley glacier tongue belonging to the late Late Glacial ice cap (Fig. 2.123, see also http://extras.springer.com: Character AD: No. 10). At 29° 0′ 56.89″N 84° 15′ 23.24″E there lies at 5570 m asl on the orographic left side a lateral moraine ledge leading downvalley and to the south. It belongs to this local ice margin. These ice margins, too, belong to the meltdown phase of Stage IV. An equally old small outlet glacier tongue adjacent to the east terminated at 29° 0′ 0.41″N 84° 16′ 56.73″E at 5220 m asl. There is a well-preserved end moraine in a small valley, also sloping down to the south. A still somewhat lower ice margin lies at a distance of 720 m farther south at the exit of this small valley. To this belongs the right lateral moraine at 28° 59′ 50.22″N 84° 16′ 38.73″E at 5320 m asl.

Up to six lateral moraine ledges and traces of ice levels form part of these ice margins and still others exist further downvalley. They lie extended, for example, at 29° 0′ 34.50″N 84° 16′ 36.79″E at 5430 m asl and further above at 29° 0′ 20.71″N 84° 16′ 22.73″E at 5510 m asl. On the opposite, orographic left flank of the small valley there are at corresponding heights also field ledges sloping gently downvalley. These are traces of the respective valley glacier levels, to which belong remnants of till covers.

In two further parallel valleys lying farther east, corresponding valley glaciers flowing south can be identified. Lying adjacent to the west from the introduced, formerly glaciated small valleys, there are—also inclined toward the south—two further reconstructable valley glacier margins (Fig. 2.123, see also http://extras.springer.com: Character AD: on the left of No. 9 and on the left of No. 8). The moraine deposits within them prove a simultaneous glacier ice filling.

These findings are important. Once more they prove the Tibetan glaciation. The highest local catchment area of the small valleys nowhere reaches an altitude of 6000 m asl. On average it reaches ca. 5700–5800 m asl. In order to reach the lowest glacier margins at ca. 5200 m asl, the ELA must have been situated at a maximum of 5475 m asl (5750 − 5200 = 550; 550/2 = 275; 5750 − 275 = 5475 i.e., 5200 + 275 = 5475). This ELA applies to the southern slope that is favorable to radiation. On the northern slope, in the Tibetan Himalaya, the ELA reaches ca. 300 m further below (ibid.: 156–161). Accordingly, this ELA at a 5475 m southern exposure corresponds with an ELA at 5175 m for a northern exposure. On the Damodar northern slope, situated only 6 km to the south (Fig. 2.123: Character AD: S of No. 3), the current ELA lies at ca. 5800 m asl. It was determined on the Lugula-North glaciers with the help of the AAR (accumulation area ratio) and also the meltdown lateral moraine (upper moraine). At a ΔELA of well over 600 (625) m, it would reach 5175 m. In this case, the N-exposed ELA would already be located at below 5500–5200 m asl (Fig. 2.123: Character AD: between Nos. 3 and 10). This is the height of the foreland level and thus of the Tibetan Plateau. Through it a piedmont (i.e., plateau) glaciation would develop, which would fill the small valleys under discussion from south to north with ice and cover their flat mountain region via the north (Fig. 2.123 north of Nos. 5 and 10). As shown by those seven ice margin positions overridden to the south, this was not the case (Fig. 2.123 north of the line from Nos. 8 to 5). Therefore, this Tibetan mountain area between Nos. 10 and 7 in Fig. 2.123 still must have been covered inevitably by glacier ice, although the ELA had already risen above the highest summits of this mountain group. The latter is an exclusive characteristic of an inland ice cover. Due to its large ice mass, it persists longer while the climatic-genetic uplift of the ELA continues. The center and later the various Late Glacial centers of this huge ice mass was/were situated further north on the Tibetan Plateau. During its meltdown phase, the ice marginally also flowed to the south, which occurred here. Obviously this took place before the young Holocene, i.e., Neoglacial Stages 'VII (ibid.: 161–163) or VI, possibly still during the next-older Holocene Stage V, but definitely not during the late Late Glacial Stage IV. For this the relevant ΔELA of ca. 700 m was definitely too large (Fig. 2.5). Probably this was the ablation phase of the inland ice after Stage IV. Due to its immense thickness, this might have continued as far as into the Holocene —perhaps even to Stage V.

Geomorphologically adequate till levels—also above the level of Stage V—are situated somewhat to the east at 29° 2′ 20.82″N 84° 23′ 35.52″E and 29° 3′ 4.45″ N84° 23′ 25.05″E. Both of them lie at ca. 5140 m asl. Below these slope positions, relatively fresh surfaces of till can be diagnosed.

A higher level of till has been preserved near the location at 29° 1′ 46.86″N 84° 21′ 19.83″E at 5500 m asl. It thus has to be classified as belonging to the glacier level of Stage IV, which had not yet melted down any further. The surrounding mountain cupolas towered above this ice level by about 400 m. Accordingly, there existed a southern Tibetan ice stream network at that time. Below an adjacent cirque, which during the Holocene was still glaciated, this late Late Glacial till was, among other things, washed away by meltwater. Remnants of till still occur here up to a height of ca. 5700 m asl.

The mountain ridges and cupolas, rounded by glacigenic abrasion, are up to 5700 and 5800 m in height, for example at 29° 3′ 48.71″N 84° 15′ 49.84″E. However, most of the mountains in this area exceeding 5600 m also show a certain sharpening. For this, two factors are responsible: first the already lowered late Late Glacial glacier level (see above); and second, a synchronous local glaciation of the more than 5600 m high mountains with the concomitant cirque development. This local glaciation of the mountain ridges and cupolas has freed them of the older remnants of till.

Possibly, the glacier ice still covered the terrain during the Late Glacial Stage III (Dhampu stage). Earlier this was certainly the case, and the summit areas of mountain ridges and cupolas were abraded up to and during Stages III and IV. At that time the ELA ran at ca. 5200 m asl, that is, at the level of the valley ground (thalweg) and up to 400 m above. It follows from this, too, that the clearly higher summits of mountain ridges and cupolas have preserved nearly no covers of till.

To sum up this digression (Sect. 2.4.7.4.5):

The two-phase condition of the past glacier movement is chronologically divided into an inland ice phase and a local glaciation phase. The first lasted up to the late Late Glacial (Stage IV). The latter started during the Holocene (in the Neoglacial Stage V) and continued into the Historical period. The first phase was glacial, the second interglacial.

This break in the geomorphological record becomes particularly clear at the NE margin of the Khumjung- (Khumjungar) North glacier valley (see above).

This upper margin of the valley at ca. 5800–5200 m asl consists of Late Glacial inland ice till breaking steeply down over a vertical distance of 100 m. The Historical to Holocene glacier stages at the valley bottom are located ca. 300–500 m lower and the pertinent glacier tongues and termini have turned towards the west.

From this the following two phases derive:

1. In the High to Late Glacial (Stage 0–IV) there existed a Tibetan inland ice cover that flowed into the northern flank of the Damodar Himalaya. The local glaciers, sloping down to the north, were adjusted to its surface level up to a height of more than 1000 m.

Fig. 2.124 Satellite image, Google Earth Pro: center 29° 31′ 52.56″N 83° 47′ 34.45″E; http://extras.springer.com: Character AE; Nos. 1–10. Character AE: At 29° 30′ 58.28″N 83° 47′ 4.05″E, 5720 m asl. No. 1: 29° 23′ 0.04″N 83° 50′ 18.75″E, 5570 m asl. No. 2: 29° 31′ 10.91″N 83° 49′ 15.42″E, 4900 m asl. No. 3: 29° 31′ 37.60″N 83° 50′ 40.76″E, about 5240 m asl. No. 4: 29° 31′ 41.37″N 83° 50′ 26.21″E, at 5340 m asl. No. 5: 29° 31′ 49.38″N 83° 51′ 20.71″E, 5210 m asl. No. 6: 29° 33′ 17.50″N 83° 50′ 15.99′E, 5245 m asl. No. 7: 29° 33′ 47.99″N 83° 48′ 28.42″E, about 5020 m asl. No. 8: 29° 26′ 52.43″N 83° 43′ 23.41″E, 5400 m asl. No. 9: 29° 31′ 38.23″N 83° 42′ 2.14″E, 5080 m asl. No. 10: 29° 31′ 56.43″N 83° 49′ 10.76″E, 4870 m asl

2. After the deglaciation of the inland ice, the Khumjung-North glacier first adjusted itself to the level of the tills of the Tibetan ice. Then its tongue followed the valley that had been developed by retrogressive erosion in soft till. This led down to the west, to the Mustang Thak Khola located 2000 m further below.

An example: this example concerns a mountain group west of the Mustang Thak Khola outlet glacier that also belongs to the Tibetan Himalaya, the North Sangda Himal. The highest summit of the currently glaciated massif is 6210 m high. The three lowest current ice margins (in 2011) are/were situated on the north slope of this massif at 5570 (29° 23′ 0.04″N 83° 50′ 18.75″E; Fig. 2.124, see also http://extras.springer.com: Character AE: No. 1), 5650, and 5700 m asl, respectively. The lowest end moraine of the former valley glacier consisting of those three joined glaciers, is located at 29° 31′ 10.91″N 83° 49′ 15.42″E (Fig. 2.124: Character AE: No. 2) at ca. 4900 m asl. Accordingly, here the ΔELA amounted to ca. 370 m ((5570 + 5650 + 5700)/3 = 5640, (5640 − 4900)/2 = 370). This ΔELA comes closest to that of the Neoglacial Stage V (Nauri stage, Fig. 2.5) or—in the Tibetan Himalaya—even to that of the younger Stage VI ("older Dhaulagiri stage") (ibid.: 159–161). Accordingly, a Medium to Late Holocene age of this ice margin position must be correct. End moraines in a lower position are lacking in this valley, which terminates at the lowest level of southern Tibet. It joins the Tsangpo main valley at 4580 m asl. The former glaciation of this main valley has already been proven (see above). There are, for example, at 29° 46′ 17.52″N 83° 46′ 36.88″E, at 5320 m asl, remnants of till lying 750 m above the Tsangpo river (see above). Likewise, at 29° 50′ 55.45″N 83° 47′ 4.64″E, at 5017 m asl, light-colored remnants of till occur ca. 450 m above the river (see above). Furthermore, at 29° 34′ 30.51″N 83° 49′ 29.65″E, at 5420 m asl, one finds glacigenic remnants of loose stones with 1–2 m sized boulders on a rounded mountain cupola south of the Tsangpo main valley. Further, at 29° 33′ 53.85″N 83° 51′ 41.07″E, at 5420 m asl, till with boulders of 1–2 m in size is found on the summit of a mountain ridge, also south of the Tsangpo. At 29° 32′ 27.64″N 83° 50′ 9.74″E erratic local moraine is observed—that with regard to its light color can only have come from farther north—at 5320 m asl on the summit of an adjacent mountain ridge south of the Tsangpo. Again to the south, at 29° 31′ 39.17″N 83° 46′ 34.72″E, at 5400 m asl, the upper margin of a till deposit can be detected. Since the deglaciation, debris flows occured within this material. It has to be classified as belonging to a Late Glacial Stage (III, more probably IV). At 29° 30′ 58.28″N 83° 47′ 4.05″E (Fig. 2.124 Character AE), on the summit of a hill, material that resembles till reaches an altitude of 5720 m. This material, rich in fine-matrix material, is light-colored and contains boulders of over 2 m in size. It here lies on rounded, dark, vertically layered metamorphic sedimentary rock. Further Late Glacial (IV–III) till with large boulders in a largely fine matrix lies at 29° 33′ 35.98″N 83° 50′ 2.86″E at 5240 m asl. Since the deglaciation it was washed away from the underlying bedrock phyllites by fresh streamlets.

Moreover, at 29° 31′ 37.60″N 83° 50′ 40.76″E (Fig. 2.124: Character AE: No. 3) at about 5240 m asl, there is a till level of light-colored fine matrix on a dark underground. Since the deglaciation it has been partly reworked downslope by rills. At 29° 31′ 41.37″N 83° 50′ 26.21″E (Fig. 2.124: No. 4) light-colored till at 5340 m asl rests on a dark summit of a mountain ridge; or, in another variation, a meter-thick till cover with a light-colored matrix lies between 5250 and 5100 m asl at 29° 31′ 49.38″N 83° 51′ 20.71″E (Fig. 2.124: No. 5) (5210 m). The latter is undercut by the valley floor and dissected by rills. Furthermore, at 29° 31′ 39.17″N 83° 46′ 34.72″E at 5400 asl lies a Late Glacial upper margin of till; on it are currently descending mudflows in the active layer. Between 29° 33′ 17.50″N 83° 50′ 15.99″E (Fig. 2.124: No. 6) at 5245 m and 29° 33′ 47.99″N 83° 48′ 28.42″E (Fig. 2.124: No. 7) at about 5020 m asl, there are in many places noticeably thick, light-colored covers of till rich in fine material and incised with deep rills. Since the deglaciation they have been transported downslope in the thick active layer. The result is multilayered flow tongues, the lower layers of which contain new permafrost.

At 29° 26′ 52.43″N 83° 43′ 23.41″E (Fig. 2.124: No. 8), at 5400 m asl, several generations of late Late Glacial till are mixed together in bulging strands of accumulation. At 29° 26′ 10.74″N 83° 43′ 20.50″E at 5450 m asl, there lies a late Late Glacial till level. It provides evidence that the ice level stagnated during its meltdown. Since deglaciation, fluvial rills have been cut into the naturally only slightly resistant till. At 29° 28′ 31.55″N 83° 38′ 41.65″E at 5300 m asl a till cover lies on top of an older surface. Due to the incised rills, its meter-large thickness can be observed. On the culmination of a hill at 5270 m asl, there lies at 29° 29′ 28.05″N 83° 38′ 54.00″E light-colored till with rich fine material up to its surface. This phenomenon is impossible to explain as a weathering cover developed in situ, because in that case the fine material should have been washed away syngenetically. On the slope, till can be found, for example, at 29° 30′ 28.43″N 83° 41′ 13.91″E at 5070 m asl. Logically it is about to be destroyed by young, extant rills cutting into its surface. At 29° 31′ 4.08″N 83° 41′ 53.00″E, there is an upper margin of till freshly cut into by narrow rills with sharp margins. It proves a late Late

Glacial ice level at 5040 m asl, possibly originating from the Holocene.

Also adjacent, i.e., at 29° 31′ 38.23″N 83° 42′ 2.14″E (Fig. 2.124: No. 9) at 5080 m asl, lies a preserved corresponding till level. The current furrowing of this loose rock here, too, requires a different previous process of development. Today, material that developed in the past is washed down and reworked. If past conditions were the same as today, this cover of loose material would never have come into being. It would have been removed syngenetically by in situ weathering. Accordingly, it must have been developed under different conditions from those prevailing at present. What might these have been? The extant permafrost conditions also existed in former times. However, the active layer must have been less thick and this also contradicts the existence of a former, deep-level (strong) weathering. A tertiary weathering cover that survived as far as into the time of the uplift of Tibet above the permafrost line is not possible. It would have been removed long ago. It could not have just been removed recently (today), that is, by the incision of fresh rills. As a last and only explanation, there remains the process of till genesis. Thus, for example, at 29° 37′ 48.80″N 83° 45′ 43.60″E at 5020 m asl, there lies one of many remnants of till on the summit of a mountain ridge, ca. 400 m above the Tsangpo valley bottom.

Tills preserved on the valley slopes prove that during the Holocene the Tibetan glaciation extended up from the north into the—even today still glaciated from the south—side valley. An arch-like end moraine, which was pushed upward from the NW, i.e., against the inclination of the valley slope, is located at 29° 31′ 56.43″ 83° 49′ 10.76″E (Fig. 2.124: No. 10) at 4870 m asl. It lies 21 km further up the side valley and 1 km downvalley from the Holocene side valley glacier terminus at 4900 m that was mentioned earlier (see above). This late Late Glacial glacier tongue flowed upvalley almost 300 (290) m. Further up the valley there are further late Late Glacial deposits from ice margins that were likewise transported upvalley; so, for example, at 29° 31′ 39.22″N 83° 49′ 17.81″E at 4916 m asl and at 29° 31′ 32.25″N 83° 49′ 38.81″E at 4970 m in an orographic right side valley, and from there further upvalley at 29° 31′ 14.15″N 83° 50′ 28.27″E at 5060 m asl. Here, we are dealing with lateral moraine ledges that are oriented upvalley but at the same time slope downward toward the valley floor.

Without the covering Tibetan glaciation (ice sheet), a ΔELA of ca. 530 m would have been necessary for the side valley glacier flowing down to the north to reach the Tsangpo valley ((5640 − 4580)/2 = 530). This is a ΔELA that in the Tibetan Himalaya was reached even during the Holocene (Stage V) and even there exceeded by ca. 40 m (ibid.: 159, 160) (see above). It follows from this, again, that the till covers south of the Tsangpo (see above) have to be classified as belonging to the late Late Glacial meltdown phase of Stage IV. Their young, partly Holocene age has to be traced back to the only slowly melting ice mass of a former inland ice sheet.

2.4.7.4.6 Summary as to the Southern Section of the Profile from the Nganclong Kangri (Northern Gandise Shan) in the North up to the Tsangpo Valley and the Tibetan Himalaya in Southern Tibet with Regard to the Late and High Glacial Ice Cover and Its Ice Level

The focus of this section is on Glacier Maps TJ21, TH23, TJ31, and TH33 in Kuhle (2004). The highest abraded mountain ridges of bedrock, with or without a cover of till, in the respective area reach 5640–6000 m asl. Of the numerous cases that could be described in detail, the following examples are selected as representative of the entire area, as they provide a suitable summary of the past ice level and thus also of the ice cover.

Beginning in the north, i.e., here in SW Tibet, we turn to the light-colored, very likely erratic till with 1–3 m sized boulders on dark bedrock at 5640 m and at 32° 46′ 28.33″N 80° 23′ 59.43″E up to more than 5800 m asl (Fig. 2.107, see also http://extras.springer.com: Character X: No. 2).

To the north, there are round, up to 6000 (6020) m high mountain cupolas. On some of their summits of abraded dark rock lie light-colored remnants of till up to 5750 m asl (32° 49′ 53.23″ N 80° 25′ 23.06″E) (Fig. 2.107: No. 3).

A little farther south at 32° 43′ 53.64″N 81° 34′ 55.13″E, there is light-colored till up to at least 5900 m asl. It lies east of the 5921 m high summit of the Dorenza (Fig. 2.103, see also http://extras.springer.com: Character V). As a precaution it has been classified as belonging to the Late Glacial Stages I or II, which indicates an ice thickness of at least 1400 m above the valley bottom. A similar height of the ice level is evidenced, among other things, at 32° 29′ 21.28″N 82° 4′ 40.28″E and up to 32° 29′ 58.85″N 82° 05′ 01.83″E by light-grey till lying on rounded mountains of dark bedrock up to 5800 m asl (Fig. 2.103: No. 1); this is also the case at 32° 22′ 41.85″N 81° 45′ 2.23″E up to 5740 m asl. Here the several-meter-thick, light-colored till cover is full of holes so that the darker bedrock is visible (Fig. 2.103: No. 3). At 32° 29′ 58.42″N 81° 31′ 13.59″E, the light-grey, very likely erratic till cover reaches 5900 m asl. It lies on darker bedrock north of the 6192 m high Yaria. There is also a light-colored till remnant lying on dark bedrock at 32° 15′ 7.32″N 82° 38′ 34.80″E up to ca. 5700 m asl (Fig. 2.125, see also http://extras.springer.com: Character AF: A) as well as at 32° 15′ 26.33″N 82° 38′ 10.31″E up to the mountain summit at 5870 m (Fig. 2.125: Character AF: F).

Again more to the south, at 32° 2′ 46.96″N 82° 26′ 43.02″E and at 5700 m asl (Fig. 2.126: Character AG), there is a light-colored erratic veil of till lying on the highest point of a dark, rounded mountain. Here, based on the topography, a feeding by solifluction can be entirely ruled out. The development of loose material in situ is also very unlikely. The same is true at 32° 3′ 26.60″N 82° 25′ 56.04″E at 5732 m asl, where till deposits cover a hill (Fig. 2.126: No. 1). At 32° 3′ 13.96″N 82° 25′ 19.71″E, at a height of somewhat below 5640 m, a perfectly preserved dark glacial knob breaks through a different-colored lighter till sheet (Fig. 2.126: No. 2). At 31° 53′ 43.58″N 83° 57′ 55.53″E, erratic remnants of till expanding across the mountain ridges and cupolas reach up to 5800 m asl (Fig. 2.127, see also http://extras.springer.com: Character AH: A). At 30° 37′ 17.73″N 82° 50′ 12.67″E, glacigenic loose rocks with 1–3 m sized boulders in a fine matrix reach up to 5850 m asl. Due to their great thickness in some places they have to be classified as belonging to the later Late Glacial (Stage III?) (Fig. 2.116: Character AA: Nos. 1–2). Abrasion forms reach a mountain cupola at 5900 m asl at 30° 38′ 38.69″N 82° 48′ 24.42″E and at 5860 m asl in the location at 30° 39′ 1.42″N 82° 49′ 14.62″E (Fig. 2.116: Nos. 2 and 4). The most southerly occurrence of till on a different-colored, abraded bedrock in this area reaches an altitude of 5800 m at 30° 17′ 13.89″N 83° 8′ 25.75″E (Fig. 2.121, see also http://extras.springer.com: Character AC: A). At this location, the continuous till sheet spans 750 m in altitude, and contains 2.7 m sized boulders.

Due to the rounded mountains, the ice level must be estimated to have been lying at least 200 m above the highest point (culmination). At minor ice thicknesses the necessary glacigenic abrasion could not have been attained. The ice level must therefore have been situated at a minimum altitude of between 5900 and 6200 m. It might well have reached even higher. The covers of till date back to the last ice cover during the Late Glacial. In this context, the older Stages I to III are the more likely candidates than the late Late Glacial Stage IV.

The mountain forms without cirques, which due to glacigenic abrasion in many places have been left with flat bi-convex profiles, make a large-scale ice cover probable. This might have existed up to the point when the ELA had reached beyond the respective summit heights. Otherwise there would have developed cirques. In places, cirques have been formed on northern exposures. Here even today small cirque glaciers do exist (Fig. 2.116: Character AA: Nos. 2 and

Fig. 2.125 Satellite image, Google Earth Pro: center 32° 14′ 7.85″N 82° 38′ 27.12″E; http://extras.springer.com: Character AF. Character A: 32° 15′ 7.32″N 82° 38′ 34.80″E, up to ca. 5700 m asl light-grey ground moraine veil lying on dark rock. Character F: 32° 15′ 26.33″N 82° 38′ 10.31″E, light remnant of ground moraine on mountain summit at 5870 m asl

4). Accordingly, these are interglacial and thus partly transmitted cirques, which for the last time only developed further after the meltdown of the Last Glacial inland ice.

As to the important question of when the Tibetan ice for the last time flowed over mountain cupolas towering high above the plateau level, the section of Fig. 2.123 is relevant. The 5800 m high, rounded mountain is the highest clearly abraded mountain situated immediately north of the Damodar Himalaya. It lies at 29° 3′ 48.71″N 84° 15′ 49.84″E at the southern margin of Tibet (Fig. 2.123, see also http://extras.springer.com: Character AD: between Nos. 7 and 10). For the last time it must have been overridden during Stage IV (see Fig. 2.5), which necessarily results from the next-younger glacier level that was already situated below the ELA and about to melt down. Yet, those parts of its ice cover still overriding the terrain could still form glacier tongues that flowed down to the south (see Sect. 2.4.7.4.5: second example). At the time when the ΔELA was still running ca. 300–400 m lower than during the Holocene, i.e., during Stage IV, the inland ice level in the area of the 5800 m high mountain must still have been situated above the ELA.

Also west of the Mustang Thak Khola outlet glacier, i.e., north of the Sangda Himal, a post-Late Glacial phase of meltdown is very likely.

Thus, the glacier tongue (Fig. 2.124, see also http://extras.springer.com: Character AE: No. 10), flowing from the north upslope towards the south, must have been older than the glacier tongue that flowed downslope in the same valley from south to north (Fig. 2.124: No. 2). Because the end of the glacier tongue that flowed

Fig. 2.126 Satellite image, Google Earth Pro: center 32° 3′ 13.29″N 82° 25′ 59.92″E; http://extras.springer.com: Character AG; Nos. 1–2. Character AG: 32° 2′ 46.96″N 82° 26′ 43.02″E, 5700 m asl, light erratic ground moraine veil on dark outcropping mountain cupola. No. 1: 32° 3′ 26.60″N 82° 25′ 56.04″E, 5732 m asl, ground moraine on mountain summit. No. 2: 32° 3′ 13.96″N 82° 25′ 19.71″E at 5640 m asl, brown roche moutonnèe breaking through anthracite-colored ground moraine; to the west are similar to still lighter ground moraine covers

downvalley toward the north proves an ΔELA of only 370 m, it belongs to the Holocene Stage V (see above). If this tongue were older, it—and its terminus—would not be preserved. That is, it would have met the upslope-flowing glacier tongue and merged with it. Accordingly, at a ΔELA of ca. 700 m (see Fig. 2.5) belonging to the next-older Stage IV (Late Glacial), the two merged valley glaciers must have filled the valley together. It follows that the glacier tongue flowing south and coming to an end upvalley must have belonged to the meltdown phase of Stage IV. During this phase, the ice located on the comparably lower Tibetan Plateau still had a greater thickness than the local ice from the over 6000 m high mountains of the Sangda Himal. This logic of the ice cover north of the Damodar proves once more the existence of the Tibetan inland ice.

Following this logic, at the time of the meltdown phase of Stage IV, the inland ice surface had already melted down below the level of the ELA. Accordingly, during Stage IV, the inland ice surface must have clearly reached higher than during the phase of its meltdown. Therefore, the abraded summits of mountains and mountain ridges covered by till up to 5660–5720 m asl during Stage IV (Sirkung stage) were still covered by the Tibetan ice sheet (Fig. 2.124: Characters A and E).

In an analogous manner, further ice margins in Tibet can be classified as belonging to the post-Late Glacial phase of meltdown: e.g., the kame ledge at 30° 38′ 28.13″N 82° 48′ 23.94″E

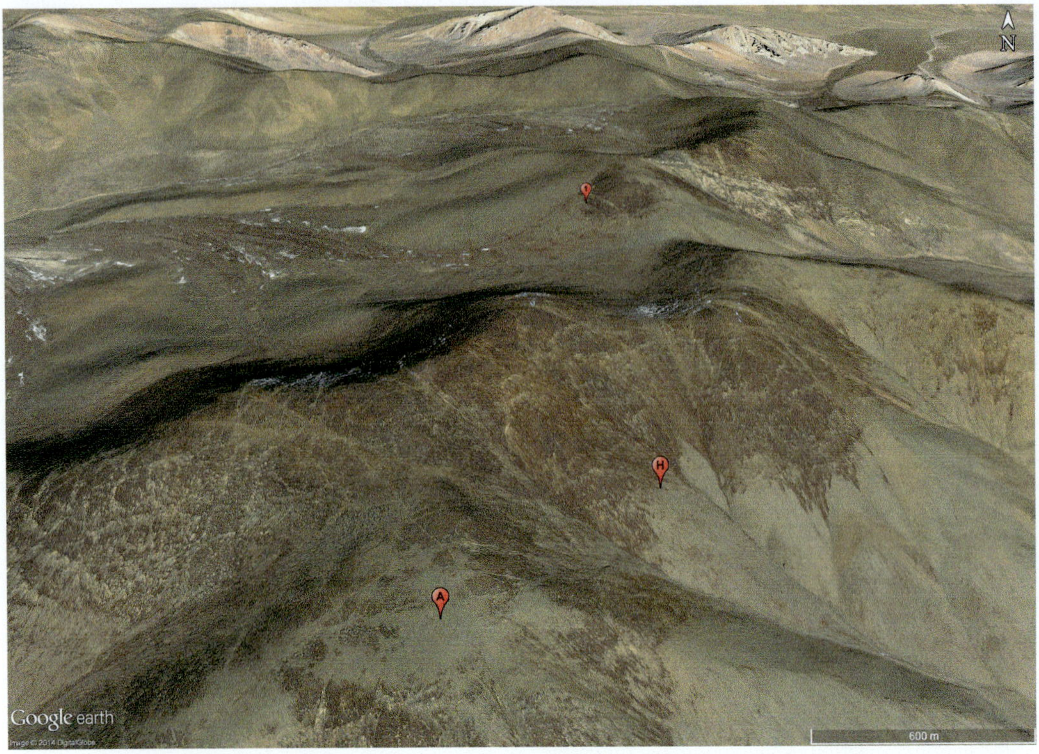

Fig. 2.127 Satellite image, Google Earth Pro: center 31° 54′ 7.61″N 83° 58′ 1.59″E; http://extras.springer.com: Character AH; No. 1. Character A: 31° 53′ 43.58″ N 83° 57′ 55.53″E, up to 5800 m asl, erratic remnants of ground moraine on mountain ridges and cupolas. Character H: 31° 54′ 1.03″N 83° 58′ 15.43″E at ca. 5600 m asl and No. 1: 31° 54′ 52.28″N 83° 58′ 11.96″ E also at ca. 5600 m asl mark a late Late Glacial ground moraine level (upper margin of ground moraine)

and 30° 38′ 31.52″N 82° 48′ 36.57″E (Fig. 2.116, see also http://extras.springer.com: Character AA: No. 3).

At 31° 54′ 1.03″N 83° 58′ 15.43″E (Fig. 2.127, see also http://extras.springer.com: Character AH: H) and 31° 54′ 52.28″ N83° 58′ 11.96″E (Fig. 2.127: No. 1), at 5600 m asl, an inland ice level is proven by an upper margin of till. Because we are here 340 km north of the southern margin of the plateau, i.e., in Central Tibet, even this high ice level might belong to the meltdown phase of Stage IV.

Above, up to nearly 5800 m asl and at 31° 53′ 43.58″N 83° 57′ 55.53″E, light-colored erratic remnants of till covers exist on dark, abraded mountain ridges and cupolas (Fig. 2.127: A). Thus, these would belong to the completely developed Stage IV. Otherwise they would belong to the next-older Late Glacial stage, i.e., Stage III.

The ice levels of northern Central Tibet (Sect. 2.4.7.3), which based on the topographic method can only at best be minimum values, must have continued across the entire region down to southern Central Tibet, as far as to the southern margin of the plateau. For the entire distance, they must have reached a similar altitude, that is, at least 5900–6200 m asl.

The ice thicknesses connected to these ice levels as far as down to the valley bottoms at ca. 4400–5000 m asl amounted to at least 1000–1600 m.

The empirical evidence for these minimum heights of the ice levels and ice thicknesses are valid for the Late Glacial. Accordingly, the ice level during the Last High Glacial, i.e., Stage 0 (Fig. 2.5), necessarily was higher still.

This ice load eventually led to depression of the underlying surface. In Tibet, the average glacio-isostatic depression was at least 650 m (Kuhle 1988c: 483–485, 505, 510; 1988d: 591; 1989; 1993: 146; 1995; 1997; 1998a: 100; 1998b: Fig. 14; 1999: 272, 273; 2013a: 199–203). The existence of glacio-isostasy in Tibet has also been confirmed by Kaufmann and Lambeck (1997) as well as Wang (2001) (cf. also Nesje and Dahl 2000: 127). This depression was basin-formed due to the inland ice geometry. This had the effect that the largest ice thicknesses were reached in the center of Tibet. The resulting largest depression in the center, in turn, caused a slowing of the ice discharge. This reduction of discharge was a consequence of the marginal counterslope that had thus developed. In this way the concentric central over-deepening in addition raised the central thickness of the inland ice. The glacio-isostatical diminution of the ice discharge in some places caused the reduced development of tills, that is, the formation of till covers of only minor thickness.

Holocene lake formations evidenced by widespread lake sediments also have to be traced back to that glacio-isostatic lowering. The ca. 1000 km long depression running from 33° 11′ 6.33″N 81° 12′ 4.17″E (4550 m asl) up to 31° 48′ 40.25″N 91° 32′ 59.19″E (4600 m asl) in a W-E direction through Central Tibet is of glacio-isostatic origins. In a topographic sense it is still over-deepened and without discharge (lacks an outlet).

2.4.8 The Past Tibetan Glaciation Between the Gangdise Shan North of the Paiku Tso in the South up to the Central Kuenlun (Kuenlun Shan) in the North

The primary focus of this section lies on Fig. 2.1: Nos. 23, 10, 12, and 3. The regional link between the areas treated in the previous chapter and the regional center of this chapter has been investigated by the present author in greater detail (Kuhle 1999: 13, 19–27, 103–116). Among other things, samples were taken from the loose rock at approximately 29° 25′ 6.80″N 85° 46′ 15.12″E (after the map ONC-H9, 1978 ca. 29° 25′N 85° 17′ 30″E) at ca. 4930 (aneroid 4680) m asl in the area of the foot of the slope. It is situated on bedrock schist and contains erratic granite boulders. The grain size spectrum of its fine material matrix (with 13% clay) and the bimodal course of its cumulative distribution function are typical characteristics of moraine (Kuhle 1998b: Fig. 6). The same characteristics are exhibited by a sample of till of the Laurentide Ice Sheet in Canada chosen here for comparison. Its material belongs to the so-called long-distance-transported “Canmore Till” from the Bow Valley in the Rocky Mountains (Locality: ca. 100 km west of Calgary near Banff and Lake Minnewanka in the eastern mountain foreland) (Kuhle 1999: cf. Figs. 15 with 16). Despite of its position in a mountain valley and the reworking of the Tibetan tills (ibid.: Fig. 5, sample 28.08.96/1), the Canmore till material still shows a 3% higher clay portion than the Tibetan sample. In the same vicinity, ca. 9 km further north, at 29° 29′ 21.54″N 85° 44′ 56.20″E at 4970 (4800 aneroid measurement) m asl, Fig. 2.128 was taken. The photography proves geomorphologically a Late Glacial (Stage IV) inland ice landscape with features of both glacigenic erosion and accumulation.

From another representative Late Glacial till landscape, samples were taken farther north at ca. 29° 59′N 84° 26′E (map ONC-H9, 1978) at 4780 (aneroid measurement) m asl (ibid.: 21, Fig. 17, 109–112). Here, a clay fraction as nearly as high as 20% was recorded. The curve for the fine-grain portions traces the bimodal course that is characteristic of moraines. A secondary maximum lies in the clay portion and reaches nearly 20%.

This typical matrix of a Late Glacial inland ice moraine in the Gangdise Shan (Tibet) corresponds with the till matrix of the till plain between the Bow River and the South Saskatchewan River ca. 32 km northwest of

Fig. 2.128 View facing NE across a ground moraine landscape at ca. 29° 29′ 21.54″N 85° 44′ 56.20″E at 4970 (4800 aneroid measurement) m asl. (■) mark ground moraines covering the entire mountain foreland of currently non-glaciated mountains in the background. (▲) are glacigenically abraded hills, and the glacial band-polishing of the outcropping edges of the strata ("Schichtkopfstreifenschliff") is visible through the ground moraine covers. The second ■ symbol from the right shows ground moraine that in this location since deglaciation has eroded diagonally to those outcropping edges of young flushing-rills. (□) is a flat valley bottom of ground moraine that has been glaciofluvially flushed. (- - - - white) is the High Glacial glacier level (Stage 0, LGP, LGM, see Fig. 2.5) of the Tibetan inland ice above the mountain group in the background. It runs ca. at or above 6000 m asl. These mountains are similar to an (during glacial times) also glacier-filled Scandinavian mountain landscape (e.g., Rondane in the Norwegian Scandes). (- - - - black) points to the Late Glacial (Stage IV, see Fig. 2.5) ice level in the mountain foreland. Analog photo 28/08/1996 by M. Kuhle

Medicine Hat (Alberta, Canada) of the Wisconsinian western Laurentide Ice Sheet (ibid.: Fig. 18). The respective glacigenic character of the landscape is shown by Fig. 2.129 at ca. 30° 13′ 21.02″N 85° 19′ 18.52″E at about 5230 (4837 aneroid measurement) m asl.

This former inland ice landscape continues farther to the north and SW (Figs. 2.130 and 2.131) up to west of the Zhari Nam Co (Zhari Nanico, Zhari Co, Taje Tso lake) (Fig. 2.132). The broad bottoms of the box-shaped valleys also consist of tills (see above) modified slightly by glacio-fluvial activity. Here, too, the valley flanks are lined with remnants of till covers up to the summits of the mountain ridges and cupolas.

Along the S-N profiles just described, on the flanking mountain ridges and cupolas, the following remnants of till covers are to be found: at 30° 10′ 48.13″N 85° 19′ 43.16″E at 5790 m asl a light-colored till cover on an abraded, dark summit of a mountain cupola; at 30° 30′ 28.40″ N 85° 3′ 50.83″E a light-colored till at 5880 m asl on dark bedrock falling away to the SE to 5500 m with an increasing thickness of the slope cover; at 30° 31′ 37.14″N 85° 14′ 41.57″E remnants of light-colored till that from 5460 m asl downward along the slope are preserved on the abraded dark bedrock of the mountain; at 30° 34′ 32.18″N 85° 15′ 46.20″E a light-colored till level at about 5440–5460 m (Stage IV or III?), frayed by denudation, on a dark, abraded mountain ridge; at 30° 37′ 32.41″N 85° 10′ 59.73″E, a light-colored till cover mantling the 5520 m high culmination of a rounded rock mountain whose dark bedrock is visible through several till windows; at 30° 43′ 52.57″N 85° 7′ 50.48″E at 5380 m asl lighter-colored remnants of a till cover on the summit of a dark, abraded mountain ridge with windows of loose rock, identified as loose rock material at 30° 43′ 14.18″N 85° 5′ 49.37″E at 4940 m asl by means of rills (see Figs. 2.130 and 2.131); and at 30° 45′ 2.68″N 85° 6′ 21.69″E, in places, i.e., at 5230 m asl, preserved lineaments of a thick-layered, light-colored, coarse-blocky till on a dark underground.

Immediately south of the western Zhari Nam Co, at 30° 43′ 36.26″N 85° 27′ 23.20″E, at 5580 m asl, there lies light-colored till on a dark rock ridge as far as up to its peak; moraine like this lies also at 30° 42′ 35.62″N 85° 29′ 53.46″E up to 5750 m asl as far as nearly up to an

Fig. 2.129 Looking downvalley at ca. 30° 13′ 21.02″N 85° 19′ 18.52″E about 5230 (4837 aneroid measurement) m asl facing roughly south, toward polished mountain ridges of dark, metamorphic sedimentary rock. The rounded orographic right mountain spur towers above the brook about 40 m and is 5270 m high; it is steeply undercut by the brook in the thalweg, so that weathering has developed (⊠). (☗) signify glacigenically abraded hills, and the glacial band-polishing ("Schichtkopfstreifenschliff") of outcropping edges of the strata is visible through a thin cover of loose rock. (↓) is a large exaration rill, preserved in the overlying ground moraine. (■) shows the important ground moraine cover in the valley-bottom, which near the thalweg has been glacio-fluvially flushed and washed out (□). (- - - -) indicates a Late Glacial (Stage IV or III, see Fig. 2.5) ice level. Analog photo 29/08/1996 by M. Kuhle

abraded, dark summit of hill cupolas. A final local example is the 800 m high slope rising from the southwestern bank of the Zhari Nam Co: at 30° 47′ 39.50″N 85° 26′ 17.76″E it reaches the extensive, light-colored, meter-thick till cover at the summit of a mountain ridge at 5410 m asl. The slopes below are also covered by a patchy, broken-through till cover, with dark rock lying underneath (Fig. 2.133).

A further research area for the extension of till covers can be found to the east. It will be discussed later in more detail in connection with the TCN data by Kong et al. (2011) of lake terraces on the Tangra Yum Co and Zhari Nam Co (Fig. 2.134a, see also http://extras.springer.com: Character AJ; Fig. 2.134b) (see Sect. 2.5 and Fig. 2.145). The field here is mantled by extensive covers rich in fine material and large, light-colored boulders (Figs. 2.135 and 2.134a: locality F). Such very likely allochthonous boulders are also present in the broader vicinity (Fig. 2.134a: D, E, F, G, H, I). Should they be taken for erratic boulders, these deposits are further proof of an extensive inland glaciation overriding the entire landscape. In further support of this claim, no local catchment areas exist through which any former glaciers might have reached the respective positions under consideration.

The question concerning the former glaciations of the northern Central Tibetan Plateau is especially important insofar as the area is

Fig. 2.130 At ca. 30° 42′ 15.07″N 85° 9′ 10.73″E from ca. 4760 m, looking west across a glaciofluvially surficially flushed ground moraine level (■ in the foreground). (▲) show up to their summits glacigenically abraded mountain ridges. (■) meter- to decameter-thick lighter remnants of a ground moraine cover up to heights of ca. 4940–5380 m asl overlying dark rocks underneath. The in places great thickness of the ground moraines can be evidenced by the ravines that have been incised since the deglaciation (■ on the left). (►) mark postglacial debris-flow fans developed from moraine rills such as these. (- - - -) is the past inland ice surface that—according to glacigenic erosion- and accumulation indicators—must have been situated at over 5880 m asl. Analog photo 29/08/1996 by M. Kuhle

relatively arid. Accordingly, most authors assume that in this area the glacier advances were only minimal during the global ice ages (Li et al. 1991; Kirchner et al. 2011). As to the Purog Kangri (Puruogangri; Purog Kangri I; 33° 54′N 89° 08′E) (Fig. 2.133 between L and N), which is the massif with the largest extant plateau glaciation in Tibet, Yi et al. (2002) have reconstructed the advance of an outlet glacier on the western slope. It had a length of only 6 km. This advance they have classified as belonging to the LGM, though the ΔELA of this (glacier) stage only comes to ca. 130 m (calculation by Kuhle). Based on the till deposits from glacier stages in the still more arid Zangsar Kangri, with an ΔELA of 90–180 m, which due to their interlocking with lake terraces must be dated to the Early to Medium Holocene (Kuhle 2013a: 146–152), this LGM classification is doubtful (see Sect. 2.7).

In addition, the findings by Kuhle in the plateau areas adjoining on the west, south, and east, such as for example the Tanggula Shan (Kuhle 1991), make clear that here too an extensive glaciation had taken place. In this way it was possible to calibrate the analyses of satellite images for this area by means of the terrestrial analyses of till covers in the plateau regions lying to the SW, SE, and NE. The characteristic surface texture of tills can be found, for example, in the vicinity of 35° 35′N 93° 18′E (south of the Kuenlun Shan). In this area, originally these large-scale deposits were considered to be tills. More specifically, one deals here with a debris–clay matrix cover with a characteristic bi- or

Fig. 2.131 At ca. 30° 44′ 8.95″N 85° 9′ 35.90″E, at 4750 m asl facing west, looking into an orographic left, small side valley, which is mantled by mostly light-colored erratic ground moraine (■). In places it overlies, at several meters in thickness, dark, metamorphic sedimentary rock (phyllites). It covers these steeply rising (nearly vertically standing) rocks sporadically up to the abraded hill cupolas (◓). (- - - -) indicates the surface of the glacial inland ice. Analog photo 29/08/1996 by M. Kuhle

tri-modal distribution of grain sizes in the matrix with maxima in the clayey-sandy section. At the same time, this fine material even contains polymictic, erratic boulders (Kuhle 2003: 76–79 and Fig. 6). On the northern margin of their stratigraphic range, at the Kuenlun pass, at 35° 41′N 94° 02′E, these till covers also occur and here contain erratic massive crystalline (e.g., granite) boulders, lying on metamorphic sedimentary bedrocks (phyllites) (Kuhle 1987b: Fig. 7; Kuhle 1987c: 269). A sedimentological analysis of till material was also carried out on samples obtained from a Central Tibetan profile from north to south between the Kuenlun and Tsangpo, including the Geladaindong group (Kuhle 1991: 134–167, 229, 230, Fig. 43; 2003: 77–86), and also along a transect between east of Gertse (32°N 85°E) and Aksai Chin (36°N 78°E), in Central and western Tibet (Kuhle 1999: 6, 7, Fig. 2, 13, 22–2.44).

To attain further certainty, additional comparisons of the material were carried out vis-à-vis the Laurentian till substrate from the former permafrost areas in North America (ibid.). On this basis, it is possible to identify by means of the analysis of satellite imagery in the broad area surrounding the Zangsar Kangri as well as the Purog Kangri (Puruogangri), localities in which the characteristic key features appear (see Sects. 2.7 and 2.8).

In order to provide such evidence, some localities on the Central Tibetan section of Fig. 2.133 (see also http://extras.springer.com: Character AI) were chosen. At the localities A, B, H, M, N, O, and P there are covers of a reddish matrix of loose material. Within them are

Fig. 2.132 At ca. 30° 47′ 30.49″N 85° 8′ 1.98″E from ca. 4720 m asl, looking east across the bottom of a main valley consisting of ground moraine (foreground). Location is south of the Coquen settlement and WSW of Zhari Co. Near the thalweg, the moraine has been glaciofluvially flushed and thus reshaped (□). (■) mark several further occurrences of the omnipresent ground moraine covers extending upslope. The culminations of the relief are rounded ridges and cupolas (☗). (- - - -) marks the approximate altitude of the accompanying inland ice surface. The glacigenic landscape situated farther north in the valley chamber near Coquen has been introduced in Kuhle 1999 (Photos 72, 73). Analog photo 29/08/1996 by M. Kuhle

embedded light-colored, isolated boulders, probably consisting of granite or quartzite (Figs. 2.137, 2.138, and 2.139). For example, Fig. 2.136 (the locality in Fig. 2.133: Character AI) also shows loose rock with a till texture that overlies sedimentary bedrock. The erratic material of the covers takes on the shape of small flow tongues that extend downward over the escarpment and the outcropping edges of the strata of solid rock.

Above, at the culmination of the rock threshold, at between 5010 and 5100 m, where it is still intact, a glacigenic lineation from west to east is shown. Somewhat further south, in a till window below dead-ice holes in the moraine, glacigenic abrasions of the layer edges ("Schichtkopfstreifenschliffe") are visible. Somewhat farther west, at 32° 45′ 04″N 87° 22′ 44″E, several generations of erratic till can be diagnosed.

In addition, one can observe small polished depressions ("Schliffwannen") in covers of loose material or in bedrock, as they are known from deglaciated *cold-based* areas of northern Greenland or Scandinavia (Fig. 2.140). Especially interesting are specific phenomena that occur on a very large scale. They are probably connected with the special conditions of an inland glaciation under *cold-based conditions*. Here we observe a striking stratification and glacial

Fig. 2.133 Satellite image, Google Earth Pro: center 33° 43′ 27.92″N 87° 34′ 26.58″E; http://extras.springer.com: Character AI. Overview across the northern area of Central Tibet between Zangsar Kangri in the west and Puruogangri (Purog Kangri I) in the east. This area is declared by numerous authors (see Li et al. 1991; Kirchner et al. 2011) as having been non-glaciated during the global ice ages. Already the analyses of satellite images, however, show the nearly exhaustive existence of glacio-geomorphological indicators. Here, one has to consider that this past glaciation definitely must have been a *cold-based glaciation*. Due to the Nordic glaciations it is confirmed that under these conditions typical landforms of flatland glaciations such as drumlins, eskers, and "ribbed moraines" *could not have developed*. The lack of these lowland keyforms in the area of the Tibetan Plateau therefore cannot be presented as evidence for the lack of a past glaciation, as has been done by Kirchner et al. (2011: 249) and Seong et al. (2009: 360). In addition to classic indicators of glaciations such as erratic boulders (positions A, B, H, M, N, O, and P; for details see Figs. 2.137, 2.138, and 2.139), polish-forms of the solid outcrop (position I; Fig. 2.135), ground moraine covers with glacigenic polish-basins (positions E, K, and L; Fig. 2.140), as well as large-scale covers of ground moraines that only locally are interrupted by bedrock (position J; Fig. 2.144), one can find here new keyforms that might be *typical of cold-based conditions* under subtropical-semiarid conditions, such as, for example, exaration rills in frozen moraine material (position M; Fig. 2.139), and broken plates of ground moraine (positions L and G; Figs. 2.141 and 2.143) (Kuhle 2014a)

scouring of what appear to be covers of till (Figs. 2.141, 2.142, 2.143, and 2.144; cf. Figs. 2.156, 2.193, and 2.190). The covers have been broken and fractured by compressions (Figs. 2.141 and 2.143).

According to measurements based on the satellite images, we are dealing with a ~400 × 450 km extended plateau area that is up to 90% covered by till. These till covers can be up to several decameters thick and most likely are of multi-glacial origin. Even today, the whole area is still situated in the region of continuous permafrost (Fig. 2.139) (Kuhle 1985b, 1987b, 2014a). Therefore, in this basin-and-plateau

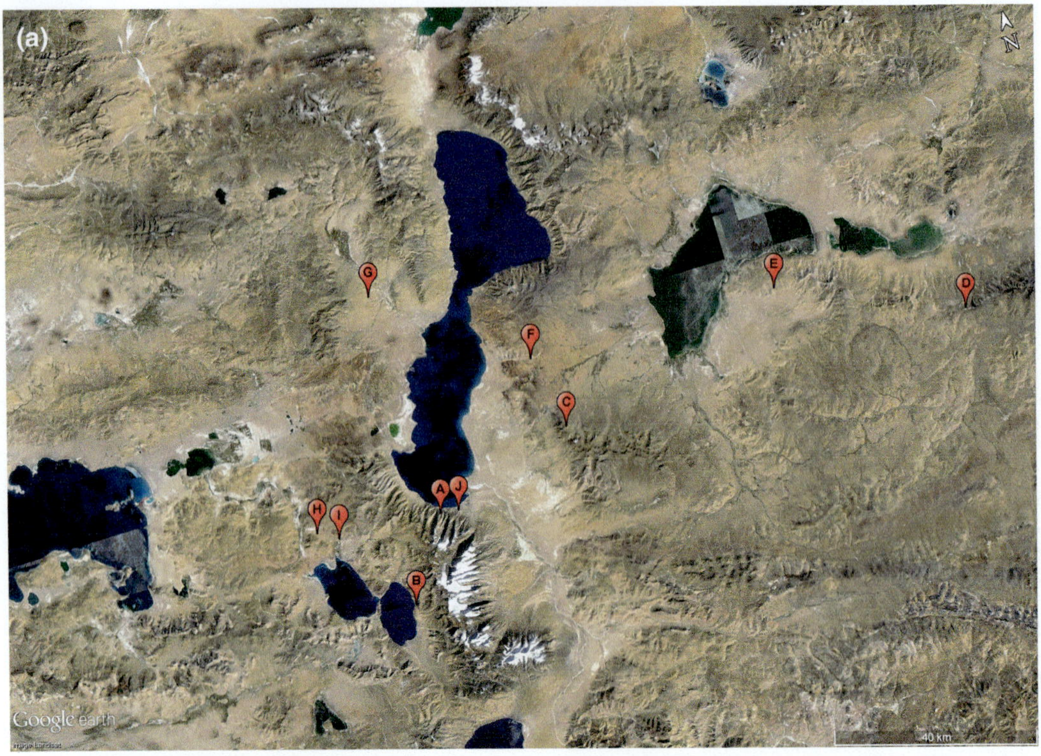

Fig. 2.134 **a** Satellite image, Google Earth Pro: center 30° 55′ 51.16″N 86° 40′ 41.26″E; http://extras.springer.com: Character AJ. The picture shows the area around Tangra Yum Co and Zhari Nam Co. Positions A and B show the three past terminal moraines that are interlocked with lake terraces (see Fig. 2.134b). Further detail concerning this interlocking is shown in Fig. 2.145. In position C (see detail in Fig. 2.135), on the flanking summits at 5760 m asl (1200 m above the recent lake level of Tangra Yum Co) there is an exposure of the presumed moraine covers that mantle the area on both sides of the lake on a large scale. Further localities where possibly erratic block material could have been identified via satellite image analysis include the positions D, E, F, G, H, and I. **b** Satellite image, Google Earth Pro: center 30° 45′ 55.56″N 86° 28′ 26.24″E. The image shows an enlarged section of Fig. 2.134a between Characters A and J, providing a more detailed picture of the interlocking of past terminal moraines with lake terraces of the Tangra Yum Co at its southern end (see also Fig. 2.145)

topography one generally has to proceed from the inventory of forms associated with a cold-based glaciation.[4]

This form of high-altitude glaciation has no parallel in the Nordic glaciation areas. This is why the potential degree and the distribution pattern of inheritance would **in a first step** have to be analyzed and determined for this high-altitude region (Sect. 2.5.2.1). Only then, if at all (see below), **in a second step** attempts could be made to determine local stages of retreat or lake terraces based on TCN dating. However, even then, the failure of such an exercise continues to be likely. This is because the method invariably has to deal with the multiple stages of rearrangement of the material in polyglacial till covers. This means that this special type of *cold-based glaciation on permafrost is probably the reason* for the regular overestimation of the age of Tibetan deposits based on TCN data in the

[4] Therefore, the statements of Kirchner et al. (2011: 249) as well as Seong et al. (2009: 360) that no plateau glaciation existed in Tibet, because forms typical of Nordic glaciations (drumlin swarms; eskers and ribbed moraines) are lacking, is *irrelevant*. The reason is that all of these phenomena *only* come into being under the conditions of a *wet-based-glaciation* or at the transition from *cold-based-* to *wet-based* conditions (see Kleman and Hättestrand 1999).

Fig. 2.134 (continued)

literature (Kuhle 2014b). At present, no proper solution to this dating problem has been found.

2.4.8.1 The Holocene Glaciation of the Zaqungngomar Massif and U-Series Datings of Two Interglacials in Central Tibet

At the eastern exposure of the Zaqungngomar massif (6290 m) there occurs loose material at 33° 2′ 33.53″N 88° 38′ 43.37″E (Fig. 2.133 at Character O) at 5020 m asl. Both its geomorphology and sedimentology point to a typical till cover. It has been dated by Blisniuk and Sharp (2003) using the U-Th method (U-series datings). According to these datings there occurred three phases of pedogenic carbonate rind formation: at ~233, ~131, and at ~16 ka. These ages coincide with three phases of global interglacial periods. In the event the older dating is correct, the material was covered twice by inland ice, for the last time during the LGP (Stage 0–IV or III). Because it was embedded within permafrost, it was protected against reshaping.

The younger dating is confirmed by the glacio-fluvial undercutting. That is, it must have occurred at a ΔELA of ca. 240–300 m. The accompanying glacier margin lay at 5050 m asl exactly in the valley exit close to the relevant terrace base. The two relevant ice margins of the largest extant glaciers are situated at 5500 and 5560 m asl. From this, the ΔELA of 240 m ((5530 − 5050)/2 = 240) on the eastern side of the Zaqungngomar massif can be derived. It corresponds with the Nauri stage (V, Fig. 2.5). Accordingly, it is of a Holocene age and younger than ca. 16 ka.

Ages younger than 10 ka are shown by Early Holocene lake level stages (see Sect. 2.4.6) cited above. They have been partly reworked by Neoglacial glacier advances that correspond to

Fig. 2.135 Satellite image, Google Earth Pro shows at 30° 48′ 38″N 86° 44′ 30″E the locality of Point C in Fig. 2.134a. Light-colored, fine material covers are exposed, into which light block material has been embedded. The darker material of the outcrop becomes visible in the surrounding uplands. The exposure lies at 5760 m asl (1200 m above the lake level of Tangra Yum Co). This material covers on a large scale the flanking summits west and east of Tangra Yum Co. This is—if the satellite image analysis is confirmed—further proof of a glaciation extending across the entire terrain

the Nauri stage. Thus, this is a repeated confirmation of that younger date of Blisniuk and Sharp (2003) as well as the glacier stratigraphy of the present author and his own observations on the same location.

2.4.8.2 Summary of Sect. 2.4

This chapter has dealt with the current to Last Glacial glacial cover of the west-central, northern, and southern parts of Central Tibet (Fig. 2.1). The analysis makes possible a more or less complete reconstruction for the individual Central Tibetan mountain groups, ranging from the current and only a few decades old glacial margins of Stages XIV, XIII, and XII (see Fig. 2.5), to the five Historical Stages XI, X, IX, VIII, and VII (Fig. 2.5), the three Neoglacial (i.e., Holocene) Stages 'VII, VI, and V (Fig. 2.5), as well as the four Late Glacial Stages IV, III, II, and I (Fig. 2.5) up until the High Glacial, i.e., Last Glacial or Würm Glacial LGM Stage 0 (Fig. 2.5).

In the neighboring mountain massifs that rise above the Tibetan Plateau, the progression of the glaciation was very similar. It was possible to synchronize it with the help of 13 or 14 gradually decreasing glacial advances that interrupted the deglaciation process from the LGM to the present. This took place between the Glacial and Interglacial, on the basis of similar ΔELAs (snowline depressions). The methodological correlation lies in the physically almost identical climatic conditions of the adjacent mountain areas.

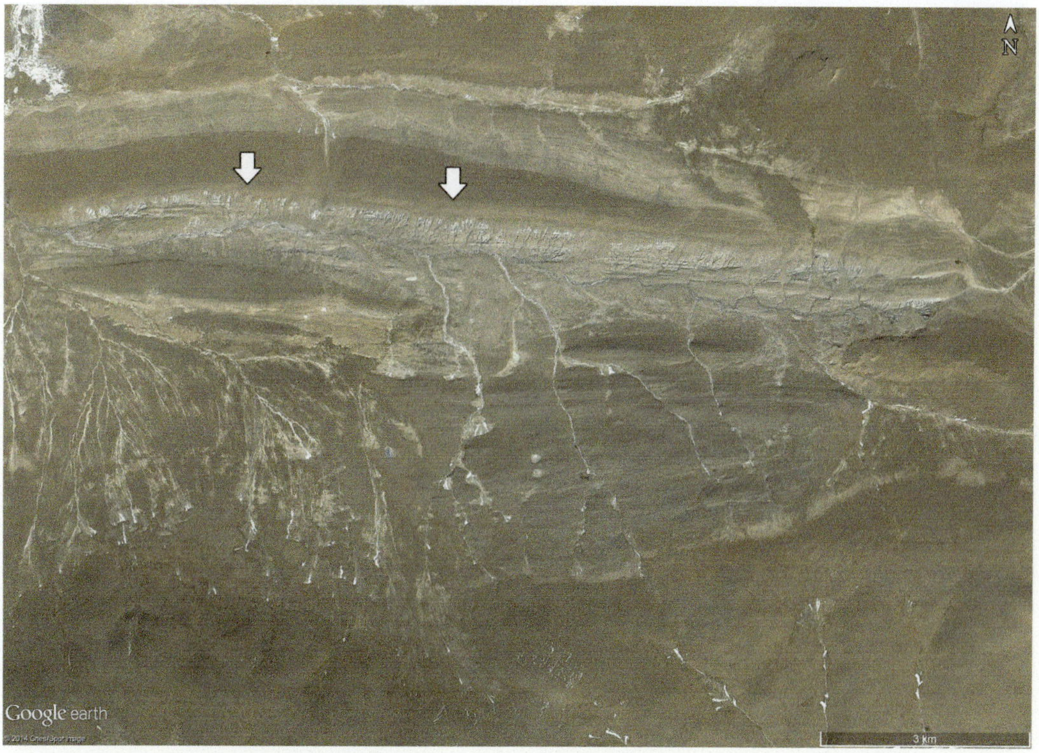

Fig. 2.136 Satellite image, Google Earth Pro shows at ca. 32° 43′ 50.66″N 87° 25′ 7.78″E and 32° 44′ 27″N 87° 24′ 30″E the locality of Point AI in Fig. 2.133 in detail. Loose rock with a typical ground moraine texture overlies outcropping sedimentary rock here at ca. 5005 m asl. The foreign material of the covers runs in the form of small flow-tongues across the outcropping edges of the strata (arrows). In the lower half of the picture, parts of a glacigenically lineated rock surface, by glacigenic scouring, are visible. Furthermore, several generations of erratic ground moraines are evident at 32° 45′ 04″N 87° 22′ 44″E. They too are partly lineated by scouring

With regard to older, pre-historical (i.e., Holocene to Late Glacial) stages, it was important to take into account the persistence of the previously existing, very considerable masses of inland ice. As long as a large, relief-dominating inland ice cover existed, the comparatively insignificant postglacial snowline depressions (ΔELAs) could not have had a significant impact on the relevant small glacier tongue advances. An inland ice sheet reacts—in comparison to small mountain glaciers—very slowly and filters small climatic oscillations out of the ice margin behavior.

An exemplary mountain massif that is representative of the large-scale study area of Sect. 2.4 is the Nyainquentanglha Shan (Fig. 2.1: Nos. 5 and 10). In its 10 to almost 20 km long valleys and connecting hanging valleys, all Historical stages are represented together. They prove the snowline depressions depicted in Fig. 2.5, or only modify them insignificantly. Even during the Neoglacial (Holocene) Stages V and even VI, at least some of the local glaciers still reached the 14–18 km wide, NE-SW oriented main valley that terminates the Nyainquentanglha on the SE. Their end moraines are dovetailed with the Late Glacial tills of the main valley glacier on the orographic right side. This was a Late Glacial outlet glacier from Central Tibet (see Ice Sheet center I2 in Fig. 3.1) that flowed down from the NE towards the Tsangpo at around 4300–4000 m asl. During the High Glacial (LGP and

Fig. 2.137 Satellite image, Google Earth Pro shows at 34° 11′ 58″N 85° 37′ 58″E in the western Zangsar Kangri the locality of Point B in Fig. 2.133 in detail. Light-colored boulders are several meters in diameter; reddish rock outcrops below. Whether these are boulders from quartz veins, local quartzite boulders, or erratic material transported over large distances must be determined from field research

LGM, Stage 0), the main valley glacier level reached up to between 5800 and 6200 m asl, and the glacial terminus reached down to the Tsangpo River. The Nam Co (Tengri Nor) lake basin lying NW of the Nyainqentanglha was also filled with glacial ice and communicated flow-dynamically with the just-mentioned main valley glacier level via transfluence passes. The Neoglacial Stages V–VI (ca. 2000–5500 YBP; Fig. 2.5) of the glaciers on the NW slope of the Nyainqentanglha Shan still reached the bottom of the lake basin. All subsequent minor advances (Stages 'VII–XII; Fig. 2.5), also marked by end moraines, took place during the Late Holocene and Historical time.

The Late to High Glacial (Stages IV–0; Fig. 2.5) ice cover between the Nganclong Kangri (the northern Gandise Shan) in the north and the Tibetan Himalaya had the following reconstructed characteristics:

The till deposits reach 5640–6000 m asl in the area under discussion and are located on rounded mountain backs and hilltops. The ice level lay between at least 5900 and 6200 m asl. This applied to both the northern and southern parts of Central Tibet on the relevant N-S profile.

For the Late Glacial Stages I or II, an ice thickness of at least 1000–1600 m above the valley bottom has been reconstructed. With slightly less but still considerable probability, an inland ice cover—with arguably the same ice thickness—still existed during the Late Glacial Stages III and IV (Fig. 2.5). Thus for example, stratigraphic evidence exists for an ice level at 5900–6000 m asl, which still existed during the late Late Glacial Stage IV in southern Tibet north

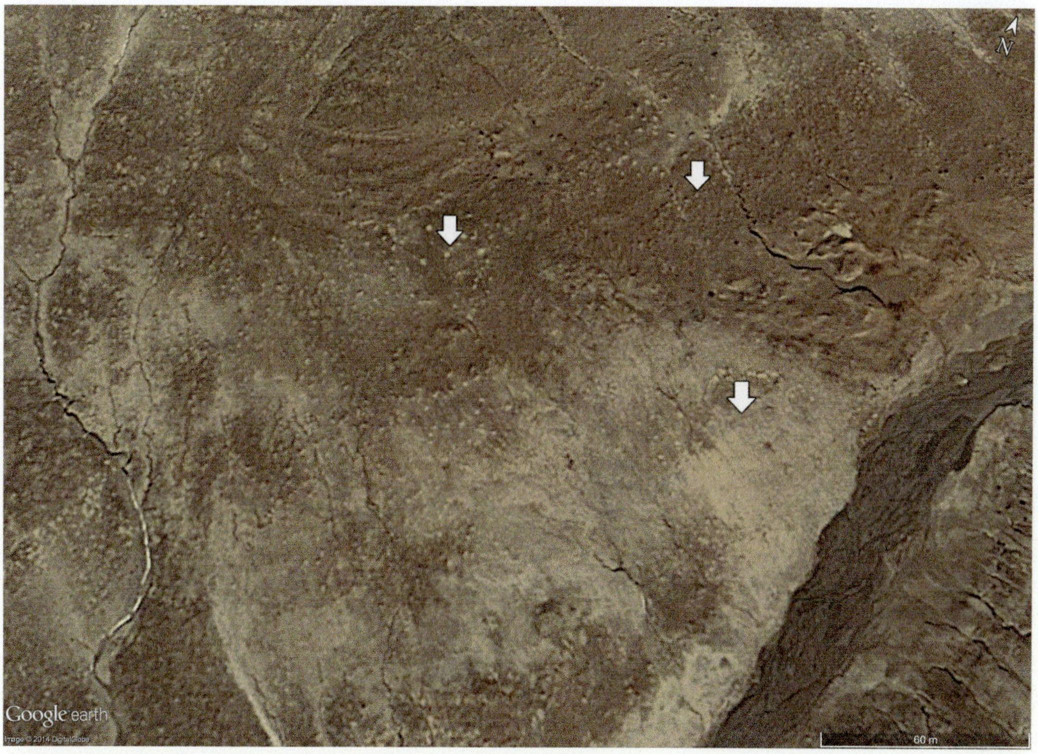

Fig. 2.138 Satellite image, Google Earth Pro shows at 33° 28′ 33″N 89° 04′ 50″E the locality of Point N in Fig. 2.33 in detail. In extensive covers of loose material, numerous light-colored boulders can be detected within a fine material matrix (arrows); 5104 m asl; 30 km south of Puruogangri (Purog Kangri I). See also Fig. 2.139

of the Damodar Himal. Corresponding evidence of a Late Glacial persistence of inland ice was given for the region north of the Sangda Himal. Here, a Late Glacial inland ice surface at above 5660–5720 m asl and a post-Late Glacial melting phase were proven. Further corresponding evidence also applies to the Central Tibetan areas 340 km farther north. The High Glacial (LGM) surface levels of the inland ice consequently must have reached considerably higher than 6200 m asl.

The average glacial-isostatic lowering in Tibet amounted to 650 m at a minimum and led to the formation of a Central Tibetan depression. This bowl-shaped depression in turn led to an increase in inland ice thickness and a reduction of ice runoff at the glacier bottom. The Holocene lakes and their very large areal extensions, which emerged after the deglaciation, are a result of this glacial-isostatic lowering.

2.5 Excursus to the Contradictory TCN Datings as Well as to the Lake Ages on a ^{14}C Basis Taken from the Published Literature—All in Regard to the Past Glaciation of Tibet

As a high-altitude heating surface of the atmosphere in a subtropical latitude, the Tibetan Plateau plays a central role for the global climate (Kuhle 1985a, c, 1986b, c, d, 1987a, 1988d, 1989, 1998a, 2001b, 2002a, b, 2011c, 2013a). The Quaternary climate history of this region, however, is controversial in several respects.

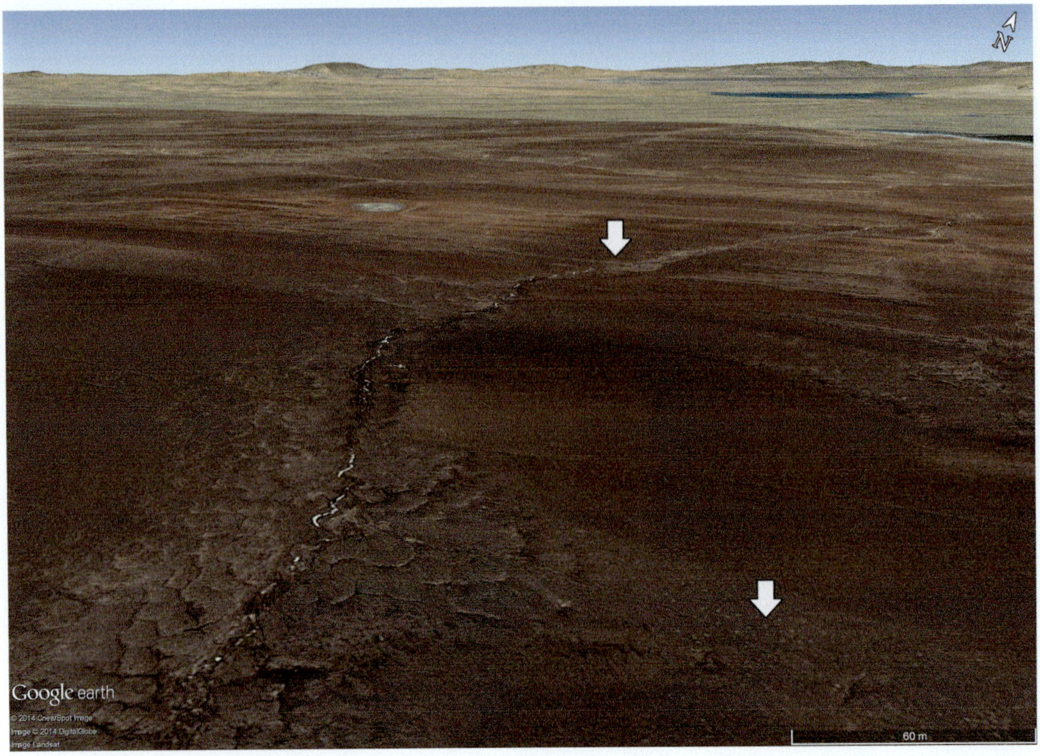

Fig. 2.139 Satellite image, Google Earth Pro shows at 33° 40′ 46″N 88° 46′ 22″E SW of Puruogangri (Purog Kangri I) the locality of Point M in Fig. 2.133 in detail. In the foreground (arrow) are reddish covers of loose material with embedded light-colored block material (cf. Fig. 2.138). The ice wedges point to the existence of continuous permafrost. The deposits are situated at 5180 m asl in the flow-shadow (lee side) of a 5540 m asl high rock ridge. Accordingly, they cannot directly have been reached by a foreland glaciation of the Puruogangri (Purog Kangri I). They thus require a glaciation prior to formation of the current landscape. In the background (arrow) keyforms of glacigenic exaration rills developed by scouring are visible (see Figs. 2.153, 2.157, 2.160, 2.178, and 2.190). They were formed in subglacially frozen moraine material (Kuhle 2014a)

This concerns, first, the extent of the past glaciations (von Wissmann 1959; Kuhle 2013a; Hövermann and Süssenberger 1986; Kuhle et al. 1989; Li et al. 1991; Shi et al. 1992) in respect to which all authors with the exception of the present author (Kuhle) followed the opinion of von Wissmann and thus contradicted a Tibetan inland glaciation. However, also controversial is its chronological progression (Owen et al. 2008; Owen 2009; Lehmkuhl and Owen 2005; Kuhle and Kuhle 2010; Chevalier et al. 2011; Kirchner et al. 2011; Kuhle 1986a, 1987b, 1998a, 2004, 2011a, b, 2012a). Pertinent to this question are the experiments performed in recent years with regard to the absolute dating of past glacier fluctuations. In one of the most extensive studies of this sort (Chevalier et al. 2011), 241 ^{10}Be-datings were gathered and statistically evaluated from the entirety of Tibet. The result, however, did not at all show a correlation of the glacial advances—neither among each other, nor in relation to the global minima in temperature or maxima in precipitation. This negative result was attributed by the authors of the study to the still-existing uncertainties of the dating technique (Chevalier et al. 2011; also Putkonen and Swanson 2003). Other authors such as Owen et al. (2012) and Kirchner et al. (2011) also came to the conclusion that there appears to be no recognizable correlation of the glacier

Fig. 2.140 Satellite image, Google Earth Pro shows at 34° 48′ 20.11″N 89° 16′ 49.84″E and 34° 48′ 54″N 89° 16′ 25″E the locality of Point K in Fig. 2.133 in detail. The intact surface of a ground moraine and a glacigenically abraded basin (with lake) in ground moraine material are visible north of Puruogangri. This type of deposit can be identified as a keyform. It extends as broad, nearly all-encompassing band over a distance of ca. 800 km at the northern rim of the Tibetan Plateau—from the massif of Kukushili in the west to the Kuenlun pass in the east (85–94°E)

fluctuations in Tibet. However, they consider this to be a fact that in the meantime has been confirmed by TCN datings (see below). This is not a plausible assumption, however, because the past climate and glacier oscillations necessarily must correlate over space. It must be considered common knowledge that they even correlate on a large (global) scale.

The reconstruction of a Quaternary inland glaciation by Kuhle since 1973 (Kuhle 1980–2014b), carried out on the basis of ca. 40 extensive field trips into the respective areas of the Tibetan Plateau and its marginal mountain massifs, is rejected by many authors (e.g., Schäfer et al. 2002, 2008; Lehmkuhl and Owen 2005; Owen 2009; Kirchner et al. 2011). This rejection is based on the TCN datings alone, while the glacial-morphological indicators on which the positive evidence of an inland glaciation rests, are not disproven.

According to the logic of the TCN datings, the following characteristics may be reconstructed for the former glacier oscillations on the Tibetan Plateau:

(a) they are not synchronous with each other and with the global ice ages (see end of chapter);
(b) during the LGM the extension of the ice was only insignificantly larger than the recent glaciation;
(c) the ice extensions of particular older glaciations are increasingly greater than that of the LGM, though the history of uplift points to a lower height of the Tibetan Plateau;

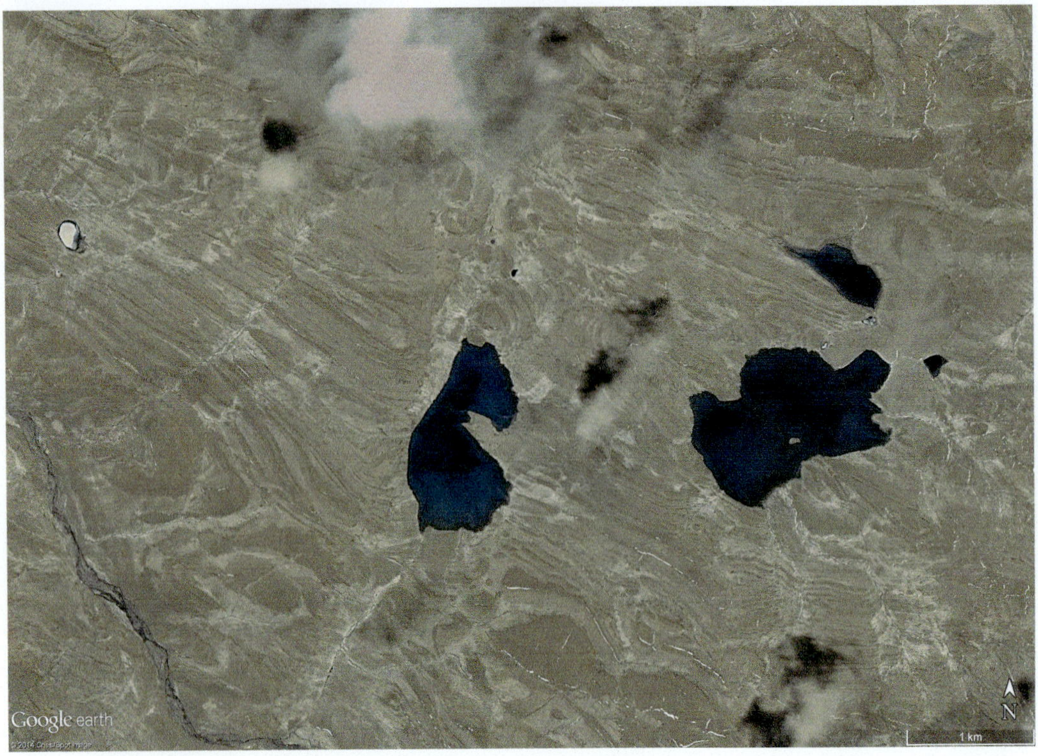

Fig. 2.141 Satellite image, Google Earth Pro shows at 34° 07′ 50″N 88° 59′ 54″E and 34° 7′ 53.11″N 89° 0′ 4.88″E the locality of Point L in Fig. 2.133 in detail. It shows broken and scoured ground moraine in the immediate vicinity of the northern Puruogangri (Purog Kangri I) foreland at 5185 m asl, 9 km from the recent glacier tongue. This phenomenon occurs also in the foreland of the current Zangsar Kangri plateau glaciation (Fig. 2.142). In addition—far from the local glaciations in the plateau area—it can be encountered in numerous places (Figs. 2.143 and 2.144)

(d) by application of the Global Circulation Models (GCMs) available today, the glacier chronology of Tibet reconstructed by TCN datings cannot be simulated (Kirchner et al. 2011: 259, 263).

These new GCMs, which reliably simulate the Nordic inland glaciations, regularly bring about large-scale inland glaciations on the Tibetan plateau during the LGM, and thereby correspond with the findings of Kuhle (1980–2014b). However, they are in a striking contrast to the opinion of Schäfer et al. (2002, 2008), Owen (2009), Heyman et al. (2010), and others (Kirchner et al. 2011: 259).

A further data record relevant for the assessment of a Quaternary glaciation of Tibet are the absolute datings of the stages of lake levels and lake infillings in the area of the plateau. By now, ^{14}C datings of drill cores from lake sediments of the entire plateau area are available. Here, maximum Late Glacial ages of <19 ka for Lake Donggi Cona are reached (Opitz et al. 2012). The other lake infillings, however, are of a clearly younger age: Lake Naleng <17.7 ka (Kramer et al. 2010); Koucha Lake <16.7 ka (Herzschuh et al. 2009);

Fig. 2.142 Satellite image, Google Earth Pro shows at 34° 12′ 25″N 86° 10′ 17″E and 34° 12′ 19.65″N 86° 10′ 28.55″E the locality of Point F in Fig. 2.133 in detail. Typical ground moraine cover in the eastern Zangsar Kangri foreland. The deposits lie up to 35 km from the recent glacier tongues and are exposed here in a thickness of only just 100 m. The lamination and compression of ground moraines by glacigenic scouring (see Figs. 2.141, 2.143, and 2.144) are evident here as well (arrows)

Lake Kuhai <14.8 ka (Wischnewski et al. 2011); Siling Co <14 ka (Morinaga et al. 1993) and <12 ka (after Gu et al. 1993); Sumxi Co <12.7 ka (Gasse et al. 1991); Lake Zigetang <10.8 ka (Herzschuh et al. 2006); Cuoe Lake <10.5 ka (Yanhong et al. 2006); Bangong Co <9.5 ka (Gasse et al. 1996); and Nam Co <8.4 ka (Zhu et al. 2008; >7.2 ka after Mügler et al. 2010). The highest lake level, however, of all the lakes, including the Donggi Cona, is coincidingly only reached in the Early to Middle Holocene (Opitz et al. 2012; Wischnewski et al. 2011: 94).

It is significant that drill cores of lakes situated *outside* of the maximum inland ice extension of the LGM as reconstructed by Kuhle, are providing clearly *older* chronologies. Examples are Lake Luanhaizi with <45 ka (Herzschuh et al. 2005), Lake Qinghai with <69 ka (Madsen et al. 2008), Lake Tengger with <42 ka (Zhang et al. 2002), Lake Manas with <37 ka (Rhodes et al. 1996), and Lake Tsaidam with an age of 35–47 ka (Kuhle 1987d, 1998a).[5]

[5]The only older ^{14}C date from the area of the Tibetan Plateau belongs to the lake terrace 70 m above the current lake level of the Pangong (Bangong) Co, and lies at ~39 ka B.P. (Rhodes et al. 1996: 118; Li and Zhu 2001: 39). This dating originates from the "Kunlun-Karakorum Sino-French team" and has not yet been published (Rhodes et al. 1996: 118). The reason for this is that the Bangong (Pangong) Co has a spillway at 4270 m asl, i.e., 24 m above the current lake level, so that the lake level of +70 m requires an artificial damming of the lake. While there are no indications of a geological cause of this damming, a valley-damming ice stream network has been positively proven for the connected Shyok Valley (Kuhle 1988a, 1999, 2001a, 2011a, 2013a). Accordingly, should the date be correct, this glaciation would indirectly have to be dated to the LGP (Stage 0, see Fig. 2.5).

Fig. 2.143 Satellite image, Google Earth Pro shows at 34° 38′ 01″N 86° 39′ 16″E and 34° 38′ 11.84″N 86° 39′ 20.00″E the locality of Point G in Fig. 2.133 in detail. The image shows characteristic detail of a probably broken and scoured ground moraine at 5055 m asl in the NE Zangsar Kangri foreland—comparable with Fig. 2.141 from the Puruogangri (Purog Kangri I) foreland. This feature was possibly created during the Late Glacial melting of the inland ice, and seems to be connected to the development of an active layer, i.e., a partial thawing of the subglacial permafrost

The TCN datings on past lake terraces, on the other hand, provide very high ages for lakes in the area of the plateau. Thus, for example, one finds >216 ka for Siling Co or >271 ka for Tangra Yum Co. Several maximum lake levels, however, also show ages of 3.8 ka (Drolung Co) or <15 ka (Zhari Nam Co) (Kong et al. 2011). In the same way that the TCN datings for glacier fluctuations show considerable variation, so also do the TCN data for highest lake levels in Tibet not result in a synchronous chronology (Kong et al. 2011: 272).

The ^{14}C datings of lakes lying outside of the limits of the maximum glaciation of the Tibetan Plateau, as reconstructed by Kuhle, uniformly point to high stages of lake levels at between ~35 and 22 ka B.P. (Kuhle 1987d, 1998a; Rhodes et al. 1996; Herzschuh et al. 2005; Madsen et al. 2008; Zhang et al. 2002). The question thus raised is why this optimum as to the lake sediments can be found nowhere else within the limits of the inland glaciation. The correlation with a maximum ice extension, i.e., with lake development on the Tibetan Plateau only occurring in the course of the Late Glacial deglaciation, seems to answer this question very well. Up until now, at any rate, no alternative explanation has been offered.

TCN datings of lake-level stages in the area of the Tibetan Plateau do not fit with the partly much younger ^{14}C datings of drill cores in the respective lakes (e.g., Siling Co >216 ka TCN vis-à-vis <14–12 ka ^{14}C age), nor do they correlate with the ^{14}C chronology of the lakes

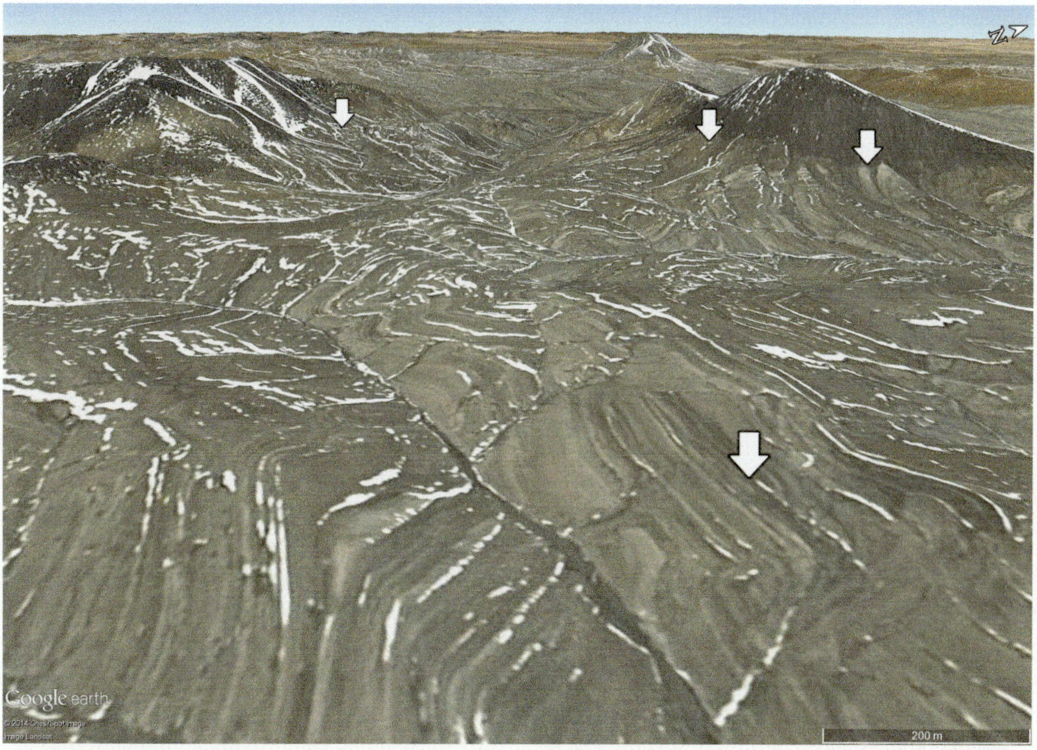

Fig. 2.144 Satellite image, Google Earth Pro shows at 33° 48′ 19″N 88° 00′ 59″E and 33° 48′ 17.72″N 88° 1′ 30.65″E the locality of Point J in Fig. 2.133 in detail. Large-scale "banded" and scoured ground moraine deposits (arrow in the foreground; cf. Figs. 2.141 and 2.142) in the area between Zangsar Kangri and Puruogangri (Purog Kangri I). Here, the light-grey ground moraines have been shifted upward to half the height of the piercing summits of darker bedrock (arrows in the background). There is no possibility of explaining these deposits by local slope (i.e., mountain) glaciations. One has to assume that they are of multi-glacial origin and came into being during the Early- and Late Glacial periods

outside of the plateau area. Accordingly, TCN datings do not even allow a chronological and much less a genetic relation between lake-level stages within and outside of the Tibetan Plateau area.

A possible reliability of TCN datings mainly relies on two factors:

First, it depends on the correct registration of TCN primary production rates at great altitude and subtropical latitude. Until now, this registration has not yet occurred in High Asia (see Sect. 2.5.2).

Second, it depends on the certainty with which an earlier exposure to radiation of the rock surfaces (= inheritance), or a shielding during certain periods in between, can be ruled out.

The problem of scaling and the problem of inheritance will lead toward an overestimation of the real ages of deposition; shielding, on the other hand, must lead to an underestimation.

Because for TCN datings in Tibet there are still practically no control values from ^{14}C age determinations, the problem of inheritance shall be explained by an example from Alaska. For the MP-I moraine of a past Muldrow glacier stage, Dortch et al. (2009) received TCN ages of between 2.5 and 146 ka (ibid.: 241). From several confirmed ^{14}C datings, however, it became clear that the age of this stage lies at 28 ka (ibid.: 239). Because it is assumed that from a data set of erratic boulders the oldest ages show the minimum age of deposition (Putkonen and

Swanson 2003; Heyman et al. 2011), without ^{14}C correction the age of the MP-I Stage would have been overestimated by a factor of 5. The boulders with the high age values probably originate from the material of an older glaciation (Dortch et al. 2009: 241). This material was later picked up and incorporated by a younger stage.

For High Asia, Kuhle and Kuhle (2010) have found a similarly large factor of overestimation. Here, TCN ages seem in many cases to be overestimated by factors of 4–6.5 (ibid.: 6–9, 12, 13).

For the area of the Tibetan Plateau the problem of inheritance is of special importance. First, plateau glaciation existed under "cold-based conditions" (Kuhle 2014a). This is already indicated by the fact that even today the permafrost line runs at only 4200–4600 m asl (Kuhle 1982a, 1985b, 1997: 243). Under these conditions, ground abrasion by the ice must have been strongly reduced or even completely absent. This means that rock surfaces in the bedrock and even of large boulders have outlasted several glaciation phases without change, and inevitably over time accumulated high ^{10}Be values. As a result, they cannot be considered as belonging to any specific glaciation phase (Linge et al. 2006; Goehring et al. 2008).

Second, the problem of inheritance is vital because in Tibet we are dealing with a plateau and basin landscape that is bordered and partitioned by many mountain groups from within and without. Therefore the inland glaciation overriding the entire terrain did not have here—unlike the Nordic flat and lowland ice sheets—an unambiguous direction of transport from N to S or W to E. Accordingly, in Tibet there must have existed different flow directions in accordance with the terrain. During the Late Glacial decay of the inland ice, they have repeatedly turned back in favor of local centers. Accordingly, the material was not shifted into the periphery—starting from a center—but multi-glacially seized and displaced. Thus, for example, on transfluence passes, the direction of transport has in part been upvalley. This back-and-forth movement has been strengthened by the glacial-isostatic, basin-shaped, ca. 650 m deep depression during the High Glacial stages (see Sects. 2.4.6.1 and 2.4.7.4.6), which was greatest in the center of the Tibetan Plateau. It additionally prevented the glacigenic transport of rock material out of Tibet, and began to form during the Early Glacial stages. It would then increase up to the point at which the maximum ice load of the High Glacial had developed, and diminish again when the load would retreat during the subsequent Late Glacial. A stronger promotion of the development of inheritance by such a to-and-fro movement as during the last 2.6–1.0 million years is hardly conceivable. Therefore, the possibility of inheritance must be considered very seriously and more than anything else.

Based on a comparative statistical study between TCN data from Tibet and the Nordic glaciation areas, Heyman et al. (2011) reach a different conclusion. Yet, the problem of this study is that for the Tibetan TCN data—in contrast to the Nordic glaciations—it does not add the independent control of ^{14}C datings. However, only these alone can provide an indication as to a possible inheritance. In fact, relevant ^{14}C data do exist, but Heyman et al. (2011) do not take them into consideration. These ^{14}C data provide significant indications of inheritance that are many times younger than the TCN data at the same glacier stages. The TCN ages are here continuously overestimated by the factor of 4–6.5 (Kuhle 1986a, 1987a, b; 1994a, 1997, 1998a, 2004, 2005a; Kuhle and Kuhle 2010: 6–9, 12, 13).

Unfortunately, Heyman et al. also did not examine the TCN ages of moraines by comparison with the accompanying ELA depressions (ΔELA) in order to test their coherence. It would have been useful and suggestive to use the ΔELA reconstructions of the respective areas published since 1982 (cf. Kuhle 1982a–2010). This neglect has led to the regrettable result that it is not possible to differentiate between inheritance and shielding. Here is an example: if one follows the TCN dating in the Bayan Har Shan (Heyman et al. 2010), then, for example, the stratigraphically oldest of five glaciation phases is the chronologically second-youngest. Now, this inversion could be explained by either inheritance or shielding (ibid.: 8–11). The authors chose the latter, arguing that otherwise the inheritance rate would lie at 53%. When viewed in comparison with the Nordic glaciations, so the authors reason, where it lies at 4%, this would seem to be an unrealistically high value. However, it is easy to see, in this way, how the argument becomes circular. To avoid this circular reasoning, it would have to be proven in the first place that the degree of inheritance in the area of the Tibetan Plateau shows identical statistical patterns of distribution to that of the Nordic glaciation areas. However, for no good reason, this identity is already taken for granted in the statistical method applied (Heyman et al. 2011).

As shown by studies in the Fenno-Scandian and Laurentian glaciation areas, the dimensions and distribution patterns of inheritance can *only* be reckoned properly based on *an already existing glacio-geomorphological reconstruction of the maximum glaciation* (see Stroeven et al. 2002; Briner et al. 2006; Linge et al. 2006; Goehring et al. 2008), and *not* the other way round. In the case of Tibet, however, the impossible attempt is made to draw conclusions by means of the TCN dating with regard to the maximum glaciations, and thus in regard to the degree of inheritance. This methodological mistake should have been avoided by Heyman et al. (2011), especially because Stroeven, who co-authored Stroeven et al. (2002) (see above), also co-authored Heyman et al. (2011) wherein he diametrically contradicts his views from 2002.

The fact that in Tibet the high basin and high plateau glaciation must have occurred under *cold-based* conditions (see above), in contrast to the Nordic flatland glaciations which mostly occurred under *wet-based* conditions (Kleman and Hättestrand 1999: Fig. 2), makes it very likely that the rate of inheritance in Tibet turns out to be significantly higher. The high TCN ages and the strong scattering of these data, as well as the absence of coherent climate trends within the TCN-dated glacier stages in High Asia (Chevalier et al. 2011; Kirchner et al. 2011), are rather clear indications of this.

In order to be able to assess this possibility, it would have been necessary to note, *before* carrying out any TCN dating, that there existed older glaciations, and to determine their extent. To ascertain the extent of past glaciations, it is without a doubt—just as in the case of the Nordic and Alpine glaciations—not necessary, nor feasible, to rely on dating. It is a question of glacio-geomorphological analysis. Thus far, with regard to the TCN datings in Tibet, this sequence of analysis has not been followed. Rather, TCN datings were simply taken without consideration of the glacio-geomorphology already in evidence for this region. This error will be explained by an example; in this way, the resulting mistakes can be made more transparent.

In several detailed studies in the areas of the Muztagh Ata, Kongar Shan, and Tashkurgan Shan, it was possible to reconstruct a High Glacial ice stream network glaciation and several Late Glacial to postglacial stages (Kuhle 1997, 1998a, 2004, 2006a). Subsequently, TCN datings were carried out on several of these stages by Seong et al. (2009) and Owen et al. (2012), but without ever taking into account the findings as to the maximum extent of the ice indicated by till sheets (Kuhle 1997, 1998a, 2004, 2006a). Thus, for example, Seong et al. (2009) have now determined for a stage of the Yangbuk glacier (NW exposure) at the Muztagh Ata an age of between 24 and 80 ka (m2C; Subaxh stage). At the Kongur Shan, situated directly opposite to the north (SW exposure), a stage of the Jangmanjiar glacier has been dated to 7.9–9 ka (m3F; Olimde stage). In the case of the younger stage, the moraines are located 8 km from the current glacier tongue, and in the case of the older, only 6 km away. The geomorphological state of conservation of the deposits is identical. The relevant ELA depression (ΔELA) in both cases uniformly amounts to 270 m (present author's calculation).[6]

Thus, the extreme difference in age between the two stages can be explained neither glacio-geomorphologically nor climatologically. On the main valley flank situated exactly opposite the Subaxh moraines, the present author has documented large-scale till covers with erratic boulders up to an altitude of 4290 m, that is, 640 m above the valley bottom. They were left behind by a past ice stream network, i.e., an inland ice sheet covering the entire high plateau of the eastern Pamir (Kuhle 1997: 29, 30, 93–95, 165–180, Photos 48–52 and Fig. 14). Thus, either way, the moraines of the Subaxh stage are completely embedded into the material of an earlier main valley glaciation. That this process inevitably led to a mixture of different accumulated material so that, accordingly, the ages of the TCN datings might have nothing to do with the true age of the respective stage where they were last deposited, is not at all reflected by Seong et al. (2009). The results derived from these TCN ages are thereby simply untenable conclusions with regard to the age of the past glaciations.

A further dating is available from the southward-connected Tashkurgan Valley. Owen et al. (2012) obtained here for the oldest stage (Dabudaer)—which they classified into the pre-last Glacial or older—TCN data with values of between 23 and 219 ka (n = 16). A moraine complex of this stage lies on the main valley flank up to more than 4800 m asl and is dated to 57–152 ka (ibid.: A1; Table 3). These moraines prove an ice thickness of the main valley glacier of more than 1600 m and belong to the ice stream network reconstructed by the present author. This network received inflow across the Kunjerab pass (4700 m asl) from the area of the northern Karakoram (Kuhle 1997, 1998a, 1998c, 2004, 2006a, 2011a). Despite the extant aridity, the accompanying ELA depression (ΔELA) lay at above 1000 m. Only 12 km downvalley from this cross-profile, near the thalweg, lies the stage of a small side valley glacier (Jialongqiete Valley) that provides a TCN age of 85–219 ka (Owen et al. 2012: Table 3, J1). Stratigraphically this stage must be clearly younger than the main valley glacier stage. However, according to the TCN dating it is older. Thus, as is shown by the comparison of the existing TCN datings, one must take into account a high degree of inheritance. Otherwise there is no explanation for such an age inversion. Young glacier stages were obviously imprinted into the material deposited by the previous and next-previous glaciations. Accordingly, the much older glacigenic material is reshaped glacio-geomorphologically. This took place without a zero position of the surface age, which however is the precondition of every dating.

Be that as it may, this was the implicit assumption made in the respective studies, albeit a faulty assumption. The age of the TCN datings has no validity as to the real age of the stages. First, the extent of inheritance ought to be clarified, which thus far has not been done.

A further example is provided by the data of Schaefer et al. (2008) from the SE slope of the Shisha Pangma, published as TCN datings. These authors, too, ignore the finding of an extensive older glaciation, proven by covers of till with erratic granite boulders. With this glaciation, an outlet glacier coming from the Tibetan Plateau flowed with a width of 25 km across the Yagru Xiong La (pass) onto the southern slope of the Himalayan main crest, and thence downward to 1600–1000 m asl (Kuhle 1988c; 1998a: 85, 86; 1999; 2001b; c; 2002b; 2004: 184, Fig. 24, Glacier Map TH33; 2005c). Schaefer et al. (2008) date—at between 8

[6]Here and all throughout the chapter, the calculation of the ELA depression (ΔELA) is performed as a variation on the Höfer method (1879) that divides the difference in height between extant and past terminal position by the factor two. This method provides good approximate values, as shown by comparison with the THAR (Toe to Headwall Area Ratio; see Meierding 1982) method that also presents a variant of the Höfer method.

and 317 ka—the lateral moraine complex of a far younger side valley glacier lying at over 3700 m asl, which bent into the main valley near Nyalam. In addition to the wide dispersion of TCN dates across the area, there also occurs an age inversion. This means that the stratigraphically youngest inner moraine shows an older date than the outer moraines (ibid.: 306). In all likelihood, the material of the younger side valley glacier was mixed with that of the older main valley glaciation that transported material from the area of the Tibetan Plateau all the way to this locality. This would at any rate explain the extreme scattering of the data. However, the authors of the study do not at all consider this possibility. As a result, the subordinate, probably Late Glacial to Holocene fluctuation of a side valley glacier is defined as a maximum glaciation supposedly reached during the pre-last or an older glaciation phase (ibid.: 295). This reasoning thus results in an extremely odd assertion: namely, it is claimed here that in this area the last global ice age, the LGP (Stage 0; Fig. 2.5) was completely omitted, i.e., did not take place.

However, not only is a Quaternary-geomorphological analysis as a necessary precondition of a dating renounced here. Also incomprehensible is the fact that these and other authors do not make use of a second, alternative for analysis: the study of ELA depressions. The TCN ages of moraines necessarily must be examined for their coherence with the ΔELA connected to them. Otherwise they remain empty numbers, that is, data with no reference. Nearby areas with the same exposure, height, and relatively recent ELAs would have to yield comparable, analogous TCN chronologies. However, this is not the case. The TCN chronologies are not at all comparable with each other, and exhibit no correspondence or agreement even in close spatial proximity. Thus, the TCN data do not stand the test of coherence with glaciological and climate-physical evidence. Chevalier et al. (2011: 536), for example, date a 420 m ΔELA stage for the north slope of Cho Oyu at 24–27 ka. Owen et al. (2009) carried out datings ca. 20 km further east for a comparable stage on the northern slope of Everest, and obtained values of >330 ka (T1). Ca. 50 km west of the Cho Oyu, in the Naisa Valley, Schaefer et al. (2008) date in a similarly inconsistent way an ELA depression (ΔELA) of 450 ± 100 m (ibid.: 308)[7] into the MIS 2 (LGP). At the same time, they assign to the Nyalam moraines lying ca. 20 km farther south, on the more humid (!) SE slope of the Shisha Pangma, a TCN age of 8–317 ka (see above). They thus do not classify them as belonging in the Last Ice Age (MIS-2), but to the penultimate or a still older glaciation (ibid.: 295), whereas, in fact, these Nyalam moraines must under any circumstances have reached an ELA depression of 450 ± 100 m (see above). Moreover, it has already been shown that the TCN data in other locations of High Asia yield similar or even more contradictory results (cf. Kuhle and Kuhle 2010).

Owen et al. (2009: 1422) deprive their statements regarding High Asia of considerable empirical value by not taking into account the usual rate of erosion. By factoring in the erosion rate, instead of the given ca. 330 ka (T1; ibid.), ages of ~600 ka or even ~860 ka (!) would be the result (Kuhle 2012a: 193, 194). One of the dated boulders of the T1 moraine has a current volume of 0.28 m^3. Under consideration of the normal erosion rate of 2.5–5 mm/ka, this boulder originally ought to have had a size of ca. 19 m^3. "Originally" means: at its former date of deposition calculated by means of TCN. Accordingly, the extant boulder would have retained only 1.4% of the original material. This small remnant of 1.4% (!) left behind by weathering has been dated by Owen et al. (2009). On this basis, it is simply not possible to recognize an empirical relation between the age of this remnant and the age of the deposition of the original boulder. Further details of exactly this TCN-dating problem are discussed in Kuhle (2012a: 193–196).

The consideration of the ELA depressions (ΔELA) points to the further grave incoherences of TCN datings. In the central area of the Tibetan Plateau, TCN datings were determined for the

[7]Schaefer et al. (2008) used the THAR method (Meierding 1982) for their calculation of the ELA depression; see Footnote 6.

Tanggula Shan. For this region, Schäfer et al. (2002) and Owen et al. (2005) obtained an age of 33–115 ka for the youngest stage with an ELA depression (ΔELA) of ~ 30 m; 46–215 ka for the stratigraphically medium stage with an ΔELA of ~150 m; and 83–181 ka for the stratigraphically oldest glaciation with a ΔELA of ~350 m. The stratigraphically youngest, middle, and oldest stages necessarily mean stages of a minor, middle, and greatest extension. Otherwise the stages would have been overridden and thus destroyed. Thus the TCN datings display a clear age inversion and can not be taken at face value.

In this context now the following is remarkable: the Tanggula Shan is surrounded by lakes for which ^{14}C dating indicates a maximum lake level in the early to middle Holocene. At the same time, this maximum level is proven for lakes in the entire plateau area (Li and Zhu 2001) (see above, Sects. 2.2.2.2 and 2.4.5.1). Now, there are several localities where moraines of past local glaciations dovetail with the highest lake terraces. This evidently came about because these glacial stages were reached at the same time or even somewhat later than the highest level of the lake (for more details, see below). These stages are connected with ELA depressions (ΔELA) of 90–180 m. All these localities are situated in areas that currently are more arid than the Tanggula Shan. Thus, they also show higher recent ELA levels than the Tanggula Shan with its stronger monsoonal humidity. These are, for example, the localities of the Nam Co (30° 32′ 15″N 90° 25′ 29″E) and Burog Co (Buruoco) (34° 24′ 22.18″N 85° 46′ 28.37″E).[8]

In a climatological sense it is completely inexplicable how, in one of the driest areas of the Tibetan Plateau, glacier advances of this magnitude (ΔELA = 90–180 m) can occur in the Holocene, while at the same time in the more humid Tanggula Shan, a glaciation with an ELA depression of >350 m is said to have *never* existed. "Never," because according to the authors of the TCN studies (Schäfer et al. 2002; Owen et al. 2005), the stratigraphically oldest glaciation (ΔELA = ~350 m) of the Tanggula Shan was also its maximum glaciation. Owen et al. (2005) even postulated for the LGM an ΔELA of only ~30 m! Both teams of authors thus come to the conclusion that the inland glaciation of Tibet reconstructed by Kuhle cannot have existed—and this solely due to the high TCN ages of the deposits (Schäfer et al. 2002: 294; Owen et al. 2005: 1408). In this context, it is erroneously maintained that Kuhle's reconstruction of the inland glaciation is not based on local indicators in the area of the plateau, but on an extrapolation from the marginal areas of Tibet (Owen et al. 2005: 1408). However, the fact is that Kuhle (1991: 135–143, 145–162, 228–230)—a treatise cited by the authors but obviously not seriously taken note of—provided very detailed descriptions of (local) glacio-geomorphological indicators such as large-scale covers of till with erratic glacial drift (meter-sized boulders) and roches moutonnées, which were also documented photographically and confirmed by sedimentological analyses.

On this basis, not only the subsequently TCN dated stages of retreat were evidenced, but beyond that also detailed evidence was given for a relief-overriding glaciation of the area between the Tanggula Shan and Nyainquentanglha Shan. The findings by Kuhle registered on the expedition in 1989 were completed during a further field trip into the area and published (Kuhle 1995: 37, 38, 40; 2003: 76–85). All these indicators are neither mentioned nor refuted by the authors of the TCN studies.

A further important fact is also that the high scattering of the TCN ages within the stages, the age inversion between the middle and older stage, as well as the high maximum ages for all three stages statistically do not at all allow the construction of an absolute chronology.

[8]In this regional context it might be of interest that north of the arid area of the Palung Co, there are light-colored erratic tills lying at 31° 04′ 38″N 83° 38′ 43″E, at between 5050 and ca. 5500 m asl. Adjacent to them, at 31° 05′ 09″N 83° 38′ 44″E, there are also preserved light-colored till covers with light-colored erratic boulders lying at between 5000 and 5160 m asl. These loose rocks lie here everywhere on dark bedrock. These deposits therefore must be classified as belonging to the late Late Glacial Stages II, III, or IV (Fig. 2.5). Here, too, they prove a (still) relief-overriding glaciation.

Exactly the same finding is also true of the age distribution of the TCN data in the Bayan Har Shan. Accordingly, Kuhle (2012a) states: "For instance in the Bayan Har Shan (E-Tibet) Heyman et al. (2010) followed the present author (Kuhle) and acknowledged till covers in the foreland as being of glacigenic origin (Fig. 1: No. 3 and 19). In contrast to the present author, who had classified these deposits as being from an Early Late Glacial stage of the LGM (LGP) (Kuhle 1982a, c, d, 1987b, 1997, 1998a, 2003, 2004), Heyman et al. (2010) put them into a separate glacial phase (BH4), which they classify as being Pre-BH 3, i.e. older than >95–165 ka." (Kuhle 2012a: 196). A lot about such a pattern of distribution speaks in favor of a multi-glacial redeposition of till material as the original cause (Kuhle 2012a: 196, 197); inheritance plays an important role here (cf. Heyman et al. 2011: Fig. 7).

This does not mean, however, that the in part extremely high TCN ages in Tibet are explained by inheritance *alone*. It is still possible also that an incorrect assessment of the primary TCN production rates—that nowhere have really been assessed (!), but that are only assumed (see Sect. 2.5.2)—contributes to this (Kuhle and Kuhle 2010). However, with the current data base it is completely impossible to separate these two factors in a clear way. At present, literally every new TCN study in Tibet contradicts the results of the previous one (e.g., see Fu et al. 2013).

This arbitrary nature of the TCN data with regard to their paleo-climatological interpretation points to the expected randomness which comes from a high rate of inheritance. This is shown in great detail with increasing density of the data. In any case, a clarification of these problems is only possible on the basis of a clarification of the maximum ice extent. Its reconstruction is supported by the documentation of its relative incremental succession. This is exactly the procedure that has been pursued in the preceding chapters and by the present author (Kuhle) since 1973. It has been evidenced as empirically stable. Accordingly, the question of the maximum extension of the ice can *only* be resolved by the methods of classic glacio-geomorphological analysis, which have existed and been constantly improved for the past 150 years. However, due to the sudden appearance of the TCN technique, it has been largely abandoned lately—unfortunately. The TCN dating technique only works within very narrow boundaries and thereby should be focused on clearly defined tongue-end positions (see above).

With this observation we mean to point to a general problem. Without exception, all the authors working with TCN datings derive from the high age of moraines—that often lie only a few kilometers away from the recent glacier tongues—that a large-scale glaciation of Tibet never existed. From this they infer, too, that there is no need to consider inheritance an important factor. However, *this argumentation is circular.* This is because the age of the boulders is only then indicative of the stage of the material in which they are found when no larger glaciation and thus no inheritance can be assumed to have taken place. The latter (no larger glaciation or inheritance), however, must first be proven. The evidence of the size of a past glaciation naturally cannot be proven by datings. It can *only* be proven *by glacio-geomorphological indicators.* These indicators point to an inland glaciation of Tibet overriding the entire landscape (Kuhle 2013a).

It is very conspicuous that the TCN datings of the Nordic glaciations are assessed quite differently than those for the Tibetan Plateau. So, for example, TCN ages between 79 and 37 ka of the rock surfaces of "tors" in the Swedish lowlands (Hättestrand and Stroeven 2002) do not lead to the assumption that they could have been non-glaciated during the LGM. Here one *knows* from classic glacial-morphological analysis that this locality was situated close to the center of the last Fennoscandinavian inland ice (Stroeven et al. 2002). If carried out in Tibet, by contrast, such TCN datings until now are *always* assumed by the respective authors to be indicators of the non-glaciation. *In consequence*, the glacio-geomorphological indicators of an extensive inland glaciation are ignored.

As has been shown in detail in the preceding sections, special indicators of this glaciation are the large-scale till covers. According to the published field results by the present author, they exist throughout the entire plateau area. However, in the meantime they can also be confirmed by the evaluation of satellite images (see above). According to assessments of the present author, in southern Tibet around 50% of the plateau surface is covered with till; in middle and northern Tibet, up to 90% of it is covered by till. At some localities, the tills are exposed with a thickness of more than 100 m (Sects. 2.1–2.4, as well as 2.6, 2.8, and following) (Fig. 2.135).

This finding is also important for the age determination of lake terraces by TCN dating, as exemplified by the Tangra Yum Co (Fig. 2.134a). On the northern shore of the Tangra Yum Co, the massif of the Bobogawu (6474 m asl) is situated. From its northern spur two valley glaciers reach down as far as to 5633 m asl (Fig. 2.134a: position A). On the western slope a further glacier reaches down to 5704 m asl (Fig. 2.134a: B). All three glaciers have developed clearly defined stages of former moraine deposits. The ELA depressions (ΔELA) pertinent to them amount to between ca. 450 and 460 m. This would be ΔELA values lying between Stage IV and V (see Fig. 2.5). In this connection, all three former glacier terminal moraines are interpenetrated by lake terraces. This dovetailing must have taken place in the following way: the moraines of the glaciers on the northern slope were overridden by the ca. 4740 m level of the past Tangra Yum Co lake level as far as down to ca. 4720–4700 m asl (Fig. 2.145; Fig. 2.134b: AJ as a section of Fig. 2.134a at AJ). The moraines of the western slope were also superimposed by the highest past level of lake terraces. In this case it was the lake terrace level of the past Zhari Nam Co (Zhari Nanico; see Kuhle 1999: 6, 7, 23; 117–120, Photos 72, 73; Fig. 2: No. 47) situated at ca. 4761 m. The moraines were overprinted down to ca. 4753 m asl. Accordingly, the glacier stages necessarily are somewhat older than the upper lake terraces that overprinted them but younger than the late Late Glacial Stage IV because of their ELA depression of only ca. 450–460 m (see above).

At the time when the lake terrace reached a level of 4740 m, the Tangqung Co must have formed a joint lake level together with the Tangra Yum Co (see above) (Kong et al. 2011: 265). TCN datings by Kong et al. (2011) on terrace debris (gravel) of this Tangqung Co stage, and thus at the same time of the Tangra Yum Co, result in an age of ~18 ka. The highest terrace level of the Zhari Nam Co, by contrast, is dated to <15 ka. To next use these datings for an age determination of the glacier stages seems to be dubious, because the data are inherently incoherent. The highest level of the Zari Nam Co at 4761 m provides an age of 14.4 ka, while a 40 m *lower level* in a contradictory manner yields a clearly *older age* of 27 ka. The present lake level provides a similarly incoherent age of 19.5 ka (ibid.: Table 1). The dating at the lake level, according to Kong et al., serves as an estimation of inheritance. For this reason, its age is subtracted from the age of the highest terrace (ibid.: 269). Accordingly, in regard to the highest level of the Zhari Nam Co, the result would be an age of −5 ka. This, however, is not a convincing result.

The chronology of the Tangqung/Tangra Yum Co is incoherent as well. The highest level is dated as 20.1 ka, a lake terrace in the rock 30 m lower to 7.4 ka; debris still 5 m lower to 22.3 ka, and the height of the present lake level to 2.8 ka. Even with several alternating lake infillings—*which naturally would call into question the intended validity of these datings altogether*—these data sets cannot be explained.

The authors of the study establish that they did not find glacial deposits in the surroundings of the lakes. Yet, they offer absolutely no Quaternary-geological, geomorphological, and sedimentological data and sampling from the field to prove their claims. But they nonetheless conclude that an influence on the dating by glacial activities can be ruled out (ibid.: 267). The glacial-geological and glacial-morphological findings by the present author (Kuhle 1999: 6, 7; Fig. 2: Nos. 26–55) from the immediate

Fig. 2.145 Satellite image, Google Earth Pro shows at 30° 44′ 22″N 86° 28′ 51″E and 30° 44′ 24.10″N 86° 28′ 49.24″E the locality of Point AJ in Figs. 2.134a and b in detail. The highest terrace level of the Tangra Yum Co is situated here at 4740 m asl (arrows). Material deposited from the older advance of the glacier (younger than Stage IV and older than Stage V; see Fig. 2.5) has been reworked by these terraces. Its deposits, which reach down to ca. 4720–4700 m asl, were destroyed by the uppermost terrace levels. Consequently, the age of these lake levels is the early to middle Holocene (see also Fig. 2.134b as a section of Fig. 2.134a at AJ)

southern, western, and northern vicinity of the Zari Nam Co prove the exact opposite.

The analysis of the satellite image (Fig. 2.134a) shows already that the lake terraces of the Zhari Nam Co and Tangra Yum Co very likely developed in the material of large-scale till covers. Material of light-colored, rounded boulders in a light-colored till matrix, for example, is well exposed in a saddle at 5760 m asl on a dark underground. It is here located ca. 1200 m above the Tangra Yum Co lake level (Fig. 2.135) and covers in the form of thick loose material the surrounding fringing heights and valley flanks down to the extant lake level. This finding would confirm that the TCN datings have nothing at all to do with the age of the lake terraces. A direct dating of the terraces, and thus also of the glacier stages interpenetrated by them, by ^{14}C samplings would be necessary here.

A further reference to the fact that the TCN datings of lake terraces have been contaminated by glacial deposits is the dating at Nam Co. Here, Kong et al. (2011) took samples of lake terraces that were formed directly in the material of a past glacier stage. This glacier stage had its terminus in the area of the current lake level (ibid.: Figs. 5d, e). From this they obtain an age of 53 ka for the highest past and and 17.2 ka for the extant (!) lake level (ibid.: Table 1). The ^{14}C datings of two independent drill cores from lake sediments of the Nam Co, however, resulted in maximum ages of 7.2 and 8.4 ka (Zhu et al. 2008; Mügler et al. 2010). Thus it is probable that the TCN datings of Kong et al. are contaminated by inheritance. This means that they do not correspond with the age of the lake-level stages. At the same time, this also means that they do not correspond with the age of the moraines into which these dated boulders were last embedded.

In this connection, the extremely high TCN ages of lake terraces in the rock at the Siling Co (>216 ka) and Tangra Yum Co (>271 ka) are also very interesting. They are no argument against an LGM glaciation (Stage 0, Fig. 2.5) overriding the entire terrain. Demonstrably, rock surfaces can persist under *cold-based conditions* through several glacial phases without being reset to zero by the glacigenic ground polishing of the subsequent ice ages (Kleman 1994; Stroeven et al. 2002; Briner et al. 2006; Linge et al. 2006).

In summary we offer the following observations:

1. the results of the TCN datings are strongly contradictory;
2. the findings of TCN datings and of lake datings also strongly contradict one another.

2.5.1 References to ^{14}C Data from High Asia, Which Show the Clear (Unambiguous) Relationship Between the Location of Glacial-Geomorphological Features and the Reconstructed Snowline Depressions (ΔELAs)

The focus of this section lies on Fig. 2.5. From the above, it can be seen that the TCN-dating studies assume a different position from that of the present author (Kuhle). The latter claims to have proven on the basis of glacial-geological and glacial-geomorphological data (Kuhle 1974–2014c), as well as with ^{14}C datings (see Fig. 2.5) (Yamanaka 1982; Kuhle 1986a: 439–454, Table 4; 1987b: 205 Table 2; 1987c: 227; 1987d: 303, 307; 1991: 177; 1994a: 141–144, 177, 178, Table 2, Fig. 36; 1997: 100, 114, 115, 127, 195–202, 253, Tables 2 and 3, Figs. 79 and 83; 1998a: 83, 85, 98, 99, Table 2; 2005a: 199–201, 217, 219, 264, 265, 266–275, 280, 288, 292, 293, 296, 299, 308, 318, 507, 508, 607, 608, Figs. 3 and 19, Tables 2 and 3, Photos 124 and 196; Grosswald et al. 1994: 282, 283) that not only in places, but across the entirety of High Asia including Tibet, there must have existed a synchronicity with the global ice ages and even with the postglacial glacier oscillations, which continues even into the present.

As recommended by the CRONUS-Earth project, ^{14}C data should be used to determine TCN 'production Rate Calibration Sites' for High Asia and thus for proper calibration (Kuhle and Kuhle 2010).

2.5.2 How Safe Are TCN Data? Remarks on the Possibility of Using TCN Data for Glacier-Historical Datings

Thus far throughout Sect. 2.5, cross-sections of glacier history and TCN datings, as they are treated in the literature with regard to High Asia, have been considered. In this context, in addition to the TCN contradictions regarding the reconstructed glacier history, similarly glaring TCN contradictions regarding the ^{14}C lake datings have been discussed. At the same time, it has been shown that the very same (^{14}C) lake datings do not contradict High Asian glacier history.

This Sect. (2.5.2) was written because the present author has geomorphologically reconstructed the glacier history of High Asia over the course of 42 years, and yet been strongly criticized for it on the basis of TCN data. This criticism will not be ignored.

2.5.2.1 To What Extent Can the Development of a Rock or Block Surface Be Known?

This section addresses the question regarding the null (zero) position of the rock surfaces to which TCN dating is applied. However, this basis of reference is lacking because of the following reasons:

1. As has been empirically proven in detail in the chapters above, *contamination through inheritance* in the (even today) permafrost-fixed block accumulations (tills) must reach a maximum. Accordingly, every past ice cover was cold-based (Kuhle 2014a: 127–139) and hardly able to dislocate the boulders that were frozen to the ground. Thus, by far most of the boulder surfaces have been irradiated poly-interglacially. Therefore they *must* show old ages.
2. At an uplifted Late Glacial ELA, side valley glaciers throughout the entirety of High Asia of the first- to n-th order reach the main valley bottoms, which have become free of ice, of a next-higher order. They pick up its poly-interglacial till and thus there is inheritance. This taking-up of material takes place from the flat active layer, i.e., from the active layer of the main valley flank as well as that of the main valley bottom (see Sect. 2.4.5; Figs. 2.76, 2.77, and 2.80) (for the example of Nyainquentanglha: Kuhle 1991: 155–161, 164–166, Photos 24, 29–32, 35; 2013a: 52–55, Photo 24; for the example of the Gurla Mandhata massif: Kuhle 1999: 42, 233, 234, Photos 168 and 169). Accordingly, a Late Glacial terminal moraine rampart *cannot be "dated" as any younger* than a few hundred thousand years. In some places, therefore, the still younger Neoglacial moraines are TCN-dated to a Late to High Glacial age (Stage IV–0; Fig. 2.5) or even still older. This is the case—to give only one of numerous examples—in Owen et al. (2010). In their TCN datings Na75–Na81, with ages up to 54.3 ka (ibid.: 1826, 1827), older till boulders were picked up—in the manner described above—from the main valley flank by the Late Glacial side valley glacier. This geomorphologically evident finding (see Sect. 2.4.7.4.1 above; Kuhle 1999) is confirmed by the scattering (age distribution) of the six TCN datings. The respective till is attached to the main valley flank up to a height of at least 5200 m. The supposedly "dated" terminus of the side valley glacier is located at 4750 m asl. In the intervening height interval of 450 m, that older main valley moraine was picked up by the side valley glacier. Thus, although it belongs to the last High to Late Glacial, it has been

contaminated (see point 1) by poly-interglacial inheritance (ibid.: 1999: 225–234). This, however has fatal consequences: an ΔELA of only ca. 435 m ((5620 − 4750)/2 = 435) belonging to a young stage in the time of transition between the late Late Glacial (Stage IV) and Neoglacial (= Holocene; Stage V) is dated here to ca. 36.4–54.3 ka (!). Thus the authors simply assume that the last ice age (= LGM, Stage 0; Fig. 2.5) did not take place in Tibet.

3. For all of these TCN tests, the date of the 0-position remains in the same way unknown as does the last exposure of the block faces on the last moraine substratum and their erosion rate. As to the exposure or burial of boulders, solifluction of the active layers on moraine surfaces would have to be taken into consideration. For instance, data from empirical measurements of moraine surfaces from the Rongbuk Valley (north side of Everest) have been available since 1988 (Kuhle and Jacobsen 1988). Yet 21 years later, when TCN data of moraine boulders were produced by Owen et al. (2009) in the immediate vicinity, they were not taken into consideration. This is inexplicable, inasmuch as they have deep relevance for the uncovering and exposure of boulders appropriate for the taking of TCN samples. In regard to the paper by Owen et al. (2009), it was thus noted in an earlier paper by Kuhle (2012a: 194): "A further point is remarkable here: at the same valley slope up-valley, at the same interval of height, the present author has established with the help of radiation-range-finders (Geodimeter) installed over a period of 40 days, at a decline of the moraine slope of 30°, current solifluction rates of 4–8 cm/a (Kuhle and Jacobsen 1988: 598). This proves an extreme reworking of the moraines. Considering these values, an in situ exposure age of a moraine boulder of ~600 to ~860 ka at least—probably in excess of 1000 ka—should be notable. Owen et al. realize that at the sight of a state of preservation like this and the position of the boulders (recently 4–8 cm/a solifluction rate) such an age cannot really be justified. Therefore the authors arbitrarily adjusted the calculation of the TCN age to their assessment of the geomorphological reality. Thus, the fact that this way the standard interpretation of the TCN-measurements produces clearly overestimated values here is suppressed."
Under consideration of this point (3) and the demanded principle of coherence, we have once again accumulated sufficient support for the claim that in Tibet the TCN age of boulders cannot be directly taken as the moraine age that is sought after.

4. Tibet is a plateau-and-valley basin landscape, fringed by marginal mountains and divided into mountain groups. When during the early Glacial the ELA dipped into this relief, the glacier flows necessarily changed depending on the ice thickness. This happened because of the numerous ice centers of minor thickness, a few centers of very great thickness must have developed (cf. Figs. 5.3 and 5.4). While the primary small centers developed on the highest terrain surfaces, i.e., the mountains, the secondary ice centers developed in field depressions. The primary glacier flow, accordingly, was of minor thickness and ran along the slopes (downward along steep slopes). The immense High Glacial ice discharge, by contrast, which led to the infilling of the terrain and transfluences, was oriented along the thalwegs. During the Late Glacial, the uplift of the ELA and its increasing uplift beyond the height of the terrain took place. From this followed the opposite development. From these processes, we can decipher the back and forth movement of the transport of tills (see Sect. 2.5), repeated time and again during the Pleistocene ice ages.
An exemplary study area for the study of this to-and-fro movement of till transport is the Rongbuk Valley. It leads from the Rongbuk Glacier below the north face of Mt. Everest to Tibet. The current Rongbuk Glacier, however, does not only flow down to Tibet. It also flows across a transfluence pass, the Lho La, down to the Khumbu Glacier, i.e., into the southern slope of the Himalaya (Kuhle 1988c:

506, Fig. 74). It has been shown by means of geomorphological indicators that during the High to Late Glacial (Stages 0–IV, Fig. 2.5) the Rongbuk Glacier overflowed the same transfluence pass with a minimum thickness of at least 450 m (ibid.: 496, 497, 502, 503, 505–507, Figs. 58, 66, and 69). Accordingly, its former polish demarcation line, which is preserved up to the present, has a slope that is contrary to its current direction of incline, namely from N to S (ibid.: 496, 497, 502, 503, Figs. 58 and 69). The LGM Rongbuk Glacier, accordingly, was a Tibetan outlet glacier, while currently it flows back to Tibet facing N. Unfortunately, Owen et al. (2009) ignore these gemorphological findings. The moraine boulders that Owen et al. TCN-dated in a valley cross-profile ca. 10 km downvalley from the current glacier terminus and ca. 300 m below it, were moved to and fro in a corresponding manner. Their determined ages occupy a range of between 32 ka (T2-complex) and >330 ka (T1). T1 is located only ca. 140 m higher on the same slope (see Kuhle 2011b: 180). The difference in age, which cannot be verified geomorphologically in this location, as well as the ΔELA of only ca. 150 m ((5220 − 4920)/2 = 150) which belongs to a Holocene stage, show that the TCN datings do not correspond at all with the age of the moraine deposits in view of their stratigraphical position.

In truth, however, their TCN datings strongly indicate that the boulders were *inherited.*

The glacial-isostatic depression caused a basin to form in Central Tibet. This lowering and its rebound during the interglacials led to changing gradients of the terrain. The back- and forth-movement of the ice flow was thus increased (see Sect. 2.5). As a result, the basin-shaped depression of Central Tibet reduced moraine transport toward the plateau margins and thus also beyond Tibet. *All these glacio-physical factors must have contributed to the predominant role of inheritance.* Accordingly, the material of tills and moraines including their boulders is polyglacial, and the TCN ages can hardly be taken at face value.

Nevertheless, the 1000–1500 TCN datings conducted in High Asia remain noteworthy—not only due to the great efforts as to finances and labor, but mainly due to their scientific value. For their pattern of distribution up to the present shows very clearly that *they do not provide a chronology* of glaciation in High Asia and thus are not yet appropriate for absolute dating in this high mountain area (see Sect. 2.5.2.3).

5. The erosion rate of moraine boulders is not solely dependent on the type of rock and the nature of its weathering, but primarily also on hydrothermal alteration as well as the strength of tectonic exposure, i.e., modification. All of these factors had already started to work in the bedrock. The existence of these primary factors has been evidenced for Tibet (Heydemann and Kuhle 1988; Kuhle 1988c: 483, 498, 499, Figs. 43, 61, 62, and 63). The later, secondary weathering rate necessarily depends on these primary factors. None of these factors, however, is known in the locally individual case. Thus, they cannot be taken into consideration, a problem that is not—so it appears to the present author (Kuhle 2014a: 195)—sufficiently emphasized by the authors of TCN studies.
6. Here, we call to mind a further aspect pointed out by Schröder (2007: 369). The structure of the atmospheric cascade of CR (cosmic radiation) makes clear that the depth profile of the isotope production rates in high mountain areas, where morainic material is earliest exposed, must be different from that in the altitude of deposition. The higher amount of particles with higher energies should cause isotope production deeper below the surface.

At the same time a relatively smaller amount of particles with lower energies reduces isotope production near the surface.

2.5.2.2 Is the Degree of Familiarity with Local TCN Radiation and Its Effect on Rock Surfaces in High Asia Sufficient to Analyze the Age of Rocks?

Whereas Sect. 2.5.2.1 is based on the professional conditions of judgement by the present author, the present section introduces quotations and arguments gathered from the specialist literature from neighboring fields of expertise. These are merely juxtaposed to one another, but naturally not judged in detail. We proceed with the aim of making certain contradictions visible. The reader is supposed to get an idea of the degree of certainty, or non-certainty, that may be achieved with regard to the preconditions of the TCN technique discussed here.

The purely technological side of surface dating is still uncertain with regard to our research area. Generally it has to be stated that up to now no physically acceptable calibration of TCN datings with reference measurements exists for High Asia. Not even the largely unknown theme complex of the geomorphological erosion rates of the rock surfaces can be estimated (see above).

Not just for High Asia, but in general the physical conditions of the TCN data are not sufficient to meet the requirements necessary for an absolute dating. This applies because "surface exposure dating" is concerned here, but must proceed without actual knowledge of the surface erosion that has taken place, because the surface erosion rate cannot be known. Also the physical preconditions demand that at most "^{10}Be-years" can be indicated (Gosse and Phillips 2001: 1508, cited in Seguin 2013). However, until now the factors influencing local TCN production rates, to a great extent, are unknown with respect to the physical fundamentals. Because of this, even the CRONUS-Earth-project fears a loss of credibility of the TCN dating (Purdue University 2008, cited in Seguin 2013). Balco (2011: 9, cited in Seguin 2013) notes that the production rate of the nuclide necessary for datings is in many places not known (Granger and Muzikar 2001: 279, cited in Seguin 2013). Lal (1991: 427, cited in Seguin 2013) adds two important factors of uncertainty: a lack of information regarding (1) the Ice Ages and (2) variations of the terrestrial magnetic field. Because most of the terrestrial cosmogenic nuclides (TCNs) can develop in the rock and the atmosphere, it remains unknown how strongly the rock samples are contaminated by the atmosphere (Gosse and Phillips 2001: 1493, cited in Seguin 2013). Merchel et al. (2010: 1183, cited in Seguin 2013) point to the impossibility of eliminating the ^{10}Be that has been developed in the atmosphere from the rock sample. Depending on the authors, the given TCN production rates vary, so that up to now there exists absolutely no reliability. In addition, Gosse and Phillips (2001: 1504, cited in Seguin 2013) have estimated that in the historical period the production rate has varied by more than ca. 15%. Thus the dating lacks a further factor of correction that up to now has not been found. It is a condition to be taken into account, because the period of time of those changed production rates can lead to extreme mistakes in evaluation. This is a fundamental uncertainty that already alone makes the whole method uncertain. The statement of Dunai (2010: 65, cited in Seguin 2013), who is an advocate of TCN dating, that "It largely boils down to the choice of how atmospheric pressure and geomagnetic field are described, and whether, and which, secular variations are taken into account" is evaluated by Seguin (2013: 11) as follows: "According to this statement there exists no claim to the complete incorporation of all parameters and secular variations, but only a personal choice which aspects shall be taken in consideration for this evaluation. The author considers this statement of a recognized TCN-specialist as being very questionable, because it makes clear that there exist no uniform standards of measurement. As a fundamental hypothesis there has to be stated at first: the TCN method provides no correct absolute age-datings, but only approximate values."

The vague situation of the data becomes still more imponderable by the result of Lal (1991: 425, cited in Seguin 2013): "A prerequisite for the application of the in situ nuclides for the study of erosional histories of surfaces is a knowledge of their production rates under different irradiation conditions; altitude, latitude, irradiation geometry and shielding." Here are listed the combinations of factors influencing the local production rates that should be known and taken into account for a dating. However, only some of these factors are actually incorporated into the calculations and then only very vaguely by means of average and approximate values. But other factors are not taken into account at all. As Seguin (2013: 14) deduces from Dunai (2010: 65), among others the thermal radiation of neutrons and muons is not included among the factors used in scaling. The neglect of these factors is confirmed by Heimsath (2006: 114, cited in Seguin 2013). In addition, the idea of Gosse and Phillips (2001: 1491, cited in Seguin 2013), that the influence of the relevant factors within the rock is greater than in the atmosphere, is disadvantageous for the datings. The radiation of muons at sea level, however, is greater than 50% of the entire cosmic radiation (Bertolotti 2013: 171, cited in Seguin 2013) and up to now its share in the whole production of TCN cannot be determined definitively (Gosse and Phillips 2001: 1492, cited in Seguin 2013). Unfortunately, opinions as to the influence of muon radiation on TCN production vary considerably in the literature. The influence of the gamma radiation that also exists in the cascade of nucleons, too, is unknown and not at all explained in the TCN-dating literature (Seguin 2013: 16). In 2009, the influence of the gamma-radiation became rather uncertain and was corrected afterwards by a global average value. Awkwardly though, this must differ widely from the values that were measured locally depending on the degree of latitude (Masarik and Beer 2009: 5 and 6, cited in Seguin 2013). Seguin (2013: 17) considers this to be a further large factor of uncertainty. In calculating the composition of the nucleon cascade, the uncertainties reached up to 25% (ibid.: 17). Therefore, the calculation following Dunai (2010: 61, cited in Seguin 2013) cannot be used directly in the determination of production rates. In addition, all simulations used for the determination of the production rate in any event are only models of approximation with uncertainties (Seguin 2013: 17).

Without being able to consider here the numerous sources of radiation, the following has to be summarized: The background radiation as well as the solar and galactic particle radiation can add to the production of cosmogenic nuclides. In addition, there exists extra-galactic radiation (Bertolotti 2013, cited in Seguin 2013) as well as the anomalous cosmic rays that also reach the earth. The latter are not shielded by the terrestric magnetic field (Bertolotti 2013: 303, cited in Seguin 2013). All of these influences of radiation are more or less not investigated, i.e., unknown as to their contributions to TCN production. In the dating literature they are not even mentioned (Seguin 2013: 18). Confusing in their contradiction is the newest literature as to the influence of solar cosmic radiation (SCR). While Dunai (2010: 3, cited in Seguin 2013) and Gosse and Phillips (2001: 1507, cited in Seguin 2013) declare it to be irrelevant for the in situ production of TCN, Masarik and Beer (2009: 3 and 17, cited in Seguin 2013) describe its influence on production rates at high latitudes N and S of 60°. An influence of SCR in the form of solar energetic particles (SEP) has been deduced by Vita-Finzi (2013: 34, cited in Seguin 2013) from the heightened ^{10}Be-content in ice drill cores in Greenland before 12.837 YBP. Seguin (2013: 19) therefore believes the influence of SCR to be more complex "than it seems to be according to the TCN-literature." Due to a lack of space, it is not possible here to describe the numerous further important uncertainties with regard to the origin of galactic cosmic radiation (GCR) (Seguin 2013: 19). Moreover it is problematic that the TCN literature has not discussed the phenomenon of supernovae, which occur in our galaxy every 25–100 years and intensify the TCN production rate on earth (Bertolotti 2013: 303, cited in Seguin 2013). Thus, the conclusion drawn by Bertolotti (2013: 313, cited in Seguin

2013) after detailed theoretical considerations of the knowledge about cosmic radiation is sobering: "… a precise vision of how and where cosmic rays are originated is still lacking, even if some of the possible mechanisms are starting to be understood." (cf. Seguin 2013: 20).

Vita-Finzi (2013: 29, cited in Seguin 2013) proves that GCR has only varied within the last 1000 Kyr about 10% and within the last 0.5–1 Gyr about 150%. Accordingly, the condition assumed for TCN datings—that galactic cosmic radiation has remained constant during the exposure durations of interest—is not correct (Gosse and Phillips 2001: 1506, cited in Seguin 2013). In addition, the production rate in the rock necessary for TCN dating is distorted by the flattening of the penetration curve at a rock depth of ca. 4.4 cm, discovered by Masarik and Reedy (Masarik and Reedy 1995 in Gosse and Phillips 2001: 1491). It leads to an underestimation of the production rate but—despite that—is ignored, whereby a further source of error is identified (Seguin 2013: 21). In addition, the depth of penetration varies with the type of rock and its moisture content. The latter can be strengthened by up to about 25% by the flow of thermal neutrons, as well as by a snow cover of minor thickness and a soil that is rich in water. Seguin (2013: 22) tried to find a publication where these central factors were taken into consideration—but in vain, although these factors would enhance the production rates.

Due to the great penetration depths of GCR of ca. 5 m (Gosse and Phillips 2001: 1492, cited in Seguin 2013), the datings of glacigenic boulders carried out by the thousands are problematic. The TCN production contained in the boulders in part must go back to the former bedrock of these boulders. Groom et al. (2001: 19, cited in Seguin 2013) and others use the term "standard rock" with regard to the penetration depth of TCN. However, according to Seguin (2013: 23), its consideration cannot be found in the relevant TCN-dating literature. While Dunai (2010: 20, cited in Seguin 2013) claims that the far-reaching radiation of muons is of subordinated importance, Lal (1991: 426, cited according to ibid.) estimates the influence of muons to reach 15.6% for the production of cosmogenic ^{10}Be.

In addition, the careful consideration of the factors influencing TCN datings is further restricted because the influence of the production rate strongly varies according to the type of nuclide (Seguin 2013: 23).

In many respects also uncertain is the influence of the terrestrial magnetic field, with its shielding effect as to the GCR. It varies in time and is dependent on the degree of latitude. The former is not sufficiently known as to an assessment of its influence on the TCN production rate (Gosse and Phillips 2001: 1507, cited in Seguin 2013). Thus, the variations of the dipole axis during prehistoric time might have had a greater influence than historical fluctuations of intensity that are a little better known (see below). And all this in the face of the fact that only 90% of the dipole field model itself is considered to be stable (Lal 1991: 438, cited in Seguin 2013).

To summarize, Seguin (2013: 24) demands that we investigate this entire complex of problems and not—as is done by most authors—ignore it. With regard to the TCN data this is especially startling since "the theory of the geomagnetic field is one of the most difficult problems of geophysics" (Cain 1987: 64, cited in Seguin 2013) and its complete modeling not possible (Seguin 2013: 24). So, for instance, in the terrestrial magnetic field local deviations of up to 30% might occur. Unfortunately, variations of this magnitude have been registered even for the past ca. 100 years (Campbell 2003: 45, cited in Seguin 2013). Accordingly, these deviations might not be correct over longer periods of time and locally might even be clearly larger. Seguin (2013: 27) refers to Klein and Gosse (1996: 162), who have shown that the fluctuations of the assumed dipole axis of the terrestrial magnetic field of more than 20° of geographical latitude have a greater influence on the galactic cosmic radiation, i.e., the TCN, than the paleontological fluctuations of magnetic field strength. At the same time, Seguin (2013: 27) notes critically that this factor of influence, too, is not discussed in the TCN-dating literature.

In addition, the influence of terrestrial magnetic storms on the TCN production rate remains unconsidered (Seguin 2013: 27). Among others, the two calculation programs available for the determination of the strength of the magnetic field at a point in time at a certain locality make Seguin (ibid.) suspicious: "the International Geomagnetic Reference Field" (IGRF) as well as the "Definitive Geomagnetic Reference Field" (DGRF) program are only uncertain theories of approximation. The variation of the local TCN production rate can be—according to the fluctuation of the magnetic field—35 to 60% (Dunai 2010: 72, cited in Seguin 2013). Thus it is of essential importance to know the exact production rate at the respective locality under investigation. For this, the terrestrial magnetic field—which is a quadrupole field—must be known. Though the models are supposed to consider the characteristics of this magnetic field, relevant descriptions are lacking in the TCN-dating literature (Seguin 2013: 28). After Gosse and Phillips (2001: 1507, cited in Seguin 2013), the equator is hit by a more energy-rich radiation than higher latitudes. The cause of this is the terrestrial magnetic field. The higher latitudes, however, are reached by a wider spectrum of radiant energy than lower latitudes (ibid.: 1488). This is a function of the angles of impact (incidence). Due to the reduced shielding effect of the terrestrial magnetic field, about double the GCR strikes the earth surface (without being slowed down) in the polar regions than at the equator. Accordingly, the production rate is doubled there (Ivy-Ochs and Kober 2008: 182, cited after ibid.). These effects must be taken into account. The consideration by Lal (cited in Gosse and Phillips 2001: 1507) that latitude-dependent alterations of the magnetic field that depend on the degree of latitude from the latitude circle of 58° upward don't have any further influence on the TCN production rate, while it increases from 0° up to 58° degrees of latitude, is also essential. With regard to the alterations between the latitude circles 0–58, this consideration must find a firm application for all local datings. Such local variations should have a profound influence. It is also problematic that the consequence of the so-called latitude shift, which implies that the slightest radiation intensity between the 25th and 41st degrees of latitude varies on the two hemispheres (Gosse and Phillips 2001: 1487, cited in Seguin 2013), has not been taken into account. However, these well-known deviations could be considered very well by corrected calculations in regard to the TCN datings (Gosse and Phillips 2001: 1488, cited in Seguin 2013: 29). Gosse and Phillips (2001: 1489, cited in Seguin 2013) argue, however: "… it was considered impractical to model all of these variables for every position on earth." The distribution model for terrestrial nuclides by Lal (1991), often used by TCN practitioners, is thought to be incorrect by Gosse and Philipps (2001: 1520, cited in Seguin 2013) and Stone (2000: 23, 753, cited in Seguin 2013). Heimsath (2006: 114, cited in Seguin 2013) even considers it to be incomprehensible. Balco (2011: 21, cited in Seguin 2013) ascertains with regard to simulation models such as these that the greatest uncertainties of dating have to be expected at great heights and low latitudes; in other words, in regions such as Tibet and the Himalaya (Seguin 2013: 30). Gosse and Phillips (2001: 1489, cited in Seguin 2013) conclude that no model exists that adequately expresses the connection between the terrestrial magnetic field and the production rates of terrestrial nuclides (Seguin 2013: 31). Apparently, this is still the current state of the art. According to Dunai (2010: 74, cited in Seguin 2013), until now none of the models as to the consideration of local magnetic field variations has been accepted in general with regard to the production rate.

Similarly disastrous as imponderables of those regional factors are the unknown former variations of the terrestrial magnetic field with respect to their influence on local production rates. The length of time of these variations also remains unknown (Seguin 2013: 31–33). Because it is simply not known of what kind and intensity these essential consequences have been, they are ignored and withheld in connection with the TCN datings.

The extreme exaggeration of the TCN ages in the Himalaya and in Tibet (see above), however, might have one or several causes. The observation that the glacial (Late Pleistocene) strength of the magnetic field was very much reduced compared with the Holocene (Gosse and Phillips 2001: 1509), and owing to this more cosmic radiation reached the surface of the earth, explains a higher TCN production rate. If this is applied in relation to actual values, this results in a value of decay that is too large and an age that is too old (Seguin 2013: 33). In connection with the overestimated TCN ages in the Himalaya and Tibet (see above, and Kuhle and Kuhle 2010), the indication (Gosse and Philipps 2001: 1509, cited in Seguin 2013) that the variations of the magnetic field in the lower latitudes (from 30°N and S and lower) are stronger than in higher latitudes, is informative. In this context, Masarik et al. (2001: 2995) have also made remarkable observations. They state that the strength of the magnetic field at between 40 and 10 ka BP become greater and since then remained the same. Due to the existing uncertainties in the production rates against the background of the also uncertain time scales after Gosse and Phillips (2001: 1508, cited in Seguin 2013), one should best speak of "^{10}Be years" in order to be correct. Though the practitioners working with TCN data may be aware of these circumstances, nearly none of them use this correct expression (Seguin 2013: 34). Instead, the majority of authors working with TCN datings claims to have available absolute age data. This, however, is definitely not the case (see Kuhle 2014b).

According to Seguin (2013: 34), Lynch (2011) is able to provide further indications in regard to the (for TCN production) also important question of pole reversal of the terrestrial magnetic field: during the last ca. 20 million years, pole reversals of the magnetic field occurred at medium intervals of ca. 200–300 ka. The last pole reversal took place ca. 780 ka ago. Seguin (2013) further remarks: "With regard to the influence of a pole change on cosmogenic nuclides, however, no literature has been found. Because the magnetic field during a phase of pole reversal is in a radical change and thus gets weakened, galactic as well as solar cosmic radiation might hit the earth stronger, so that the nuclide production would be heightened. However, up to now no complete disappearance of the magnetic field has been evidenced (cf. Lynch 2011)" (ibid.: 35). However, a change of poles such as this does not take place all of a sudden, but within several millennia (Lynch 2011, cited in Seguin 2013), which makes the influence even more complicated.

A further influence on TCN production is exerted by deformation of the terrestrial magnetic field due to the solar wind, i.e., the heliosphere (Campbell 2003: 111, 112, 115, cited in Seguin 2013). Depending on the character of incoming galactic cosmic radiation, i.e., whether it is high- or low-energy radiation, the influence of the solar magnetic field on the GCR varies (Li et al. 2008: 36, cited in Seguin 2013; see also Bertolotti 2013: 300). Furthermore, solar flares and coronal holes also have an influence on the terrestrial magnetic field (Campbell 2003: 124, cited in Seguin 2013) and thus on the TCN production rate. Seguin (2013: 38) attributes the measurement of proton output (Campbell 2003: 126, cited in Seguin 2013) to a strengthening of the nucleon cascade during such events. The TCN-dating literature, again, does not take this into consideration.

Obviously indisputable is the relationship between changes in climate and the thickness of the earth's atmosphere. At the same time, a relationship of climate with solar activities can be observed as well (Vita-Finzi 2013: 40, cited in Seguin 2013), from which Gosse and Phillips (2001: 1515, cited in Seguin 2013) infer a change in TCN production rates. In this context, the intensity of geomagnetic storms must be considered, which is dependent on the 11-year solar circle (Campbell 2003: 139 p, cited in Seguin 2013), and which influences GCR.

From references like these, Seguin deduces a strengthened nucleon cascade and, accordingly, a higher production rate of cosmogenic nuclides, e.g., across South America (Campbell 2003: 165, 167). This might be caused by the weakness of the magnetic field there. Furthermore, it establishes that "anomalies like these," leading to such excessively overestimated TCN datings, "at the moment are not taken in consideration" (Seguin 2013: 39) in the TCN-dating literature.

In addition, the factor of the surface cover and thus the shielding of the rock surface of buried materials requires consideration in connection with TCN dating. Here, the period of time of past vegetation cover is unknown. The dating of moraine boulders starts from the—however opaque—assumption that since the deglaciation classified as belonging to it there has existed no new cover or erosion. Thus, for example, even if one wished, the climate-dependent past snow cover could be recorded only very incompletely. Assessments such as these are only due to indirect, current seasonal values of snow thicknesses. The seasonal influence of snow can lead to differences in age up of to 15% (Gosse and Phillips 2001: 1523, cited in Seguin 2013). Merchel et al. (2010: 1183, cited in Seguin 2013) admit its importance and at the same time point to its indisputable unknowability. Other important factors of influence are the shadowing by surrounding mountains as well as the slope inclinations at the TCN sample locality (Ivy-Ochs and Kober 2008: 184; Gosse and Phillips 2001: 1521, 1522, cited in Seguin 2013).

The inheritance problems related to the Tibetan Plateau have already been discussed in detail (see above, and Kuhle 2013b: 8).

Masarik et al. (2001: 2996, cited in Seguin 2013) consider the influence of the absolute altitude on TCN production rate to be unimportant. Others (Balco 2011: 4; Ivy-Ochs and Kober 2008: 182, cited in Seguin 2013) assume that an increase in cosmic radiation intensity occurs with increasing absolute altitude. This interpretation is based, among other things, on the decreasing atmospheric pressure (see Lal 1991: 437, cited in Seguin 2013; Gosse and Phillips 2001: 1515; Bertolotti 2013: 83). Thus, for the TCN-dating tests in High Asia, a strong overestimation is probable (i.e., the calculated TCN ages are too old) (Kuhle and Kuhle 2010; Kuhle 2011a).

Seguin (2013: 44) remarks critically that up to the present, the chapter "Estimation of production rate" in Gosse and Phillips (2001: 1516, cited in Seguin 2013) for many TCN-dating practitioners still serves as a relevant basis for their age estimations, even though these are—even according to the information provided by the authors themselves—only *estimated* values.

Similarly, Lal (1991: 428, cited in Seguin 2013) provides problematic estimations. Seguin is especially surprised about these (ibid.: 44), because apparently Lal himself had previously observed that a small error of 10% in connection with the respective calculations (that actually are incalculable) of the nucleon cascade must lead to a error of 120% at sea level, i.e., in lowland areas (Lal 1991: 426, cited in Seguin 2013). In the context of Lal's statement that "[a]ccurate estimations of production rates are generally not possible because of lack of knowledge of the probabilities of formation of nuclides in the different reactions" (Lal 1991: 425), Seguin (2013: 44) raises the inevitable question of why TCN datings are performed at all given that the necessary production rates of the different nuclides —due to the unknown reactions in the nucleon cascade—are obviously not known. Seguin's criticism goes even further. He cites CRONUS-Earth's own project objective: "The overall goal of CRONUS-Earth is to put the systematics of terrestrial cosmogenic nuclide methods on a firm foundation. It is not intended to fund development of new methodologies or site-specific investigations" (Purdue University 2008); and in this context, he criticizes the fact that the models of CRONUS-Earth are available for age datings with ^{10}Be and ^{26}Al, free of cost online, especially since "… several basis-informations, as e.g. the exact production rate of ^{10}Be or the influence of the Myon-production are still not settled …" (Seguin 2013: 45). Accordingly, "the approach is made possible for a greater group of researchers and the results become comparable. Possibly, there could also develop more and more mistakes due to ignorance or unclean work, because all researchers again and again refer to the same models, without improving them." (ibid.). "In this way," Seguin concludes,"also participants of CRONUS-Earth as e.g. Susan Ivy-Ochs herself (see Akçar et al. 2008)…" contribute to this trend and also to new, but methodologically unproven possibilities of applications. As a result,

CRONUS-Earth indeed becomes inconsistent with its project goal cited above.

Aside from this "vicious circle" there is a further foreseeable consequence. It is based on the fact that despite of the manifestation that "it is not intended to fund development of new methodologies or site-specific investigations" (Purdue University 2008), in the meantime a large group of people collecting "absolute TCN-data" constantly reference CRONUS-Earth, relying uncritically on these "online-available models free of cost." The respective practitioners who collect "absolute age-data" naturally obtain by means of mutual quotations and examinations an exponentially increasing space for publishing their data in journals. Up to the present, however, TCN publications for High Asia rather signal the acceleration of a TCN hype.

A further problem are the requirements for a TCN sampling locality. A necessary prerequisite are extensive horizontal areas, so that mountains pose difficult terrain for TCN datings. The reason for this is the extensive shielding by the mountains themselves and their valley flanks. Similarly unsuitable is the strong erosion in steep mountainous terrain (Nishiizumi et al. 1993: 419 and Gosse and Phillips 2001: 1537, cited in Seguin 2013). Because the respective TCN-dating practitioners seek to replace these requirements—because in mountains they cannot be met—by so-called correction factors, the already existing (see above) age uncertainties increase further. Accordingly, the following generalizations apply: As long as the aim is restricted only to the production of TCN data in the form of "^{10}Be-years," a sample application in regard to the question of ages such as this would only be *useless*. However, if the ^{10}Be-data obtained are declared to be a true dating, it is *misleading*. Astonishingly enough, among Quaternary geologists one can even hear the statement: "Better wrong dates than no dates!" Unfortunately, the present author is not able to commit himself to this understanding of science.

In the glacial chronology, which within a glacier forefield is unambiguous, TCN data of "^{10}Be-years" can only be controlled with respect to their sequence, but not their absolute age (Kuhle and Kuhle 2010; Kuhle 2011a). Compared with the intersubjectively verifiable and evident arrangement of the positions of stratigraphic ice-margin indicators that need absolutely no correction factors, TCN data merely have a subordinate position in the test hierarchy. Thus, they have only minor empirical value.

Despite these briefly summarized, and to a great extent unresolved, basic methodological issues, the majority of these publications accepts the naïve application of TCN dating. However, the astrophysical conditions of the TCN production must be clarified first, as they determine the applicability of the dating. So in fact the second step has been made before the first. As Seguin (2013) concludes: "… the TCN-dating still is far away from being an exact and reliable method for the absolute dating of ages …" (ibid.: 52).

In the light of this, the following comment by Dunai (2010) seems overly optimistic: "Depending on the nature of a surface (glacial feature, lava flow etc. …), independent age constraints may be derived from diverse techniques such as radiocarbon-, luminescence- and Ar/Ar-dating, as well as dendrochronology and varve chronology … At its simplest, a cosmogenic nuclide production rate is calculated from the measured nuclide concentrations in a sample, divided by the surface's independently determined age." (ibid.: 62). In our study area, i.e., in High Asia, in the meantime TCN data for e.g. "^{10}Be-years" have been collected, but they have not been calibrated in the way proposed by Dunai. On the contrary, glacial features and radiocarbon datings always have been ignored by the dating experts in the event they did not correspond with those "^{10}Be-years." TCN data not calibrated with Quaternary geology and geomorphology have been treated as though they were absolute ages (see Kuhle and Kuhle 2010; Kuhle 2013a).

The demand for absolute age data is understandable, but it should not lead to ostentatious optimism despite a dysfunctional method. Otherwise one risks pursuit of a dead end.

2.5.2.2.1 New Results on the Terrestrial Magnetic Field Deny the Usefulness of TCN Data for the Determination of Ages

The global weakening and regional shifting of the terrestrial magnetic field reconstructed by the "Swarm" project, i.e., the instability of the field in terms of its position and intensity during the Holocene, are the cause of variations in global and local shielding effects up to a factor of ca. 100 (Hartmann and Pacca 2009; Lühr 2013). This means that currently, regionally there is the possibility that incoming cosmic radiation along a given parallel of latitude may be 100 times more intense than elsewhere along the same parallel of latitude.

In addition, the magnetic field may change its global intensity for a short period of time (Seidler 2014: 1). Thus, for example, its intensity has been weakened during the last 100 years by about 5% (ibid.). Regionally such a decrease has occurred even within only the past 10–15 years (ibid.). Regional weakenings, as well as opposite strengthenings (amplifications) have been observed within only a period of six months (ibid.).

Since 2013, the European Space Agency's three "Swarm" satellites have been measuring the deformation of the terrestrial magnetic field by "solar-storms" ("Sonnenstürme") with magnetometers (Seidler 2011; 2014: 2); additionally there is the influence on the magnetic field by the petro-variance of the earth's crust (ibid.) and feedback factors in the earth's interior by convection streams and earth rotation (ibid.). Important mass movements such as these reach distances of ca. 10 km per year (ibid.). Accordingly, this only partly known "Geodynamo" modifies the magnetic field. Thus, for example, the shifting of the magnetic pole amounts to 90 m per day (ibid.).

Already this *precisely measured instability of the terrestrial magnetic field* alone denies the possibility of a stable reference system and makes TCN datings problematic.

2.5.2.3 Conclusion of Sect. 2.5.2

We here continue the line of reasoning presented at the end of Sect. 2.5.2.1, Point 4:

The prognosis for TCN dating based on empirical facts of glacial-geomorphological research is demonstrated by the 1000–1500 TCN datings undertaken in High Asia. What is really being measured by the ^{10}Be-years are not the moraine ages. This is evidenced by the fact that these ages are often clearly too old.

1. Very young, subrecent moraines, classified as belonging to the Little Ice Age (LIA = stage X–VIII)—e.g., in the forefield of the Rongbuk Glacier (north side of Everest)—are TCN-dated to 1.6 ka (Owen et al. 2009: 1415). This amounts to an *overestimation* by a factor of ca. 4–8.
2. The TCN data from Central Tibet also are much too old, which may be accepted as necessary and explicable on the grounds of a *high degree of mixing of old and young materials and thus by inheritance.*

   (a) This prognosis is proven by the fact that *age inversions* occur (Schäfer et al. 2002; Owen et al. 2005; Heyman et al. 2010) which in the case of inheritance inevitably has to be expected (cf. Sect. 2.5).
   (b) Furthermore, the lack of correspondence between ^{10}Be-years and moraine ages is manifested by the fact that the TCN data sets identify the glaciation history of Tibet in an incorrect way, i.e., as *chronologically heterogeneous.*
   (c) In the same logical fashion, these data records in error show *no synchronicity* with the Northern Hemisphere (global) Ice Ages.

These *confirmations of the prognosis inevitably result* more and more clearly from the distributional pattern of all TCN data in Tibet.

Thereby one has to distinguish cause complexes A and B:

A1: Due to the instability of the terrestrial magnetic field, the GCR (TCN) radiation is *higher than assumed.*
A2: Due to the great absolute altitude, the radiation in Tibet transmits more energy than in lowland areas.

In High Asia both factors *must always lead* to a clearly higher ^{10}B-age than the scaling factors calculate.

B: Because on the Tibetan High Plateau an ice cover existed *continuously as cold-based ice on subglacial permafrost* (Kuhle 2014a), there exists a *very high* degree of inheritance. This is clearly higher than in the area of the Nordic glaciations, which for a great part were dominated by wet-based ice.

Thus, in the latter region there must have been an incomparably smaller degree of inheritance than in Tibet. Its influence on the overestimation of the TCN ages, therefore, must necessarily have been much more insignificant than in Tibet.

Due to the proven *instability of the magnetic terrestrial shielding*, true TCN age datings of glacier stages were in fact not possible in the first place (see Sect. 2.5.2.2.1).

Dunai (2010) has attempted to show with his publication *Cosmogenic Nuclides—Principles, Concepts and Applications in the Earth Surface Sciences* that theoretically it would be possible to apply the TCN technique in the event that the empirical conditions for it were available at all. However, this is not the case.

2.5.2.4 Summary of Sect. 2.5

It can be established that the TCN datings collected for High Asia are extremely contradictory. Moreover, they stand in stark contradiction to the geomorphological data and the glacial-geological chronologies and stratigraphies in High Asia and Tibet. The only remaining convergence between these strongly diverging TCN ages consists, if at all, in their shared contradiction to the glacial-geomorphologically proven Tibetan inland glaciation for the last ice age (LGP, LGM, Stage 0). Moreover, the TCN ages also deviate from the Tibetan lake ages determined in the surrounding areas. These lake datings were based on the ^{14}C method and in many places diverge from the TCN findings by a decimal place or even more. By contrast, the ^{14}C based ages of the lakes correspond well with the reconstructed Late Glacial Tibetan inland glaciation. In addition to these limnic ^{14}C data (from lakes), the same can be said for the other terrestrial ^{14}C data based on wood samples in moraines, etc. All these data sets correspond well, without contradiction, with the relative glacial chronology and its Late Glacial to Historical snowline depressions (ΔELAs). In classic Quaternary-geological fashion, the ^{14}C data thereby transfer their stage chronology and stratigraphy, proven geomorphologically and through their spatial-positional relationship, into an absolute chronology (Fig. 2.5).

It is uncontroversial that rock surface ages must be known in order to be able to determine the impact of local TCN irradiation. For the so-called TCN dating method, however, the determination of the rock surface ages is the actual goal. In order to bypass this contradiction, an erosion rate generated through weathering or prepared through hydrothermal disintegration is imputed, i.e., assumed in advance. Just as little is really known about the development of rock surfaces in regard to their pre-historical covering, the same can be said about boulder surfaces in regard to their covering and/or TCN irradiation exposure or nonexposure as a result of dislocation and rotation. In terms of a measuring technique, a primary basic prerequisite for carrying out TCN dating of rock boulders is thereby not given. The erosional impact of glacial transport on moraine boulders also remains unknown. In order to make TCN dating seem reasonable at all, one simply imputes that the sampled moraine boulder has exclusively received its nuclide irradiation in the very position and state of covering in which it is found. Empirically, however, this assumption is completely unfounded. Multiple exposures to irradiation of unknown length and intensity are the norm.

The second set of basic prerequisites also is not fulfilled. The regionally and hypsometrically varying TCN irradiation is unknown. Nowhere in

Tibet and High Asia has this been measured yet. Its intensity is extrapolated from completely, even climatically different geographical positions around the globe and thus from regions defying comparison, and following unknown standards.

In order to be able to measure, it would be necessary for the TCN irradiation to remain constant at least locally. This also would have had to have been measured in order to fulfill a viable third basic prerequisite. This, however, has not been done. Instead, it is uncontroversial that the irradiation of cosmogenic nuclides not only varies considerably across space but also across time, which is due to numerous modifications and shielding effects, e.g., by the terrestrial magnetic field. Even the new findings with regard to the continual shifting of the terrestrial magnetic field alone, as only one aspect of the third basic prerequisite, negate the possibility of reliance on TCN data for age determination.

That which the TCN analyses measure with their ^{10}Be-year estimations is thus definitely not the age of the moraines in High Asia. That fact at least is proven convincingly by the 1000–1500 TCN datings collected for the region.

Hence, only the geomorphological analysis alone allows a pre-historical glacial reconstruction. As this is true of all reconstructed inland-ice regions on earth, it is also true of High Asia. ^{14}C data can, for the period between the late ice age (Stage 0; LGP) and today, help determine and specify these geomorphological findings through the assignment of minimum ages.

2.6 Comparisons of Tills and Lakes from Central Tibet with Those in Arctic Glacial Regions

In a first comparison, the area of a lake neighboring the Dogai Coring, located at 34° 32′ 25″ N 88° 36′ 00″E (Fig. 2.146, see also http://extras.springer.com: Character AK), is juxtaposed with a lake representing the area of Last Glacial Scandinavian inland ice. The latter is located at the Arctic Circle near Jokkmokk (66° 36′ 52″N 19° 30′ 19″E) (Fig. 2.147). Both lakes have a similar outline. In addition they are divided into two pieces by W-E- and E-W-extending ridges consisting of loose material. Near Jokkmokk (Fig. 2.147) it is generally known that this is a till ridge. In Central Tibet we are also dealing with till. This till is exposed 10 km farther to the northeast on the southern shore of the 11 times longer, 71 km long Dogai Coring, with a thickness of 3–6 m (Fig. 2.148).

Also similar to the dead-ice pools among the tills of the small lakes in the vicinity of Jokkmokk are the small pools in the area surrounding the lake in Tibet (Fig. 2.149). The pools in Tibet are more numerous and fresher as regards the geomorphology. They have the same arrangement positionally as their Scandinavian equivalents. The glacial genesis of both of these study areas is certain.

The Last Glacial inland ice area, e.g. at 65° 11′ 42″N 19° 40′ 30″W, also shows all type characteristics of the Dogai Coring area (34° 48′ 42″N 89° 16′ 51″E) (cf. Figs. 2.150 and 2.151). Here, too, the deglaciation in Tibet is more recent than that close to the Arctic Circle. The till cover seems to be fresher, which becomes clear due to the greater number of small dead-ice pools. This side lake of the Dogai Coring occupies the basin of a glacier tongue that last advanced from the east (see Fig. 2.152). Naturally it overrode the older till cover from the east. West of the lake its surface has remained. The fresher condition of the till cover east of the lake is geomorphologically evident. The number of small pools is much larger there.

The two till areas in Central Tibet were covered with glacier ice for the last time during the Late Glacial Stage IV (see Fig. 2.5). During the High Glacial stage, the main direction of discharge of the inland ice was from west to east. The W-E extent of the longish Dogai Coring, the flank polishings on the steep slopes above its northern steep bank, and several islands in the lake in the form of roches moutonnées and glacially streamlined hills are all indicators for this. They prove the mainly W-E and then, during the Late Glacial, E-W flow direction of the ice (Fig. 2.152).

Figure 2.153 (see also http://extras.springer.com: Character AL) shows a further Central

Fig. 2.146 Satellite image, Google Earth Pro: center 34° 32′ 25″N 88° 36′ 00″E; http://extras.springer.com: Character AK. Last (High) Glacial inland ice area in Central Tibet in the area of a lake neighboring Dogai Coring (cf. Fig. 2.147). Glacigenic scouring of the ground moraine was facilitated by the hardening of the loose rock by subglacial permafrost (cf. Figs. 2.156, 2.157, 2.160, 2.178, 2.190, and 2.193)

Tibetan glacially scoured till landscape at 35° 31′ 42″N 93° 07′ 59″E (cf. Figs. 2.190, 2.178, and 2.166). It points out a scouring orientation from roughly west to east. Here, at ca. 4500 m asl, it shows the High to Late Glacial direction of discharge of the inland ice which corresponds to the direction of incline of the plateau. For comparison: Fig. 2.154 shows an equivalent glacial landscape in Greenland at 67° 54′ 08″N 52° 37′ 22″W and Fig. 2.155 in Iceland at 65° 29′ 39″N 17° 20′ 39″W.

Scourings ("Striemungen") such as these do not only occur frequently in Tibet, but on a regular basis (Kuhle 1991: 149, 150: Photo 8, 155, 156: Photo 22, 179–181: Photos 41 and 44, 183: Photo 51, 194–196: Photos 77 and 78). They occur in the peripheral areas of the highland as well as in its center. In the center, accordingly, they must have still been covered by glacier ice during the late Late Glacial to Holocene (Stage IV–V, Fig. 2.5). Only a few thousand of years ago, therefore, they (i.e., the scourings) were still in the process of being formed (Figs. 2.156 and 2.166).

Figure 2.157 (see also http://extras.springer.com: Character AM) shows a further landscape with tills at 35° 32′ 57″N 93° 05′ 49″E at about 4500 m asl. Again it shows the typical scourings from W to E and proves them (see above) to be present on a large scale. This is further proof of a covering inland glaciation overriding the entire terrain also to the north; here, too, with the large-scale and corresponding flow direction from west to east. Figure 2.158 shows as a large-scale comparison an equivalent glacial landscape in the Arctic, Novaya Zemlya at

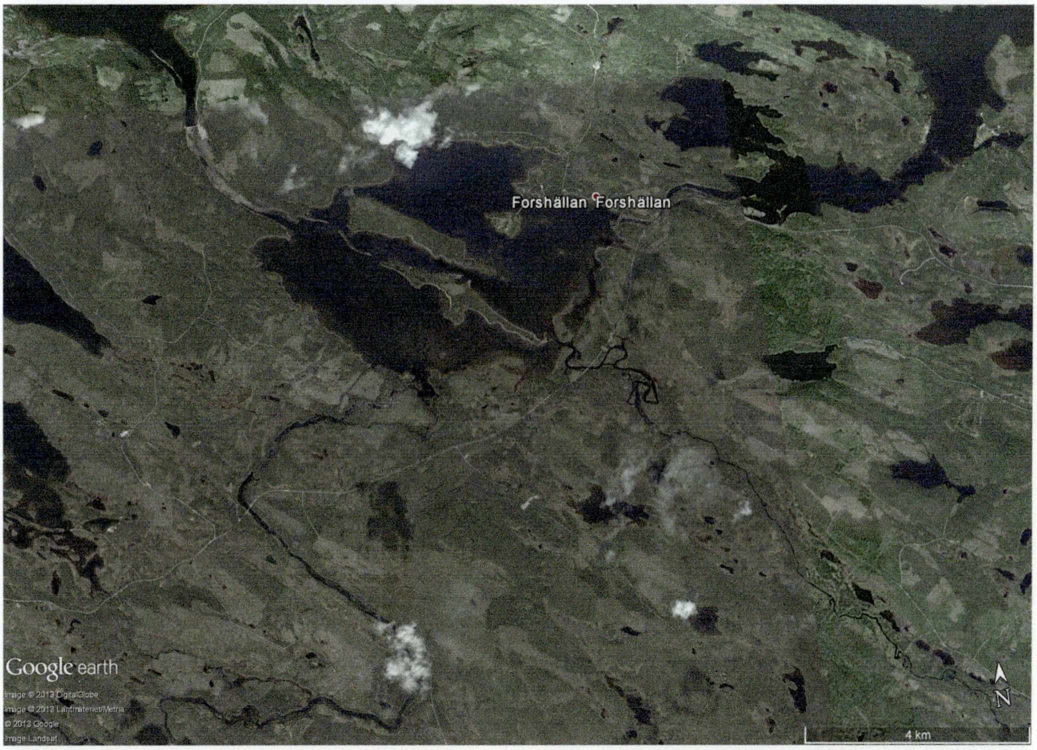

Fig. 2.147 Satellite image, Google Earth Pro 66° 36′ 52″N 19° 30′ 19″E. High Glacial inland ice region at the Arctic Circle in Scandinavia at Jokkmokk (cf. Fig. 2.146)

71° 53′ 58″N 52° 06′ 38″E. Comparable permafrost indicators such as these tundra polygons have also been proven by Kuhle for Tibet (Kuhle 2014a).

In the terrestrial sense, the nearly completely covering till surface can, for instance, also be unambiguously diagnosed glacial-geologically in the surroundings of the Dogai Coring (main lake: located in Central Tibet at 34° 45′N 89° 05′E) (Fig. 2.159).

The satellite overview completes this proof in the location 20 km WNW at ca. 34° 45′N 89° 05′E (Fig. 2.160, see also http://extras.springer.com: Character AN) and can be compared, for example, with the regions in northern Norway at 69° 05′N 22° 44′E (Fig. 2.161), the Kola Peninsula at 67° 35′N 31° 36′E (Fig. 2.162), Northwest Territories in Canada at 65° 44′N 124° 05′W (Fig. 2.163) as well as Baffin Island at 71° 39′N 81° 18′W (Fig. 2.164).

A young ablation landscape ("Niedertaulandschaft"), developed due to the disintegration of the inland ice cover, is situated in the north of Central Tibet, e.g., at 35° 34′ 01″N 93° 12′ 31″E at 4470 m asl (Fig. 2.165, see also http://extras.springer.com: Character AO). In a typical way it is marked by a larger, in its outline irregular, overdeepening of a large lake basin. Among many others, a corresponding example is provided by the lake located at 5070 m asl in the center of Central Tibet, at 33° 51′ 1.78″N 88° 36′ 2.32″E (Fig. 2.166).

Its surrounding area shows the, in the same way characteristic, small-scale pattern of dead-ice pools. These are lying in a thick till cover. The fact that the pools still exist in great numbers proves their young age. Accordingly, not all of them are silted up solifluidally. Figure 2.167 shows a corresponding example from Iceland at 65° 36′ 15″N 17° 00′ 21″W (cf. Figs. 2.165 and 2.167).

Fig. 2.148 View from 34° 37′ 53.69″N 88° 45′ 13.31″ E at 4830 (4834) m asl toward the east at the southern shore of the Dogai Coring. The surge from a strong NW wind prevails here. The shoreline undercuts the ground moraine and the surge penetrates it at a depth of at least 3 m (■ below). Rounded and subangular boulders (◉) of many different varieties of rock are found embedded in the clayey matrix. They lie isolated from one another and alongside other, edgy, fist-sized components. (■ above in the background) indicates up to decameter-sized, light-colored and in many places erratic ground moraine covers. They extensively cover a considerably darker colored bedrock landscape. (◓) mark—also on the opposite northern lakeshore—abraded mountain slopes such as glacially streamlined hills, consisting of the aforementioned darker bedrock. Analog photo 10/10/2014 by M. Kuhle

Figure 2.168 shows a somewhat older Late Glacial meltdown landscape that stems from the Tibetan disintegration of inland ice from northern Central Tibet, at 35° 21′ 02″N 92° 39′ 35″E (Fig. 2.168, see also http://extras.springer.com: Character AP). Similar, but still poorer in till and thus with a greater proportion of glacially scoured basins ("Schliffwannen") in the bedrock, is the glacial landscape presenting itself at 78° 33′ 51″N 69° 20′ 37″W in northern Greenland (Fig. 2.169). Neither the Late Glacial inland ice landscape in the Scandinavian Sareks area at 67° 08′ 22″N 16° 50′ 44″E (Fig. 2.170), nor the corresponding inland ice landscape in northern Svalbard at a distance of only ca. 650 km from the North Pole at 80° 07′ 16″N 18° 16′ 27″E (Fig. 2.171), seem to be nearly as fresh and thus young as those in Tibet (Fig. 2.168, see also http://extras.springer.com: Character AP). In northern Svalbard a till of nearly the same thickness exists as in northern Central Tibet, but the destruction of the small dead-ice holes in the smooth loose material of the till is already further advanced than in Tibet. Here, the deglaciation obviously took place only recently, i.e., during the Late Glacial (Sirkung Stage IV) to Early

Fig. 2.149 View from 33° 57′ 7.01″N 88° 42′ 23.31″E at 5100 (5103) m asl roughly toward the NW. Ground moraine (■) with erratic, black basalt boulders of up to 0.8 m length (see person at far left for scale). The conspicuous depression is a typical kettle hole, i.e., a closed, hollow form filled with a tarn. (▲) marks abraded mountain slopes and glacially streamlined hills of hard bedrock, which are also covered by striated, Late Glacial ground moraine which has been fixed by subglacial permafrost. (- - - -) indicates the upper surface of the High Glacial inland ice sheet at above 6500 m asl. The rounded mountains are covered by snow. Analog photo 29/09/2014 by M. Kuhle

Holocene (Early Neoglacial, perhaps Nauri Stage V, Fig. 2.5).

The juxtapositions and comparisons presented here for Central Tibet with former inland ice areas elsewhere on earth that are 4000–5000 m lower in elevation can be concluded with the example of Axel Heiberg. Figure 2.172 (see also http://extras.springer.com: Character AQ) shows a very young till landscape at 34° 46′ 12″N 90° 49′ 36″E at about 5050 m asl, situated only a little below the current ELA. Figure 2.173 shows the respective till landscape in the Arctic at 79° 57′ 25″N 88° 51′ 55″W. It has the same relative position with respect to the ELA. In both cases till covers the bedrock to a large extent. The tills contain countless small dead-ice depressions. In many places they are so fresh that they still contain water pools. At the same time, the large-scale branching-out of outlet rills is conspicuously visible. They can have only been developed since the deglaciation (Fig. 2.174). Accordingly, they are still younger than the fresh dead-ice forms. The cause of this rapid rill development must be the surface of the clayey material of the till. The cause of the noticeable branching-out of rills must be their shallowness. This in turn is due to the permafrost. The thick polyglacial tills would have made possible deep linear erosion. The active layer of the permafrost, however, in both areas is only decimeters to maximally 1 m thick (Kuhle 1985b, 2014a). Accordingly, these are two permafrost areas of the same geomorphological genesis. Both of these past inland ice areas were freed of ice only a relatively short time ago (during the Holocene). Only then was developed the flat and therefore branching rill system in the active layer of the permafrost.

2.6.1 Summary of Sect. 2.6

This chapter can be summarized as follows. We have applied the geomorphological method of cross-regional comparison, for which 14 Arctic regions were chosen. For all these 14 regions a former as well as a Last Glacial inland glaciation is generally accepted to have been proven glacio-geologically (sedimentologically) as well as geomorphologically. These regions were compared with the representative study areas of Tibet. The correspondence of glacial-geomorphologically relevant indicators was made evident. Accordingly, we have given complementary proofs for the inland glaciation of Tibet—also for the Last Ice Age.

Fig. 2.150 Satellite image, Google Earth Pro: center 34° 48′ 42″N 89° 16′ 51″E. The locality of Point K in Fig. 2.133 in detail. The last High Glacial to last Late Glacial inland ice region in Central Tibet in the region of the Dogai Coring. The hard bedrock faces of the central Tibetan Plateau are here completely covered by ground moraine. The land pockets located at ca. 5000 m asl exhibit silted-up kettles (depressions) as well as kettle holes filled with small lakes (tarns). These appear in varying densities. The 5 km long and 4.7 km wide lake is segmented by a ground moraine trail in an E-W direction. The lake partly fills a late Late Glacial to Holocene glacier tongue-shaped basin (Stage IV–V, see Fig. 2.5) in 4850 m asl. The respective glacier flowed down to here from the ca. 30 km distant, 5920 m high massif located to the ENE. At the time it had an ΔELA of ca. 275–300 m ((5400 − 4850)/2). A glacial-geomorphologically equivalent ground moraine and lake landscape can be diagnosed in many parts of the Arctic and Subarctic (cf. Fig. 2.151)

2.7 The Present to Holocene Glaciation of the Purog Kangri, Especially Its Western Slope in the Center of Central Tibet

The primary focus of this section is on Fig. 2.1: No. 37. The ca. 6500 m high massif of the Purog Kangri is covered by a ca. 420 km^2 large, connected glacier (Figs. 2.156, 2.175, and 2.178). Approximately 35, mostly small glacier tongues descendradially from its central nourishing area. The two southern glacier outflows on the western slope, forming slightly larger outlet glacier tongues, were investigated as examples. The lowest ice margins of these tongues are located at 33° 50′ 42.70″N 89° 4′ 30.73″E at 5438 m asl (Fig. 2.176) and at 33° 49′ 24.86″N 89° 5′ 1.17″E at 5445 m asl. Both these glacier tongues possess several-meter-high, steep margins. These are the characteristics of *cold ice* that reaches down here as far as the ablation areas. The characteristics belong to temperatures of ca. −6 to −13 °C at the ELA (Chi and Ren 1980; Ding 1987; Kuhle 1990b, 1994a). The more northernly of the two glacier tongues discussed here has built up a terminal ablation moraine ("Satzendmoräne"). This glacier tongue disintegrates into

Fig. 2.151 Satellite image, Google Earth Pro: center 65° 11′ 42″N 19° 40′ 30″W. Area of the former High Glacial inland ice in Iceland (cf. Fig. 2.150)

numerous ice pyramids (Fig. 2.176). Also the margin of the glacier tongue is strongly frayed at the edges. Accordingly, we observe here the process of retreat (ablation). Its next-lower ice margin position was situated at 5400 m asl. It lay at a horizontal distance of 160–400 m from the current ice margin. The ice margin position is proven by the lower margin of the ablation moraine ("Satzendmoraine"). It is classified as belonging to Stage XI (1920–1950). This ice margin documents an ΔELA of ca. 20 m ((5440 − 5400)/2) in comparison with today (see Fig. 2.5). The more southerly located glacier tongue under discussion points to an increase in ELA of only half as much since Stage XI, that is, to an ΔELA of ca. 10 m.

By comparison, three of the glacier tongues flowing down towards the north and four towards the east clearly show greater horizontal retreats. But only one tongue reaches an ELA uplift of two-and-a-half times since Stage XI (ΔELA of Stage XI = 50). The retreat behavior of the other tongue ends corresponds with the first-mentioned exemplary glacier tongue in Fig. 2.176.

These comparably minor fluctuations make obvious a warming of only 0.1–0.2 °C during the last 60–90 years. Due to an increase in precipitation, the thermally conditioned glacier retreat in part could have been compensated.

In the forefield of the more southerly of the two glacier tongues that were chosen as samples, a tongue basin is identifiable as far down as ca. 5400 m asl. The accompanying ΔELA lies at ca. 23 m ((5445 − 5400)/2) in comparison with today. Stratigraphically, this glacier stage has to be classified as belonging to Stage X.

The confluence of the two southern glacier tongues on the western slope occurred at ca. 5380 m asl. Their joint glacier tongue reached down to ca. 5330 m asl. As far as to this point (at

Fig. 2.152 View from 34° 37′ 24.55″N 88° 45′ 15.26″ E, at ca. 4850 m asl, from the south shore of the Dogai Coring looking toward the NNW. A strong wind blows here from NW, recognizable by the foam-capped waves. (large ■ below) marks glacially scoured ground moraine running from east to west, more than 3 m thick. (small ■) mark up to decameter-thick, light-colored and in many places erratic ground moraine cover remnants, which cover a dark-colored bedrock landscape below. (▲) marks a small island in the lake, which is a ca. 140 m long roche moutonnée consisting of bedrock. Its longitudinal axis runs in an E-W direction from left to right. It is covered by ground moraine, which breaks off nearly vertically at the lakeshore as a fresh, meter-thick cliff face (cf. Fig. 2.12). (- - - -) indicates the Late Glacial inland ice surface at ca. 6000 m asl. Analog photo 10/10/2014 by M. Kuhle

33° 49′ 56.48″N 89° 1′ 11.36″E), we find positive geomorphological indications of a tongue basin that was shut off by a terminal moraine. Its outline can be diagnosed by an end-moraine-like, bent edge in the terrain. The ΔELA of ca. 55 m ((5440 − 5330)/2) classifies this ice margin as belonging to the Early Historical Stages VIII or VII (Fig. 2.5).

Farther west and further below, the comparably flat mountain foreland lies connected to this area. It is covered with tills, which are in turn covered on both sides of the current streams by narrow, glaciofluvial gravel bands (Fig. 2.156). The youngest till cover reaches down to ca. 5160 m asl (33° 49′ 16.20″N 88° 50′ 50.83″E). Further below, i.e., more to the west, older till extends throughout the region. The demarcation of both till covers proves the ice margin of the western piedmont glacier of the Purog Kangri. Its lobe formed a many-kilometers-wide, large-scale

Fig. 2.153 Satellite image, Google Earth Pro: center 35° 31′ 42″N 93° 07′ 59″E: http://extras.springer.com: Character AL. The formation resulted from basal abrasion ("Grundschliff") by the inland ice. This glacigenic scouring of the ground moraine was facilitated by the hardening of the loose rock by subglacial permafrost (cf. Figs. 2.152, 2.157, 2.160, 2.178, and 2.190). This is a High Glacial northern Tibetan ground moraine landscape (cf. Figs. 2.154 and 2.155)

bow. On the SW margin of this past piedmont glacier older till material was pressed together and thrust up to form a ca. 30 m high terminal moraine rampart (33° 46′ 7.31″N 88° 51′ 53.30″ E). Facing ENE, it forms an open concave bow typical of terminal moraines.

This foreland glacier was nourished by all glaciers of the western slope of the Purog Kangri. All these today-separated glaciers were once united in it. Among the four largest outlet glaciers are the two glaciers described as examples above.

A naïve calculation of the ΔELA for this piedmont glacier would result in a value of ca. 140 m ((5440 − 5160)/2). This value would classify its ice margin as belonging to the Neoglacial Stage VI ("older Dhaulagiri stage"). However, this value is too high. With a piedmont glaciation such as this, consisting of cold-based glaciers, the heightening of the surface (thickness) is already considerable. This led to a positive feedback effect on glacier nourishment. Therefore, this ice margin must be classified as belonging to Stage 'VII ("middle Dhaulagiri stage") (Fig. 2.5).

Farther north, Yi et al. (2002) have reconstructed the advance of another outlet glacier on the western slope. However, it only had a length of 6 km. This advance they classified as belonging to the LGM, though the ΔELA of this stage amounts to only ca. 130 m (ΔELA calculation by Kuhle).

Of the seven glacier tongues descending to the north, two of the eastern glacier tongues were joined together. Their former ice margin has left behind—at 34° 4′ 58.68″N 89° 13′ 28.18″E, at

Fig. 2.154 Satellite image, Google Earth Pro: center 65° 29′ 39″N 17° 20′ 39″W. High Glacial landscape in Iceland (cf. Fig. 2.153)

5160 m asl—a clearly demarcated tongue basin. Against the lowest current ice margin, one can calculate an ΔELA of ca. 110 m (5380 − 5160/2). This ΔELA classifies the former ice margin as belonging to the Late Holocene Stage ‘VII (Fig. 2.5).

Two of the glacier tongues in the west on the northern slope of the Purog Kangri Massif likewise have left behind a united tongue basin at nearly the same time. At 34° 8′ 44.12″N 89° 7′ 4.18″E, it reaches down to 5100 m. The accompanying ΔELA lay at ca. 130 m ((5360 − 5100)/2), which thereby classifies this ice margin as also belonging to the Late Holocene Stage ‘VII or the next-older Holocene (Neoglacial) Stage VI (Fig. 2.5).

The Holocene glaciers of the eastern slope of the massif had—like those of the western slope (see above)—flowed together to a united piedmont glaciation during this ΔELA, i.e., the Holocene Stages ‘VII and VI. The last of these piedmont glaciers possessed a fresh till belt down to ca. 5280 m asl (at 33° 56′ 13.82″N 89° 21′ 43.85″E). This can be proven due to the numerous small pools that have been left behind. With a ΔELA of ca. 80 m ((5440 − 5280)/2) it even belongs to the oldest Historical Glacier Stage VII. This youngest piedmont glacier, accordingly, is maximally only ca. 1700 years old (Fig. 2.5) and thus still younger than the one that flowed down from the Purog Kangri to the west (see above). The causes of this are the climatically more favorable eastern exposure for the glaciation as well as the ca. 100 m higher altitude of the foreland.

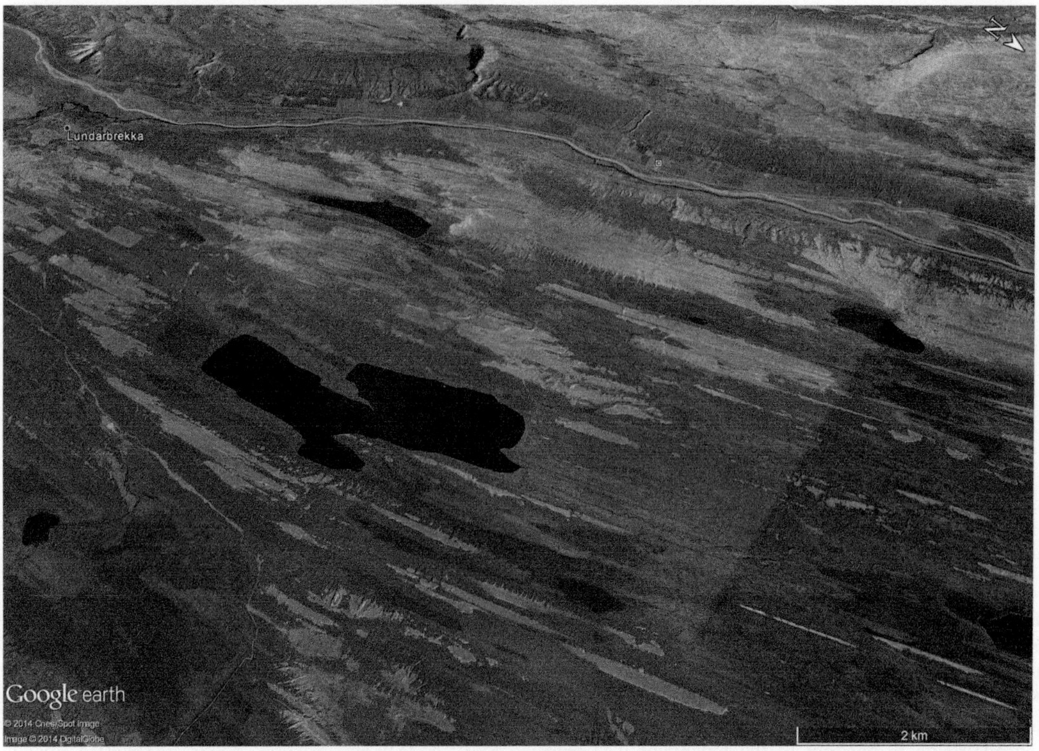

Fig. 2.155 Satellite image, Google Earth Pro: center 67° 54′ 08″N 52° 37′ 22″W. High Glacial landscape in Greenland (cf. Fig. 2.153)

At the time not all the tongues of the eastern slope had joined this piedmont glacier.

The next-younger ice margin of the central east-facing glacier tongue reached down as far as ca. 5350 m asl (33° 55′ 20.35″N 89° 20′ 33.60″ E). For this a ΔELA of ca. 45 m ((5440 − 5350)/2) can be calculated. It thus belonged to the Historical Stage VIII or IX (Fig. 2.5).

During the Middle to Late Holocene Stages ‘VII, VI, and V with their ΔELAs of ca. 300–100 m, the piedmont glacier of the eastern slope assumed the form of a hammer head. This form was caused by the watershed of the Tibetan Plateau, which is situated on the E-W axis of the Tangula Shan across the Purog Kangri and Zangsar Massif, as far as to the massifs of the Aksai Chin. As a result, it falls away along the N-S running main valley bottom east of the Purog Kangri to the south and the Dorsoing Co as well as down to the north to the Dogai Coring. Accordingly, the piedmont glacier flowed down from the culmination of the valley bottom at an altitude of ca. 5360 m (at 33° 51′ 39.81″N 89° 20′ 45.41″E) to the north and south, respectively. The diffluence pass is covered with till.

Still during the Holocene this hammerhead-shaped piedmont glacier also had inflow from the east. The largest glacier tongue of the 6130 m high massif (at 33° 54′ 59.34″ N89° 28′ 5.01″E) retained ice contact the longest. Today it terminates at ca. 5700 m asl. At an ΔELA of 80–150 m, i.e., during the Late Holocene Stage ‘VII, it flowed ca. 160–300 m lower downward than today. Thus it is likely during this stage that it still reached the surface of the eastern piedmont glacier of the Purog Kangri at ca. 5400–5550 m asl.

Fig. 2.156 View from 33° 43′ 42.76″N 88° 56′ 20.08″ E at 5315 m asl, from the piedmont region looking toward the NE onto the west flank of the Purog Kangri Massif (No. 1; ca. 6500 m). In the western piedmont lies an at least meter-thick ground moraine cover (■) that extends down to the foot of the mountains. (Δ) mark striations in the ground moraine, which were formed by exaration rills on subglacial permafrost (see Fig. 2.141). (☗) marks a glacially abraded mountain slope and truncated mountain spur. (□) marks a glaciofluvial gravel field (sandr) that is currently intersected by meltwater outflows from the Purog Kangri glacier. (- - - -) indicates the surface level of the High Glacial inland ice sheet at ca. 6800–7000 m asl. Analog photo 28/09/2014 by M. Kuhle

Even further distant, the lowest reaching glacier tongue of the more southerly 5990 m high massif (33° 50′ 1.09″N 89° 28′ 9.11″E) probably still reached the piedmont glacier during the Middle to Late Holocene. Today it terminates at ca. 5500 m asl. Accordingly, it must have flowed down to 5340–5200 m even still during Stage ‘VII. This is low enough for a confluence with the respective piedmont glacier, which at the time was also a ca. 14 km wide main valley glacier.

2.7.1 The Lowest Ice Margin of the Oldest Holocene (Neoglacial) Stages on the Western Slope of the Purog Kangri

The next-lower piedmont glacier margin west of the Purog Kangri Massif is situated ca. 60 m lower than that of the Young Holocene Stage ‘VII (see Sect. 2.7 above). There, at ca. 5080 m asl, at 33° 51′ 40.99″N 88° 41′ 50.60″ E, it folded together the older till into a terminal moraine. Figure 2.177 shows the till 7 km farther east within the pertinent tongue basin. Here, the clay-rich till cover has been exposed at a depth of 3–5 m by the current meltwater river. It is the tongue basin of Stage VI. At this location, 25 km from the mountain foot of Purog Kangri, the piedmont glacier tongue had a breadth of ca. 16 km. It lay at between 33° 54′ 32.75″N 88° 46′ 14.96″E and 33° 46′ 35.14″N 88° 49′ 26.70″E. This position proves the N-S extension of the pertinent homogenously colored till. On its NNW margin it contains fresh, still water-filled, small dead-ice holes. The ΔELA of this foreland glacier tongue amounted to ca. 180 m ((5440 − 5080)/2).

This foreland glacier tongue reached the lake basin of the Linggo Co (Figs. 2.166 and 2.178)

Fig. 2.157 Satellite image, Google Earth Pro: center 35° 32′ 57″N 93° 05′ 49″E; http://extras.springer.com: Character AM. High Glacial Central Tibetan ground moraine landscape (cf. Fig. 2.158)

during the oldest Neoglacial Stage V (Nauri stage). The lake level lies at ca. 5068 m asl (33° 50′ 41.85″N 88° 35′ 27.65″E). The depth of the 15 km long and 9 km wide lake basin is not known. Thus, a ΔELA of >186 m ((5440 − 5068)/2) for Stage V can only be determined as an approximate value.

2.7.2 Summary of Sect. 2.7

The Purog Kangri Massif shows the largest current contiguous glaciation in Central Tibet. It has the characteristics of a cold-based glacier that outflows on permafrost. Ice fringes of some of the ca. 35 outlet glacier tongues prove a ΔELA of 10–50 m for Stage XI (Fig. 2.5), which indicates a temperature rise of ca. 0.1–0.2 °C during the last 60–90 years. For a next-older ice margin (Stage X), for example, a ΔELA of 23 m was established. For the Historical Stages IX, VIII, or VII, a ΔELA of 45–80 m was established. During the Neoglacial Stage 'VII, the glacier tongues were united into a western piedmont glaciation down to an altitude of 5160 m asl. The ΔELA lay at about 140 m. The next-older glacial advance of Stage VI still reached 5080 m asl, which meant an ΔELA of 180 m. Another ΔELA of Stage 'VII or VI reached into the foothill region to the north and amounted to 130 m. From the Purog Kangri downward to the east a foothill glaciation already resulted during Stage VII. Even during Stage 'VII the eastern piedmont glacier probably still received glacier inflow from the adjacent mountain massifs to the east.

At a ΔELA >186, the western foothill glacier tongue of Stage V reached the Linggo Co (lake). This is the lowest terrain section so it is not possible to determine the exact ΔELA. Because the ΔELA of Stage VI already amounted to 180 m, the ΔELA at this time of the oldest Neoglacial Stage V (Fig. 2.5) must have been

Fig. 2.158 Satellite image, Google Earth Pro: center 71° 53′ 58″N 52° 06′ 38″E. High Glacial ground moraine landscape in Novaya Zemlya, Russia (cf. Fig. 2.157)

considerably larger than 186 m. The existence of an extensive ice cap or of remnant inland ice is therefore very likely.

2.8 The Last Glacial (Late Glacial and Stage 0 = Würm Ice Age, LGP, LGM, MIS 3-2) Glaciation of the Purog Kangri and Its Immediate and More Distant Surroundings in the Center of Central Tibet

The primary focus of this section is on Fig. 2.1: No. 37.

2.8.1 Relief-Specific Snowline Considerations

The youngest pre-Holocene stage was the late Late Glacial Stage IV (Sirkung stage). The period from this Stage IV to the Holocene Stage V (Nauri stage) experienced the strongest global warming between two glacier advances. It amounted to ca. 2.0 up to 2.75 °C and was the step from the Glacial to the Interglacial, with ELA uplifts of about 400–550 m (Fig. 2.5).

Viewed in a reverse-chronological sequence, this means the following for the area of the Purog Kangri: here, the ΔELA of >186 m in Stage V had dropped further during Stage IV of the late Late Glacial by ca. 400–500 (514) (see above). At this

Fig. 2.159 View from 34° 37′ 33.83″N 88° 46′ 21.68″E, from ca. 4843 m asl toward the west. The Dogai Coring lake is located in the right background. (■ bottom and top middle) mark meter- to decameter-thick ground moraine covers with widely distributed erratic boulders (e.g., basalt boulders, with hammer for scale). This ground moraine is bisected by a stream, together with its up to 100 m wide gravel bed (◘). (■ top on far right) marks light-colored, erratic ground moraine cover remnants that extensively cover a low mountain range landscape of dark-colored sedimentary bedrock above 5000 m asl. (- - - -) indicates the High Glacial inland ice surface at 6500 m asl. Photo taken 10 October. Analog photo 10/10/2014 by M. Kuhle

ΔELA of at least 186 m in Stage V, the lowest points in the terrain were almost reached. This is because in the area under discussion, the surface of the terrain nowhere reaches below 5000 m. Likewise, 100 km farther north of the watershed of this centerpiece of Tibet (see above), as well as 100 km south of it, the terrain nowhere falls below 4800 m asl. This means that the ΔELA during Stage IV was no more than ca. 50 (exactly 34) m larger than during Stage V in the highest area of the watershed region, and thereby was sufficiently large to build up an inland ice cover. In the somewhat lower peripheral area of the center north and south of the watershed, a 150 m larger ΔELA was necessary for this in comparison with Stage V. This amounts to a ΔELA of ca. 350 m (200 (>186) + 150 = 350) which during the Late Glacial Stage IV must already have led to an inland glaciation. However, the ΔELA that is characteristic of High Asia during Stage IV amounts to twice as much, namely 700 m (Fig. 2.5).

This calculation makes a completely terrain-covering inland glaciation even during the late Late Glacial very likely.

2.8.2 Glacial-Gemorphological Findings

Till covers (see Sect. 2.7.1) are located both to the south and north of the Holocene piedmont glaciation of the Purog Kangri Massif. They can

Fig. 2.160 Satellite image, Google Earth Pro: center 34° 45′N 89° 05′E; http://extras.springer.com: Character AN. High Glacial, polished, Central Tibetan ground moraine landscape at Dogai Coring (cf. Figs. 2.161, 2.162, 2.163, and 2.164). Ground moraine furrows alternate with small ground moraine ridges. Their formation was the result of basal abrasion ("Grundschliff") by the inland ice. This glacigenic scouring in the ground moraine was facilitated by the hardening of the loose rock by subglacial permafrost (cf. Figs. 2.141, 2.146, 2.153, 2.157, 2.165, 2.178, and 2.190)

be proven nearly everywhere. Their large-scale expansion shall first be traced towards the north. Their expansion here involves the highland between Linggo Co and north of the Dogai Coring.

We are dealing here with a more than 100 km long profile. Apart from the analysis of sedimentological samples with granulometric and morphoscopic analysis, erratic boulders were also recorded and analyzed. Figure 2.179 shows ground moraine with polymictic centimeter- to decimeter-sized components at 33° 54′ 54.21″N 88° 45′ 54.67″E. Basalt components transported long distances are preserved in isolation from one another. Sedimentary rock is located in situ underneath. This till sheet is here located ca. 30 m above the valley bottom as well as the current glacier outflow of the Purog Kangri Massif. Farther west, there are also till sheets with up to meter-sized basalt boulders in the vicinity of dead-ice depressions with tarns at 33° 56′ 23.52″N 88° 44′ 14.98″E and 33° 56′ 49.41″N 88° 43′ 47.57″E (Fig. 2.180). At times, this till surface even reaches 10 meters below the glacier outflow at 5086 m asl. Accordingly, it forms an even wider expanding glacigenic depression area than that of the small tarns (water pools).

Further till covers with erratic rubble, at times including basalt components transported long distances, are located at 33° 56′ 22.92″N 88° 44′ 12.40″E at 5089 m asl, at 33° 56′ 21.87″N 88° 44′ 31.72″E at 5089 m asl,

Fig. 2.161 Satellite image, Google Earth Pro: center 69° 05′N 22° 44′E. High Glacial ground moraine landscape with lakes in northern Norway (cf. Fig. 2.160)

33° 57′ 12.65″N 88° 44′ 55.78″E at 5115 m asl, 33° 57′ 33.55″N 88° 45′ 8.48″E at 5140 m asl, 33° 57′ 38.95″N 88° 43′ 56.60″E at 5110 m asl, and 33° 58′ 20.60″N 88° 44′ 24.54″E at 5140 m asl.

Ca. 2.6–7 km farther west, at 33° 57′ 9.97″N 88° 42′ 23.12″E at ca. 5100 m asl, there are meter-thick till covers with up to 1 m large erratic basalt boulders (Fig. 2.181). Embedded into this till is a water-filled dead-ice depression (Fig. 2.149). It is located on a flat slope 6 m above the valley depression line. Even so, it is a long way removed from being reached by the backward (retrogressive) erosion. This is an indicator of its youth. All of the tills mentioned here were last covered by glacier ice during the late Late Glacial of Stage IV (Fig. 2.5). It was then still an inland glaciation (see Sect. 2.8.1). Because of its large ice volume, the ablation process must have lasted up until the Holocene.

A further till sheet with erratic basalt boulders is observed at 5160 m asl at 33° 58′ 33.51″N 88° 43′ 28.37″E. They lie here on light sedimentary rock (calcareous sandstone) (Fig. 2.183) and are distributed across a large area. They were also found 1.6 km (Fig. 2.184), 2.2 km (Fig. 2.185), and 2.8 km (Fig. 2.186) farther to the northwest, as well as 5.2 km farther west (Fig. 2.187).

The accompanying till cover in between is preserved without spatial interruption. It thereby covers a relief of ca. 170 m of vertical distance. The highest positions of erratic basalt boulders observed here are at 5260 m asl.

The till cover can be observed to an altitude of up to 5640 m on the highest ridges WNW of the Purog Kangri Massif.

Moreover, the completely rounded form of these mountain ridges (e.g., Figure 2.190) also verifies an ice thickness of more than 600 m

Fig. 2.162 Satellite image, Google Earth Pro: center 67° 35′N 31° 36′E. High Glacial ground moraine and lake landscape on the Kola Peninsula, Russia (cf. Fig. 2.160)

above the lowest neighboring plateau level. This minimum value was still reached during the late Late Glacial Stage IV (Sirkung Stage). During the LGM (Stage 0, see Fig. 2.5), a minimum ice thickness of more than 1500 m can be estimated. This fact is proven by way of the rounded summit ridges of the Purog Kangri Massif (Figs. 2.156, 2.175, and 2.178), which reach heights of up to ca. 6500 m.

A further example of erratic boulders and their mixing with polymictic boulder material is shown by Figs. 2.188 and 2.192 (detail). Here at ca. 5174 m asl the boulders are, as in the case of the previously mentioned occurrences, positioned in isolation from one another in a pelitic matrix. Their position is located on a very flat slope with an average inclination of only 2°. For this reason also solifluidal transport of these boulders must be excluded.

Again for example, 20 km farther north is the flat north slope of a 140 m high ridge that is covered by till (Fig. 2.189). It contains polymictic, partially erratic boulders, among them erratic basalt boulders transported over long distance. The 8 km wide valley (or basin) bottom, which lies north from here at ca. 5060 m asl, is completely composed of till, on which flow a few narrow rivulets.

On flat slopes of such basins the ground moraines often display spacious furrowed structures, or glacigenic scourings and abrasions, respectively (Fig. 2.190; cf. Figs. 2.166 and 2.178); this is the case, for example, near 34° 8′ 46.06″N 88° 37′ 50.44″E at 5065 m asl. The surrounding hills and ridges exhibit, without exception, the glacially abraded forms of glacially streamlined hills and roches moutonnées (Fig. 2.190).

Fig. 2.163 Satellite image, Google Earth Pro: center 65° 44′N 124° 05′W. High Glacial ground moraine and lake landscape in the Northwest Territories, Canada (cf. Fig. 2.160)

Ca. 40 km farther north, at 34° 28′ 47.65″N 88° 39′ 42.39″E, the till cover is already located north of the highest elevation point of the plateau surface of Central Tibet. Accordingly, its surface is slightly inclined toward the north. The continuous till sheet here therefore is found "only" at 4890 m asl. It contains an increasing number of calcareous boulders. Many of them are rounded at the edges, i.e., subangular (Fig. 2.191). Here also it displays spacious abrasions with glacigenic scouring as well as closed-off cavities. The much smaller dead-ice depressions are only in part dried up. Numerous preserved tarns also again prove the young age of the till (cf. Fig. 2.146; satellite image Google Earth Pro: Center 34° 32′ 25″N 88° 36′ 00″E; see also http://extras.springer.com: Character AK). At this location, the inland ice cover was so thick that it must have outlasted the late glacial ablation process by thousands of years. Remains of it therefore must have been preserved up until the Holocene.

The till sheet, at times reaching a minimum thickness of 3–8 m, can also be traced 20 km farther north up to the Dogai Coring (Fig. 2.148). Across this distance, it covers ridges that are up to 5000 m high and consist in situ of sedimentary rock, partially limestone. This mountain area has the relief of a low mountain range. The surface of its rises and depressions, or small valleys, also displays the same glacigenic abrasion (cf. Figs. 2.190 and 2.191). This glacigenic scouring is characteristic for the Tibetan tills that have developed under cold-based ice. We are dealing here with typical tills of subglacial permafrost (Kuhle 2014a).

The till sheet here descends to the Dogai Coring down to ca. 3830 m asl. It contains

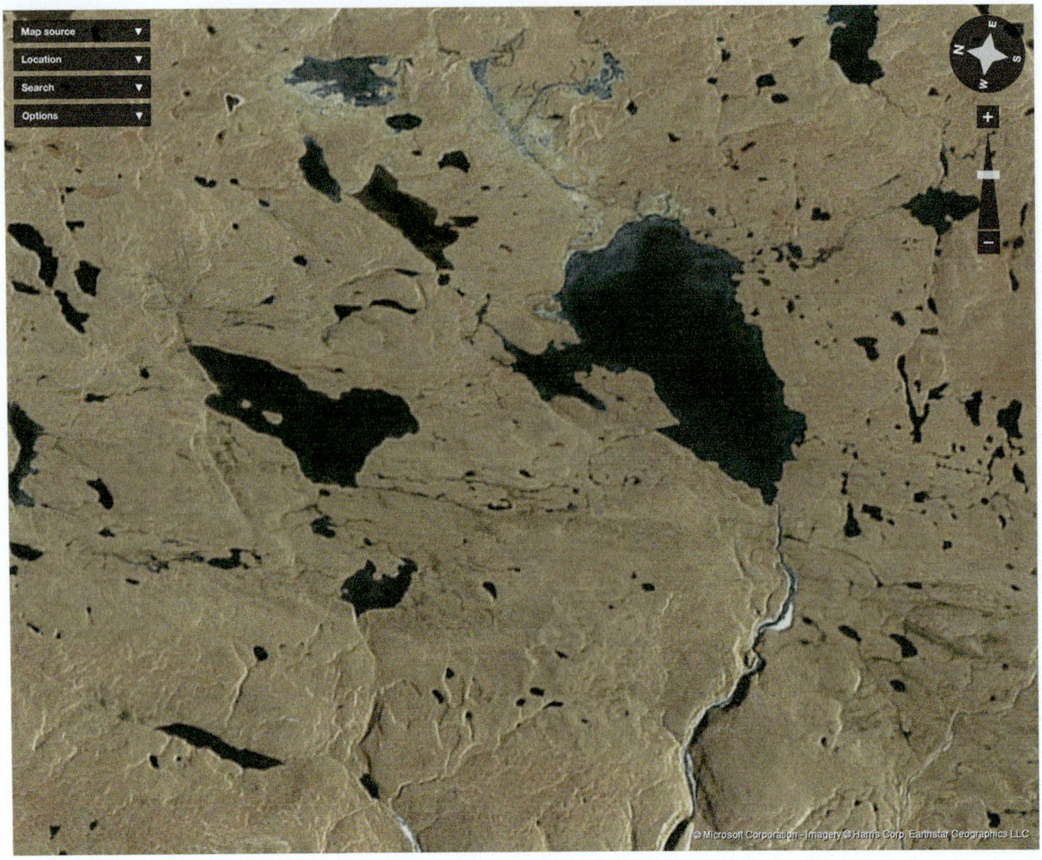

Fig. 2.164 Satellite image, Google Earth Pro: center 71° 39′N 81° 18′W. High Glacial ground moraine and lake landscape in Baffin Island, Canada (cf. Fig. 2.160)

polymictic, angular and subangular boulders (Figs. 2.148 and 2.159). Among them are erratic basalt boulders up to 1 m in size. These have been transported long distances to the Dogai Coring.

The same typical glacigenic abrasion that is shown in the Tibetan low mountain terrain between Fig. 2.191 and the Dogai Coring is also an indicative formation in the Arctic. On Novaya Zemlya at 74° 01′ 44″N 55° 24′ 03″E and 74° 02′ 53″N 55° 18′ 34″E, the same glacigenically striated ground moraine exists as in the aforementioned Tibetan area at 34° 33′ 18″N 88° 45′ 26″E (Google Earth). In both cases we are dealing with similar conditions of development under subglacial permafrost.

The till of several meters in thickness, which is exposed at 3 m (Fig. 2.148) and 8 m depth, reaches the Dogai Coring from both north and south. The elongated island in the center of the lake, as well as a roche moutonnée that forms a 140 m long island (Fig. 2.152), is covered by till (Fig. 2.12). The young age of this loose material is supported by the fact that it has been at all preserved on the continuously eroding steep bank of the roche moutonnée. This assumption is valid because a renewed supply of loose material can be excluded. It is always windy here and the wind velocity is high. The surge of the lake is accordingly strong. The ablation of the steep bank material, i.e., till, is therefore very high. Such till cover has been recorded continuously both on the flat in the west as well as the steeper, more actively eroding slope on the east (see Sect. 2.6). This indicates its young, in this case late Late Glacial age (Stage III–IV, cf. Fig. 2.5).

Fig. 2.165 Satellite image, Google Earth Pro: center 35° 34′ 01″N 93° 12′ 31″E: http://extras.springer.com: Character AO (cf. Figs. 2.146, 2.153, 2.156, 2.157, 2.160, 2.178, 2.190, 2.192, and 2.193). Late Glacial ground moraine and lake landscape in northern Central Tibet, the formation of which has taken place by down-thawing and decay of the inland ice cover (cf. Fig. 2.167)

The young till cover continues on the northern shore of the Dogai Coring. It is even preserved along steep slopes with a thickness of at least one meter. A quickly advancing abrasion is evident on these 100–180 m high mountain slopes. Slope gullies have been formed there from which moraine material has been and still is being transported away. The material is re-deposited at the slope bottom in the shape of mudflow cones and fans as well as alluvial fans (Fig. 2.152).

Underneath the till remains on the slope are rock outcrops belonging to the in situ sedimentary rock. They were subject to the typical streamed abrasions on top of the ridge. On these rocks, between ca. 5100 and 4870 m asl, the inland ice created abrasion basins and abrasion thresholds. Embedded within those basins are lakes and tarns. The rock surfaces alternate with till surfaces. About 10–20 km farther north there exist spaciously interconnected till covers. They show numerous dead-ice tarns in rather close vicinity to one another. Therein lies an indicator of their young, late Late Glacial age (Stage III–IV; cf. Fig. 2.5). However, there are also bigger lakes of 5–15 km extent that are preserved on top of the till covers (Fig. 2.150). The glacial-geomorphological evidence and chronological classification of one of those larger lakes has been presented in Sect. 2.6. (cf. Figs. 2.150 and 2.151).

After having discussed the downward slope from the culmination of Central Tibet toward the north, we now turn to the slope in a southward direction. Again we reconstruct the extent of the LGM to Late Glacial inland glaciation.

For the western piedmont glacier of the Purog Kangri to expand into this till landscape, a ΔELA (snowline depression) of at least ca. 140 m is

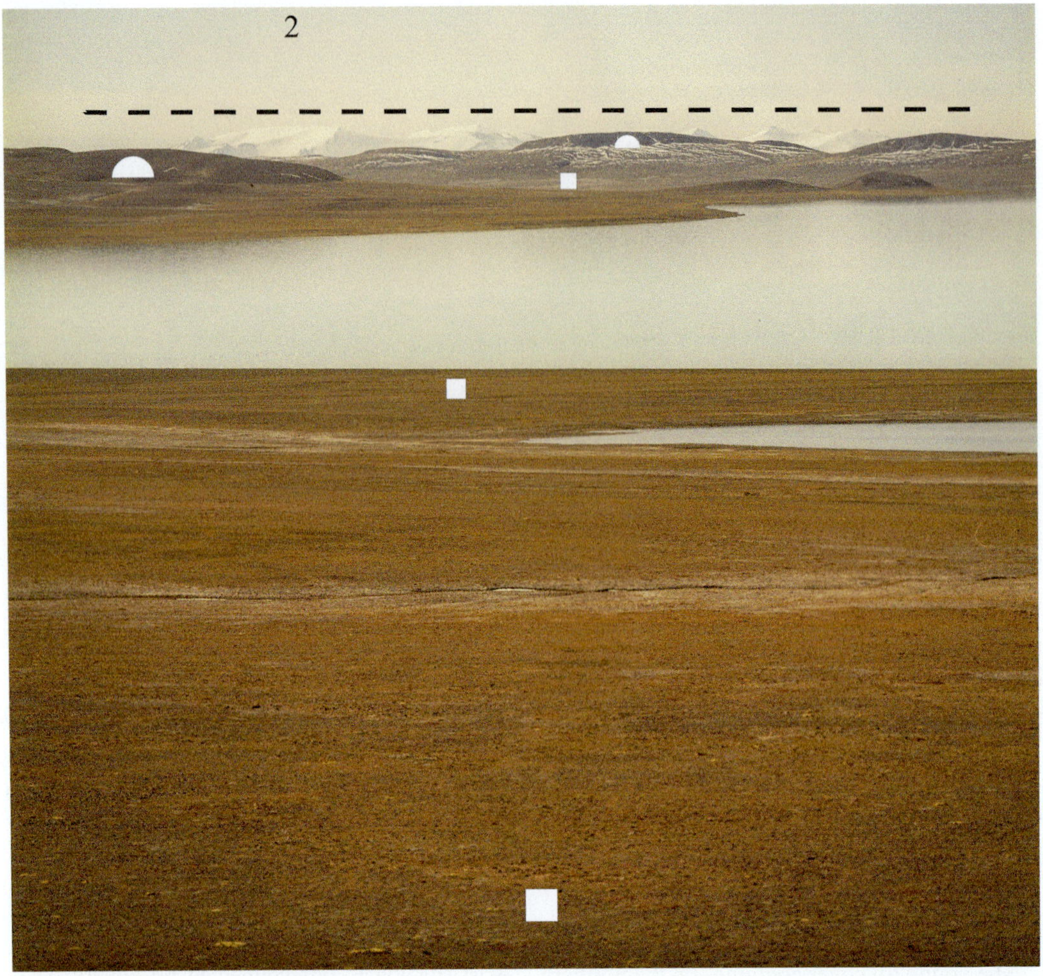

Fig. 2.166 Looking southward, in the center of Central Tibet at 5070 (5068) m asl and 33° 51′ 1.78″N 88° 36′ 2.32″E, onto the Linggo Co (lake). It was photographed from the northern shore from ca. 5110 m asl. In the background, at a distance of 64 km, a glaciated 6120 m massif is visible (No. 2). (■) mark meter- to decameter-thick ground moraines, on top of which the current tongue-shaped basin-lake has been formed by a glacier tongue that descended from the Purog Kangri Massif which lies to the east. (▲) mark hills of sedimentary rock that were once abraded and polished into a round shape through basal abrasion. (- - - -) indicates the High Glacial inland ice surface at over 6500 m asl. Analog photo 06/10/2014 by M. Kuhle

necessary. The current glacier tongue endings are located at 5450 m asl. Accordingly, this value of ca. 140 m is a minimum value (5450 − 5170/2). The ΔELA value of 350 m that was calculated for this region in Sect. 2.8.1 (see above) is therefore estimated very generously. It is with certainty, therefore, that one can posit a late Late Glacial glaciation cover, which correlated with Stage IV (Fig. 2.5). Because the ΔELA still reached to up to 700 m in High Asia during this Stage IV, there was still inland ice during the late Late Glacial Period. There was thus not only a local piedmont glacier but also a widely interconnected Central Tibetan ice cover.

The till sheets in the southern foreland are also marked by characteristic glacigenic scouring. These subglacial abrasion structures can be diagnosed in a SSW direction up to the ca.

Fig. 2.167 Satellite image, Google Earth Pro: center 65° 36′ 15″N 17° 00′ 21″W. High Glacial ground moraine and lake landscape on Iceland, the shaping of which has taken place by down-thawing and decay of the inland ice cover (cf. Fig. 2.165)

5390 m high peripheral hills framing this wide valley. The in situ rocks of these glacially streamlined hills are covered by the late Late Glacial till cover. They were previously, during the LGM (Stage 0, cf. Fig. 2.5) and up until the early Late Glacial (until Stage II, cf. Fig. 2.5), abraded and polished and thereby rounded over.

An exemplary location for observing the effects of the late Late Glacial (Stage IV) ice cover can be found in the hilly mountain area SW of the Purog Kangri. Figure 2.193 shows a typical till-covered landscape at 33° 39′ 27.76″N 88° 56′ 46.37″E at 5260–5400 m asl (Fig. 2.156).

The streamlining of the landscape surface evident here in many places follows the geological structure. The in situ bassets (rock outcrops) of vertical sedimentary rocks have here been exposed through the inland ice by glacigenic scouring. The bassets form fine-boned ridges. This has occurred in many places in Tibet where the striated till surface appears. The harder bassets in many places persist within similarly fine-boned till ridges. The glacier bottom overriding the ridges thus formed, as a continuation of the basset, the till into a similarly shaped small ridge. The additional weight of the inland ice pushed the ductile till material into the furrow of the glacier bottom. This furrow in the ice had been carved out by the in situ rock ridge of the basset. At the spot where the ridge was interrupted and where the till started forming the undersurface of the glacier, the till was pushed into the furrow of the ice that had previously been formed by the small ridge. In this way it is comprehensible why the glacigenic abrasion of the layer edges (“Schichtkopfstreifenschliff”) is today being continued by a streamlined till surface.

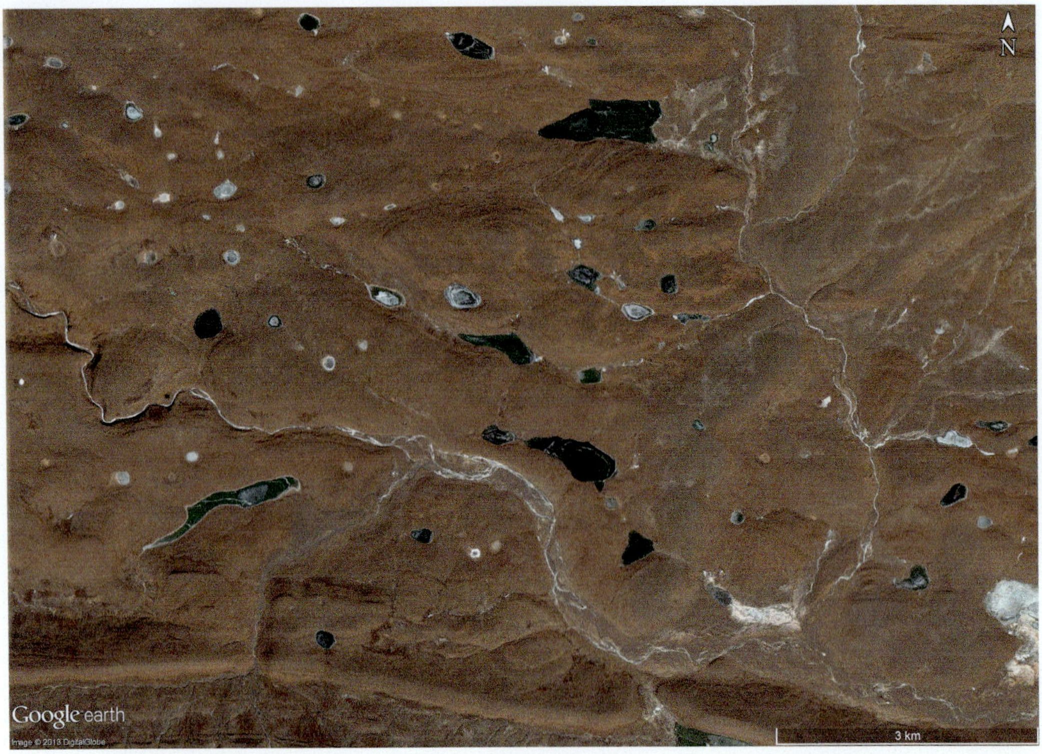

Fig. 2.168 Satellite image, Google Earth Pro: center 35° 21′ 02″N 92° 39′ 35″E: http://extras.springer.com: Character AP. Late Glacial ground moraine and "down-thawing" landscape of the Tibetan inland ice in northern Central Tibet (cf. Figs. 2.169, 2.170, and 2.171)

Mountain ridges, ca. 5200–5300 m asl, in the adjacent, western hill country have been glacigenically rounded. Simultaneously they exhibit small tor-like outcrops of the in situ limestone (Fig. 2.194).

From here farther to the south, along and past the eastern and western sides of the 6120 m high mountain massif (33° 20′ 22.18″N 88° 45′ 10.64″E), there are round-streamlined hill profiles and a till covering the entire landscape. We are dealing here with a total N-S distance of ca. 51 km. The base area of the Tibetan Plateau, i.e., the lowest surface level of any size, descends here from SW of the Purog Kangri from ca. 5150 m down to ca. 5050 m asl. This barely perceivable descent is thus already located slightly south of the culmination of the highest plateau section of Tibet (see above).

2.8.3 Summary of Sect. 2.8

The glacial-geomorphological findings of the discussed area from the center of Central Tibet are located within a N-S profile of ca. 50–60 km width and ca. 180 km length. The region ascends from its base level at ca. 4900 m asl in the north to ca. 5200 m west of the Purog Kangri Massif in the south. From here it descends toward the south to ca. 5050 m asl. The geomorphological as well as sedimentological data prove a complete glaciation. They prove that this decakilometer-wide study area was covered by inland ice beyond the highest-elevation part of the plateau surface of Central Tibet. At a minimum, the thickness of the ice amounted to 1500 m. Based on the young age of the glacigenic assemblage of glacial formations the

Fig. 2.169 Satellite image, Google Earth Pro: center 78° 33′ 51″N 69° 20′ 37″W. High Glacial landscape in northern Greenland (cf. Fig. 2.168)

findings further demonstrate that this ice sheet must have persisted throughout the Late Glacial. Even during the late Late Glacial during Stage IV (Sirkung Stage, Fig. 2.5), such an ice cover still existed. For this to happen, a ΔELA of even less than 350 m was needed (Sect. 2.8.1). In many places in High Asia, Stage IV even reached the doubled value of this snowline depression (Fig. 2.5). In this particular region, the melting process and disintegration of the ice cover outlasted the late Late Glacial. Presumably, it persisted until the Holocene.

For reference: during the LGM the snowline depression in High Asia was approximately three times as high as Δ350 m (Fig. 2.5: Stage 0 = Würm = late High Glacial maximum).

2.9 Further Terrestrial Evidence from the Center of the Plateau Regarding the Inland Glaciation of Tibet During the Pleistocene Ice Ages Including the Last Time During the LGP (Last Glacial Period, Stage 0–IV, MIS 3-2)

The primary context of this section is provided by Fig. 2.5.

Fig. 2.170 Satellite image, Google Earth Pro: center 67° 08′ 22″N 16° 50′ 44″E. High Glacial inland ice landscape in the Sarek area, Scandinavia (cf. Fig. 2.168)

2.9.1 The Late Glacial Glaciation Between the 6120 m High Massif (33° 20′ 22.18″N 88° 45′ 10.64″E) as Well as the Zaqungngomar (6290 m) and the 5400 m High Mountain Range West of Baingoin (31° 24′ 7.10″N 89° 54′ 46.46″E)

The primary context of this section is provided by Fig. 2.1: No. 37. On the southern slope of the 6120 m high massif (33° 20′ 22.18″N 88° 45′ 10.64″E) there are meter-sized erratic boulders at 5250 m asl (33° 15′ 23.49″N 88° 43′ 23.46″E). They lie in and on a pelitic till matrix above hard rock consisting of reddish sandstone (Fig. 2.195). In order to transport these boulders with a local valley glacier, i.e., with a mountain glacier from the north to this location, a ΔELA of ca. 220 m ((5570 − 5130)/2) is necessary.

In order for the valley glacier margin to reach the location of the till at all, which has been preserved 2.4 km farther east and ca. 100 m higher at ca. 5335 m asl, an ΔELA of ca. 270 m ((5570 − 5030)/2) is necessary. The form of this even, decameter-thick, and spacious till cover at 33° 15′ 0.18″N 88° 44′ 43.61″E, however, is only explicable through a large-scale, all-encompassing glaciation (Fig. 2.196). The till contains polymictic boulders that also include granite boulders. The corresponding latest continuous glacial cover existed during the late Late Glacial Stage IV. Afterwards, during the Holocene, the aforementioned local mountain glacier incised itself into this late Late Glacial till cover. This southward-flowing valley glacier reached down to 5030–5130 m asl, corresponding to a

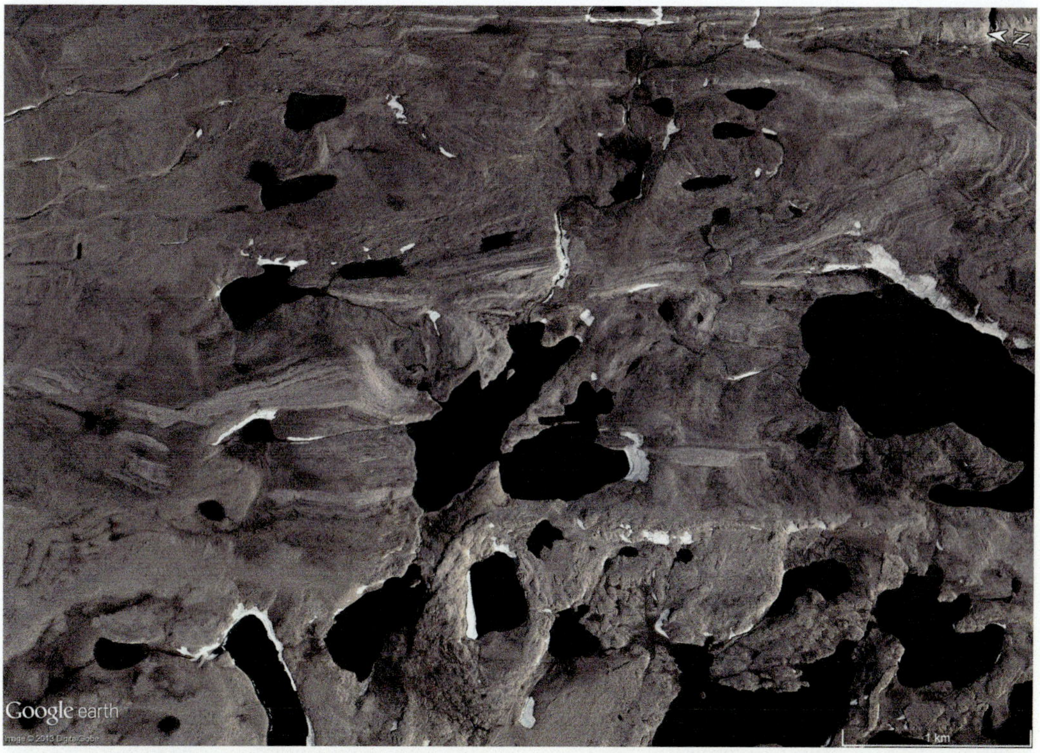

Fig. 2.171 Satellite image, Google Earth Pro: center 80° 07′ 16″N 18° 16′ 27″E. Glacial landscape, left behind by the High Glacial inland ice in northern Spitsbergen (Svalbard; cf. Fig. 2.168)

ΔELA of ca. 250 (245) m. This ELA depression existed during the Holocene Stage V (Neoglacial in Fig. 2.5).

On the pass at ca. 5317 m asl there is a meter-thick till cover with up to meter-sized polymictic boulders. Figure 2.197 shows one of the basalt boulders on this pass at 33° 14′ 52.45″ N 88° 41′ 13.60″E at ca. 5325 m asl. The boulder is accompanied by loose rock material of other petrographic compositions. Because on both sides of the pass, east and west, the terrain descends equally down to a little below 5000 m, it is not possible to reconstruct the direction of flow (“Transfluenz”) of the glacier ice across the pass. In particular, SW of the transfluence pass and up to 5550 m asl, the adjacent slopes display glacigenic flank smoothings and abrasions. These also reach up to the mountain tops. The inland glaciation that covered this terrain with great probability persisted until Stage IV of the late Late Glacial (see above). During the previous late glacial stages, Stage III and Stage II, the inland ice surface completely covered and surmounted the 6120 m high massif. The corresponding ΔELAs still amounted to ca. 1000–800 m (Fig. 2.5). The even greater inland ice thicknesses during the Late Glacial Stage I (Ghasa stage; Fig. 2.5) and the LGM (Last High Glacial maximum = Stage 0; Fig. 2.5) have abraded even the highest peaks of the 6120 m high massif (at 33° 20′ 22.18″N 88° 45′ 10.64″ E) and thereby rounded them even further. In general, these abraded mountains are polyglacial formations. Like the fells in Scandinavia, they were abraded during earlier Pleistocene glaciations and thus reshaped repeatedly by various inland ice sheets. The small mountain glaciation that currently still exists in this area consists of glacier tongues of no more than 2–3 km length. They flow from the peaks in all directions down into the peripheral small valleys. They flow transversely with respect to the High Glacial

Fig. 2.172 Satellite image, Microsoft Bing (flashearth.com): center 34° 46′ 12″N 90° 49′ 36″E; http://extras.springer.com: Character AQ. Late- to Post-Glacial Central Tibetan ground moraine landscape (cf. Fig. 2.173)

glacier abrasions and thereby destroy those features. In fact, they even further sharpen these slopes because of the numerous ice margins of today's glaciers. Along their black-and-white margins, strong frost weathering of the rocks occurs, which did not take place underneath the inland ice. This interglacial sharpening process of mountain terrain can also be proven on the Geladaindog Massif located 225 km farther east (Kuhle 1991). The same is true of the adjacent Zaqungngomar Massif located SSW of the current study area.

Till covers with polymictic rubble have been sampled WNW of the described pass region up to 33° 16′ 57.82″N 88° 36′ 34.44″E at 5100 m asl. They further extend broadly from here towards the west.

East of the 6120 m high massif and the adjacent, 36 km distant, ca. 6290 m high Zaqungngomar Massif (33° 3′ 27.21″N 88° 32′ 30.05″E) are almost completely uninterrupted till covers. They are at least several meters thick in many places and have been sampled and analyzed more closely along a W-E distance of ca. 35 km. They consist of polymictic material that also includes large and coarse rubble. These biggest components are, in isolation from one another, embedded in a clayey fine material matrix. Local moraine material, which contains a certain amount of rock material from the bedrock, is naturally mixed together with nonlocal, moraine material transported great distances. The bedrock below consists of tectonically steeply (vertically) layered sandstone and limestone.

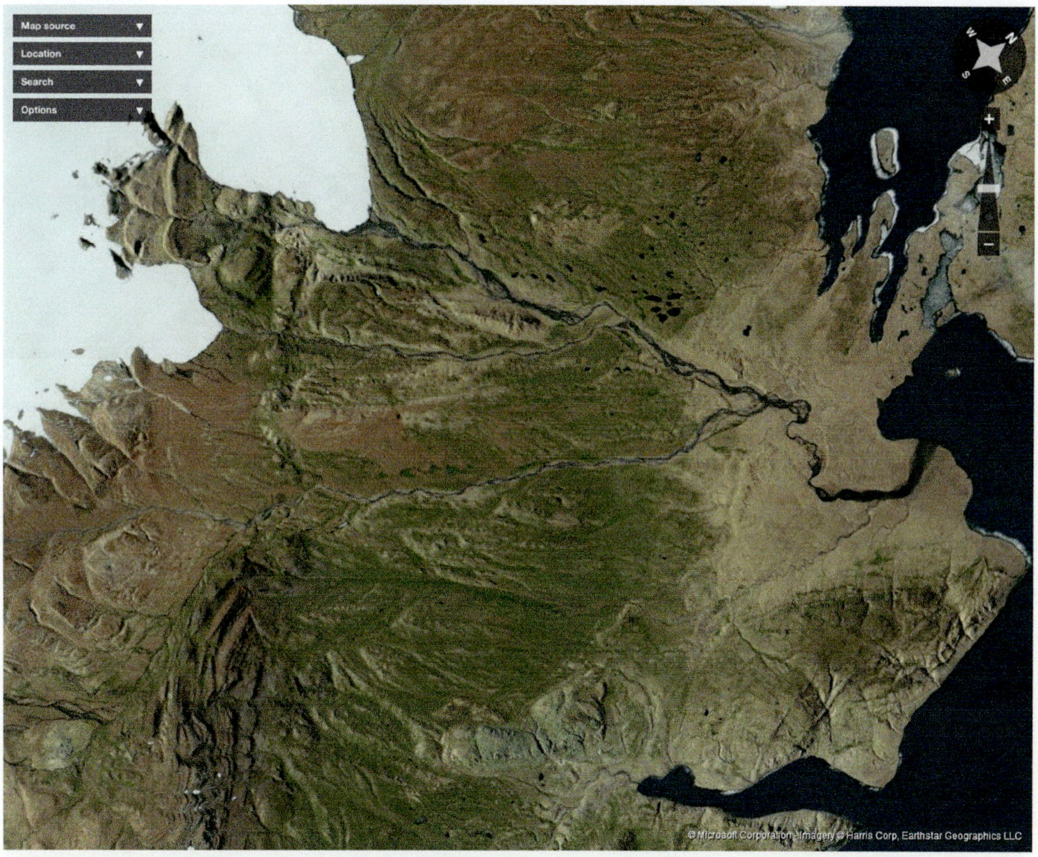

Fig. 2.173 Satellite image, Microsoft Bing (flashearth.com): center 79° 57′ 25″N 88° 51′ 55″W. Late- to Post-Glacial ground moraine landscape in Axel Heiberg, Nunavut, Canada (cf. Fig. 2.172)

Fig. 2.174 View from 33° 49′ 38.26″N 89° 0′ 10.26″E, from ca. 5385 m asl, looking across the middle-western piedmont of the Purog Kangri I (No. 1, ca. 6500 m). The center of the photograph is oriented toward the NE. (■) mark meter- to decameter-thick ground moraines (in the middle left, a person is standing to provide scale). (▲) mark mountain slopes and truncated mountain spurs abraded by the High to Late Glacial inland ice. (□) marks a historical, glaciofluvial gravel field (sandr) that is currently intersected by an outflow of meltwater from the Purog Kangri glacier. (■ in the right background) is a historical ground moraine surface that displays striations (exaration rills and small crests) typical of subglacial permafrost in Central Tibet; this striation is now being traced (transformed) by the current snow ledges. In the middle foreground, (- - - -) indicates the Late Glacial inland ice surface at ca. 6000 m asl. Analog photo 28/09/2014 by M. Kuhle

Fig. 2.175 View from 5092 m asl, at 33° 54′ 14.91″N 88° 39′ 31.93″E, toward the east and the Purog Kangri Massif (No. 1). Most of the up to 6500 m high peaks of the massif were rounded by prehistoric inland ice. At present, they rather tend to be steepened by the current (in comparison thin) layer of interglacial ice; i.e., they are being sharpened rather than rounded. (■) marks meter- to decameter-thick, striated ground moraine. (↓) marks an isolated, ca. 1 m sized, long-distance-transported basalt boulder that is "swimming" in a fine-grained ground moraine matrix. (☗) marks a perfectly roche moutonnée-shaped mountain slope of sedimentary rock, which was abraded by High to Late Glacial inland ice. Finally, (▣) marks the shore of a current meltwater outflow from the western slope of the Purog Kangri. Analog photo 06/10/2014 by M. Kuhle

Fig. 2.176 View from 33° 50′ 38.81″N 89° 5′ 14.25″E, at 5500 m asl, looking onto the western piedmont of the Purog Kangri I Massif. The middle of the panorama is oriented toward the NW. It shows the second outlet glacier tongue south of the Purog Kangri's ice-capped western side. The ice of the glacier tongue descends with a nearly decameter-thick, steep margin (↓). It ends at a hilly terminal moraine ("Satzendmoräne") (■ white). Its ending is fringed, disintegrated into firn ice pyramids, and is in the process of melting down. In the remaining piedmont the historical ground moraine is striated by exaration rills and separating crests that run diagonally to the orientation of the current glacier tongue (above ■ white). (■ black) marks the stony-gravelly-sandy-clayey ground moraine surface, which is characteristic of the majority of surfaces in Tibet, i.e., of nearly 2 million km^2. In the current location, the surface was only freed from glacier ice very recently, i.e., only ca. 65–180 years ago. (↓) points down at a part of the glacier with a ground moraine covering, which is turning into dead ice, and also at a meltwater outlet close to the steep margin of the outlet glacier tongue. (☗) marks mountain slopes once abraded by High to Late Glacial inland ice, as well as mountain spurs in bedrock that were truncated last by the Holocene and previous outlet glacier tongues of earlier interglacial periods. (- - - -) indicates the High Glacial inland ice surface above 6500 m asl. Analog photo 28/09/2014 by M. Kuhle

Fig. 2.177 View from 33° 52′ 0.62″N 88° 46′ 19.38″E, from ca. 5115 m asl toward the east and the Purog Kangri I Massif (No. 1). (□) marks the ground moraine, whose surface is here expanding extensively because of glaciofluvial activities of the glacial meltwaters. We are here dealing with the many hundred meters wide, strongly fluctuating meltwater outlet of the outlet glacier tongues of the western slope of the Purog Kangri. It has undercut the primary ground moraine cover (■) mainly by removing some of the pelites (mostly clay), which led to the stripping of the ground moraine to depths of up to 3–5 m. Analog photo 28/09/2014 by M. Kuhle

Among the most prominent erratic rubble components are, again, the basalt components lying on the surface. Such long-distance tills with basalt components were charted, for example, at 33° 1′ 37.11″N 88° 51′ 7.51″E at 5000 m asl as well as at 33° 1′ 32.72″N 88° 51′ 18.75″E at 4980 m asl and from there slope upwards. The thus proven glaciation for the currently described area thus again overrode the entire landscape. This is demonstrated by the glacial, streamlined hill profiles, which rise up to ca. 5300 m asl, as well as the large whaleback-like formations. These were the result of glacigenic abrasions of the layer edges (“Schichtkopfstreifenschliff”). The reconstructed inland ice sheet persisted up until the late Late Glacial. This was at the time of Stages III and IV (Fig. 2.5). Presumably, remnants of inland ice also lasted here until the early Holocene.

The glacier advance that then followed took place during the Middle Holocene, i.e., during the Neoglacial. As already mentioned above, it was reconstructed along the eastern slope of the Zaqungngomar Massif (Sect. 2.4.8.1) and reached a ΔELA of ca. 240 m.

Glacial till covers were also found south of the Zaqungngomar Massif as well as south of the 5010 m high pass (south of 32° 59′ 32.48″N 88° 30′ 52.73″E) between two nameless lakes, up to 20 km long. They are here positioned on the slopes of hills that rise to ca. 5200 m asl. They alternate with glacigenic ground polishing

Fig. 2.178 View from 33° 54′ 8.44″N 88° 39′ 29.15″E, from ca. 5090 m asl, toward the east and the Purog Kangri Massif (ca. 6500 m, No. 1), across the estuarial area of the massif's glaciers' meltwater outlet into the Linggo Co (Tso, lake). (□) marks a small, temporary, further inflow area that stems from the active permafrost layers that begin to melt during summer. (■) designate decameter-thick glacially scoured ground moraine that was still covered by glacier ice during the Holocene. (☗) marks a roche moutonnée or otherwise a glacigenically abraded mountain slope. (- - - -) indicates the High Glacial (LGM = Stage 0, see Fig. 2.5) inland ice surface at ca. 6700–7000 m asl. Analog photo 07/10/2014 by M. Kuhle

Fig. 2.179 View at 33° 54′ 54.21″N 88° 45′ 54.67″E, from 5137 m asl, looking across a late Late Glacial, glacial landscape. The center of the panorama lies in SSW direction. In the background, halfway to the left, lies the 6120 m massif (No. 2) at a distance of 65 km. (■) designates an at least several-meter-thick ground moraine that extensively covers the bedrock landscape underneath; it contains erratic basalt components. (☗) are roches moutonnées and abraded mountain slopes, i.e., glacially streamlined hills consisting of bedrock. They are also coated in ground moraine, although in some places this cover is only decimeter-thick and interrupted. Analog photo 01/10/2014 by M. Kuhle

Fig. 2.180 View from 33° 56′ 49.41″N 88° 43′ 47.57″ E, from ca. 5086 m asl, looking ENE across a kilometer-wide, slightly undulating ground moraine plain (■). (⭘) marks one of the ca. 1 m sized basalt boulders; these are erratic because sedimentary rock forms the bedrock underneath the moraine cover. Beyond the kettle hole, which is filled with a frozen tarn, the ground moraine (■) (cf. Fig. 2.141), which is abraded in a N-S direction, nestles against the slowly ascending slopes of glacially streamlined hills (▲). In the background, these glacially abraded mountain slopes are covered with fresh snow. (- - - -) demarcates the High Glacial (Stage 0, Fig. 2.5) surface of the inland ice at ca. 6700–7000 m asl. Analog photo 30/09/2014 by M. Kuhle

Fig. 2.181 View from 33° 57′ 9.97″N 88° 42′ 23.12″E at ca. 5100 m asl toward the WNW across a ground moraine cover (■) consisting of meter-thick loose rock material. Ca. 1 m thick basalt boulders are, in isolation from one another, embedded in pelitic moraine material. (▲) indicates a rounded slope of a mountain that reaches an altitude of more than 5400 m asl. Its glacigenic form closely resembles that of the fells and glacially streamlined hills of the Scandes in Northern European Scandinavia. (- - - -) demarcates the Late Glacial (Stages I–III, Fig. 2.5) inland ice surface at ca. 6000 m asl. Analog photo 29/09/2014 by M. Kuhle

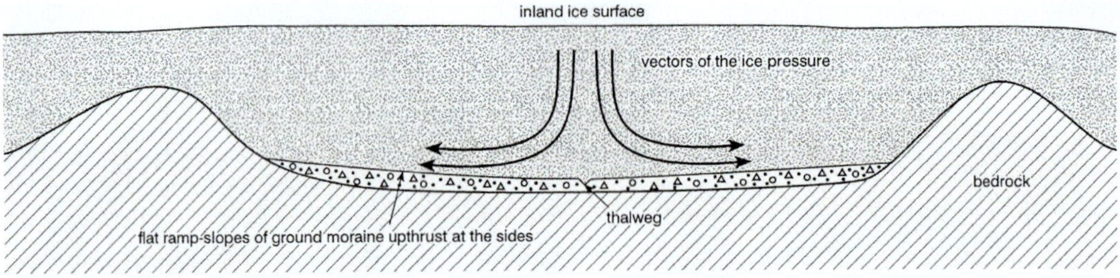

Fig. 2.182 Depicts the genesis of flat, glacial ramp-slopes in the High Glacial to Late Glacial (Stages 0–IV) ground moraine in Central Tibet (after Kuhle 1999: 23, Fig. 20)

Fig. 2.183 View from 33° 58′ 33.51″N 88° 43′ 28.37″ E at ca. 5160 m asl across a slope covered with ground moraine (■), looking toward the NW and upslope. The ground moraine surface carries polymictic boulders, among them limestone and basalt boulders (**O**). The ground underneath consists of calcareous sedimentary rock (◇). The several-decimeter-thick basalt boulders (**O**) lying on top are erratic (a person in the photo provides a sense of scale). (- - - -) indicates the Late Glacial inland ice surface at ca. 6000 m asl. Analog photo 29/09/2014 by M. Kuhle

on bare bedrock. On the basis of fresh slope gullies, these tills can be identified geomorphologically as meter-thick unconsolidated rock material that is resting on the surface.

Another, nearly uninterrupted till cover of the valley and basin bottoms is found between the till covers at ca. 33° 13′ 12.37″N 89° 0′ 38.06″E, north of the ca. 22 km long, banana-shaped lake, and up to and beyond 32° 51′ 8.71″N 88° 52′ 26.27″E in a southward direction. The hill and mountain slopes lying in between, which rarely exceed 5400 m in height, have been

Fig. 2.184 View from 33° 59′ 18.90″N 88° 44′ 0.43″E at ca. 5200 m asl downvalley toward the south across a moderately inclined ground moraine surface (■). Isolated, up to several-decimeter-sized subangular and rounded basalt boulders (hammer provides a sense of scale) lie embedded in the fine-grained matrix of the ground moraine. (◓) marks a mountain slope rounded by abrasion. (- - - -) demarcates the Late Glacial inland ice surface at ca. 6000 m asl. Analog photo 29/09/2014 by M. Kuhle

Fig. 2.185 View from 33° 59′ 38.58″N 88° 44′ 1.73″E, from 5236 m asl, across a moderately ascending ground moraine plain (■), looking toward the NNW and upslope. (**O**) marks one of the up to meter-sized, subangular basalt blocks that have been outwardly roughened by weathering. (◓) marks roche moutonnée-shaped, abraded mountain slopes and rock faces that reach an altitude of up to 5410 m asl. Analog photo 29/09/2014 by M. Kuhle

Fig. 2.186 View from 33° 59′ 53.91″N 88° 44′ 14.61″E, at 5260 m asl, across a ca. 17°-inclined slope covered with ground moraine (■), which is rising toward the north. The ground moraine cover reaches up to the tops of the ridge and the hill at 5400–5410 m asl (the two ■ at top left). (⭘) mark decimeter-sized subangular basalt boulders that have been partially fragmented through frost weathering and thereby sharpened on one side; they lie isolated or in accumulations and are basally embedded in the pelite-rich matrix of the ground moraine cover. (- - - -) indicates the Late Glacial inland ice surface at ca. 6000 m asl. Analog photo 14/10/2014 by M. Kuhle

Fig. 2.187 View from 33° 58′ 3.67″N 88° 40′ 54.87″E, from 5150 m asl, across a gently inclined, slightly bulged ground moraine surface toward the WSW. On the right next to the person lies a ca. 80 cm long basalt boulder—one of many large boulders on the surface of the ground moraine. (▲) marks a flat, roche moutonnée-shaped, abraded mountain slope; it is located to the right of glacially streamlined hills that reach an altitude of up to ca. 5380 m asl. (- - - -) indicates the Late Glacial inland ice surface at well over 6000 m asl. Analog photo 29/09/2014 by M. Kuhle

Fig. 2.188 View from 33° 58′ 35.13″N 88° 40′ 7.56″E, from 5174 m asl, toward the north across a flat upper-valley-end depression ("Talursprungsmulde"). The slopes are covered with ground moraine (■). Its surface is polygenetically transformed, since it displays a combination of both periglacial stone- and deflation pavement consisting of coarser moraine components. The larger boulders visible in the foreground (to the right of the person) are several decimeters in size; they are edgy and—despite frost weathering—partly subangular or rounded. (▲) mark the accumulations of considerably larger, up to several-meter-sized boulders. (Δ) marks glacigenic scouring (cf. Figs. 2.146, 2.166, and 2.190) in ground moraine hardened by subglacial permafrost. (◓) marks an abraded mountain back in the shape of a flat, glacially streamlined hill that reaches ca. 5320 m asl. (- - - -) indicates the Late Glacial inland ice surface at well over 6000 m asl. Analog photo 02/10/2014 by M. Kuhle

Fig. 2.189 View from 34° 8′ 43.79″N 88° 38′ 14.94″E, from 5080 m asl toward the SE across a very flat, rising slope of ground moraine (■). Apart from basalt boulders, the ground moraine also contains sedimentary rock fragments which, for example, consist of limestone. The large components are mostly angular in shape, although a few are subangular. The light-colored stripes in the middle-ground are of fluvial origin; they were generated by seasonally outflowing meltwater and therein flushed out ground moraine material. (◓) marks an abraded mountain slope (5220 m asl), which at a later stage became coated in striated ground moraine; the striations were the result of glacigenic scouring. (- - - -) indicates the Late Glacial inland ice surface at well over 6000 m altitude. Analog photo 13/10/2014 by M. Kuhle

Fig. 2.190 View from 34° 8′ 46.06″N 88° 37′ 50.44″E, from 5065 m asl, toward the NE, onto an up to 5250 m high mountain slope. (■) marks ground moraine that completely covers the glacigenic landscape. It contains polymictic boulders, among them the erratic basalt boulder in the foreground (below lowermost ■). The texture of the ground moraine is well recognizable in the foreground and up to the terrain of the low mountain range in the background. It consists of rills or furrows that run from NW to SE. They are oriented transversely with respect to the orientation of the photograph. Ground moraine furrows alternate with small ground moraine ridges. Their formation resulted from the basal abrasion ("Grundschliff") of the inland ice. This glacigenic scouring in the ground moraine was facilitated by the hardening of the loose rock by subglacial permafrost (cf. Figs. 2.141, 2.146, 2.166, 2.178, 2.192, and 2.199). (- - - -) indicates the Late Glacial inland ice surface at ca. 6000 m asl. Analog photo 11/10/2014 by M. Kuhle

rounded (Fig. 2.198). In some places, the covers contain isolated, notably bigger boulders of several decameters to a meter in size, which are either subangular or rounded. In most cases, however, the surface of the till consists of pebbly or rocky polymictic components, which sometimes reach the size of a fist.

In the locations mentioned here, there exists a ca. 8 km-wide, basin-like depression that is filled with such a till sheet nearly up to level ground. At its margins are mudflow and alluvial fans, some of which reach a breadth of 1 km (e.g., at 32° 47′ 9.75″N 88° 53′ 26.68″E, 5000 m asl). They are oriented toward the till covers in the

Fig. 2.191 Taken at 34° 28′ 47.65″N 88° 39′ 42.39″E, 4890 m asl, looking across a ground moraine cover (■) toward the NE. The ground moraine cover was furrowed by glacigenic scouring (cf. Figs. 2.152, 2.166, and 2.190) in a NW-SE direction (transversely to the orientation of the photo). In the foreground lies a ca. 1 m sized, subangular, limestone boulder (above the lowermost ■) that has been transported over a long distance. Since the deglaciation, the surface of the ground moraine cover has been flushed out and redeposited in a ca. 30–50 m wide meltwater streambed (below uppermost ■). (▲) marks an abraded, ca. 5036 m high mountain slope. Its longitudinal axis follows the NW/SE-oriented direction of flow of the High Glacial inland ice (Stage 0 in Fig. 2.5). (- - - -) indicates the High Glacial inland ice surface at more than 6500 m asl. Analog photo 11/10/2014 by M. Kuhle

center of the basin. These fans originate from small side valleys out of which the till material has been transported since the time of the deglaciation. The dislocated moraine material was then redeposited in the aforementioned fan-like formations.

Such decakilometer-large, evenly leveled till fillings of basins—often together with gradually silted up lakes—are characteristic for the whole of Tibet. For example, the basin east of the Zaqungngomar Massif is ca. 20 km wide in a W-E direction from mountain margin to mountain margin, and ca. 40 km long in a NNW-SSE direction. Accordingly, this represents a classic glacigenically abraded basin landscape ("Schliffwannenlandschaft") that is segmented by abrasion thresholds and more or less medium-sized mountain ranges.

South of two smaller, basin-like depressions, which are separated from one another through low mountain range thresholds and hills of a decameter up to several hundred meters in height, there is another, similarly large basin. It is located at 32° 33′ 0.18″N 88° 56′ 2.56″E. Its

Fig. 2.192 View from 33° 40′ 3.85″N 88° 51′ 22.49″E, from 5170 m asl, toward the WSW, onto an abraded hillslope (▲) that rises up to 5220 m in the north. It is a glacially streamlined hill whose reddish sedimentary rock is partly covered by ground moraine. (■) mark further ground moraines that extensively cover the underlying bedrock landscape. (↓) mark localities with up to several-meter-large, polymictic boulders on and in the ground moraine, among them erratic boulders. (Δ) indicate characteristic furrows (scourings) in the ground moraine surface caused by glacigenic scouring (cf. Figs. 2.139, 2.146, 2.152, 2.166, 2.178, and 2.190). (- - - -) indicates the Late Glacial inland ice surface at ca. 6000 m asl. Analog photo 03/10/2014 by M. Kuhle

NE-SW expanse reaches ca. 50 km and its NW-SE expanse ca. 30 km. Its till filling is interspersed with flat, glacially scoured moraine hills. In many places, it leans onto the slopes of local rock hills (Fig. 2.199). The till sheet is also striated in these positions. Some of the rock hills are streamlined and have thus been rounded by the inland ice into large roches moutonnées with lengths of between 1 and 2 km and elevations of up to ca. 200 m (Fig. 2.199). The nearly leveled till filling of the basin shows evidence of a predominantly prehistoric fluvial transformation. It is transformed close to the surface by numerous, hardly more than meter-deep fluvial incisions. These are now vegetated by dwarf shrubs in the climax state (Fig. 2.199).

It is very characteristic of Tibet that the till plains merge with the slopes at their margins and, furthermore, in many places gently ascend from the center of the valley or basin. Figure 2.182 tries to capture the glacio-dynamic cause of this arrangement. The imposed load of the

Fig. 2.193 View from 33° 39′ 27.76″N 88° 56′ 46.37″E, from 5230 m asl, toward the NE, across a furrowed, striated ground moraine landscape (■). The furrows are traced by snow ledges. This characteristic striation pattern, caused by glacigenic scouring of the inland ice, is particularly prominent in places where it has led to the tracing or further abrasion of rock outcrops in the sedimentary bedrock (🡅) (cf. Fig. 2.141). Analog photo 29/09/2014 by M. Kuhle

kilometer-thick inland ice compressed the till material transversely with regard to the relief-dependent direction of flow gently sideways and onto the valley slopes. The very same glacier dynamics is also held responsible for the formation of trough valleys. It has been demonstrated by means of mountain glaciers and the shovel-shaped abrasion areas that ascend sideways along the glacier-ice margins (e.g., von Klebelsberg 1948/1949).

Between 32° 47′ 15.95″N 88° 52′ 36.16″E and 32° 33′ 0.18″N 88° 56′ 2.56″E, on top of rock faces up to 5200 m high, there are glacigenic abrasions of the layer edges (“Schichtkopfstreifenschliffe”). These are preserved on local sediment bedrock ridges. An at least meter-thick till cover is found at 32° 45′ 37.83″N 88° 52′ 52.53″E as well as between 32° 42′ 44.18″N 88° 54′ 23.60″E and 32° 41′ 29.75″N 88° 57′ 22.16″E at 4910–4950 m asl. Its considerable thickness can be diagnosed on the basis of permafrost formations that are known to derive from extremely large expanses of tills in other areas such as, for example, northern Canada. Thus, one also finds here similarly large ice wedge net formations or tundra polygons with expanses of 10–40 m in diameter.

Further detailed investigations, data collection, and sampling have been made in the regions lying farther SSW along the profile of Central Tibet and extending ca. 170 km from the previous study area until reaching the mountainous region of Baingoin (31° 24′ 7.10″N 89° 54′ 46.46″E). In the context of this book, however, we need only summarize and provide examples for the most important empirical facts: Between 32° 33′ 0.18″N 88° 56′ 2.56″E and the lake of Duomaxiang (32° 23′ 29.34″N 89° 11′ 5.87″E), five areas lying at altitudes of between ca. 5200 and 4680 m have been studied and shown to indicate prehistoric, glacigenic transformations. The current landscapes originate from glacigenic ground polishing and/or meter-thick till deposits and their subsequent transformation since the deglaciation period. The profile toward the south continues at almost the same altitude. It only descends ca. 20 m to 4660 m asl from the lake of Duomaxiang to the locality at 32° 12′ 32.17″N 89° 8′ 51.66″E. In the north, the mountain ridges, which are at maximum ca. 350 m high, still consist of yellow-brownish sandstone. Then follow light-colored limestone hills rising 200–300 m above the plateau, and hills consisting of lime-marl. On the basis of 10 test areas, glacigenic forms and sediments have been documented along a distance of ca. 23 km. Till covers are found here that have been incised along the slopes by numerous, very fresh gullies with a depth of no more than a few decimeters to meters. The limestone hills and mountains show

Fig. 2.194 Taken at 33° 39′ 8.68″N 88° 52′ 25.44″E, from ca. 5150 m asl, looking toward the NE, showing a glacially streamlined hill (▲). This hill is composed of steeply (vertically) standing sedimentary rock layers. With some inclination toward the more resistant (harder) rocky outcrops, gate-like rock castles have been formed (left and right of ▲). (■) marks a snow-covered ground moraine cover carrying erratic boulders. Analog photo 15/10/2014 by M. Kuhle

two things: On the one hand, they were incised or dissolved by subglacial meltwater erosion or corrosion. On the other, traces of ground polishing such as basset edge abrasions and the remains of roches moutonnées have been preserved. An example of this is provided by the hill group rising ca. 75 m high above the level of the plateau at 32° 15′ 18.08″N 89° 6′ 41.93″E. The meltwater gullies are partly filled with ground moraine. In many places, convexly preserved loose rubble cushions have no higher local catchment area from the slope. Purely on the basis of this it is possible to conclude that they must have been deposited here by glacier ice. The deposits thus necessarily consist of till.

Another N-S section of our study area is located between 32° 12′ 32.17″N 89° 8′ 51.66″E and the NE 3shore of the huge Siling Co, the second-largest lake of Central Tibet (after the Nam Co). The Siling Co is 80 km long and 68 km wide. Although it is hardly smaller than the Kuku Nor at the NE margin of the plateau, it is located 1350 m higher. The N-S section that is now described lies between ca. 4780 and 4550 m asl. It is a particularly interesting region because hardly any bedrock is showing anywhere near the surface, which is composed of three different kinds of loose rock material: (1) till covers that mantle the hill chains; (2) glaciofluvial gravel belts that were washed out along the depression line from the till during deglaciation; and (3) limnic sediments of postglacial tarn and lake formations in glacigenic depressions of different sizes.

Fig. 2.195 Taken at 33° 15′ 23.49″N 88° 43′ 23.46″E, from 5250 m asl, looking toward the north and across more than 2 m sized erratic granite boulders (⭘). These lie on and in a several-meter-thick ground moraine cover; below the moraine is reddish sandstone bedrock. The ground moraine covers the ground of the entire 3.4 km wide valley chamber (■). On the opposite mountain slope it reaches an altitude of 5200 m asl (uppermost right ■ in the background). As erratic ground moraine it there also covers the reddish sandstone bedrock in some places. (▲) marks striated ground moraine, caused by glacigenic scouring, on the valley slope. (☗) mark glacigenically abraded mountain spurs. (- - - -) indicates the Late Glacial (Stage I–III; cf. Fig. 2.5) inland ice surface at ca. 6500 m asl. Analog photo 04/10/2014 by M. Kuhle

The most crucial considerations concerning prehistoric landscape formation relate to the question regarding the extent and dynamics of glaciation during the ice age. It has already been evidenced for the area north of here (see above). For this region also the glaciation must be proven. This is possible here solely on the basis of till covers. The ground-polish forms of the inland ice described for the area further north do not reach the surface any more in this region. Instead, they are covered by the youngest till sheets, which were deposited here shortly before the onset of deglaciation. The till must therefore be classified as being of Late Glacial age.

The formation of gullies is essential for the diagnosis of tills. For example, one notes branching gullies in loose rock at 32° 0′ 24.91″N 89° 12′ 24.60″E at ca. 4630 m asl. They have been incised here with decameter-large spaces between them by means of fluvial (retrogressive) erosion. Their mutual isolation, i.e., wide-spacedness of their occurrence, is indicative of their young age. They only just reach depths of decimeters or meters. As manifestations of an only recently initiated fluvial transformation, along slopes that are nearly devoid of vegetation, they mark a geomorphological change of regime. Previously, this region must have been covered by the inland ice; this is also proven by the above-mentioned sedimentological indicators. Fluvial incisions are inevitable on slopes without vegetation. If these are current and recent, i.e., young, there should have existed a protective cover of vegetation before. Remains of plants,

Fig. 2.196 Taken from 33° 15′ 0.18″N 88° 44′ 43.61″ E, from 5335 m asl, across a Late Glacial ground moraine (Stage IV; Fig. 2.5) (■ in the fore- and middle-ground), looking toward the ENE. It shows a decameter-thick, erratic ground moraine on a rock pedestal consisting of reddish sandstone. Distributed across the moraine surface lie meter-sized granite boulders (above ■ in the fore- and middle-ground, respectively; the yaks allow for a size comparison). (■ in the background) marks the uninterrupted, Late Glacial ground moraine covers from Stage III at between ca. 4940 and 5100 m asl. (☗) mark low mountain range crests and slopes that were polished by glacigenic basal abrasion ("Grundschliff"); they reach an altitude of up to 5400 m asl. (- - - -) demarcates the High Glacial (Stage 0, Würm, LGP, MIS 3-2; Fig. 2.5) inland ice surface at more than 6500 m asl. Analog photo 16/10/2014 by M. Kuhle

however, are entirely absent. Another alternative that must be considered and disproven is solifluction. This process would have to be stronger than in prehistoric times. The reason: today, the depth of the active layer is larger than it was prehistorically, i.e., during the ice age (Kuhle 2014a). Because of this method of elimination, a glacial inland ice cover, along with its recession during the Holocene, must be taken as the necessary prerequisite and starting point for the juvenile gully formation.

Because of the aridity and the formation of gullies in unconsolidated material, there exists a latent danger of confusing this kind of landscape with that of arid mountain areas, for example, in parts of the Sahara. However, this danger only exists if one does not differentiate the ages of gullies into juvenile and mature ages. Persistent deserts, semi-deserts, and steppes exhibit mature gullies. The difference becomes very evident at the northern margin of Central Tibet down toward the Tsaidam (Qaidam) Basin, where a mature gully landscape exists among catchment areas with peak altitudes of below 3700 m asl. Its type can be clearly diagnosed based on the strongly snow-covered ridges that separate the gullies. This process of gully formation is mature, i.e., old or very old. It is not possible to further distinguish geomorphologically between old and very old because the highest level of denudation of the ridges separating the gullies has been reached. The difference between juvenile and mature, on the other hand, is typologically evident.

In the case of mature gullies, we are thus dealing with a fundamentally different type of gullies than that of postglacial, juvenile gullies. The former have reached their climax state; the

Fig. 2.197 Taken at 33° 14′ 52.45″N 88° 41′ 13.60″E, at 5325 m asl, looking toward the west and downvalley from a transfluence pass covered with ground moraine. On this very pass lies a basalt boulder (hammer is shown for scale). It is one of many boulders consisting of different types of stone that are embedded in a pelitic ground moraine matrix. The boulder is subangular and its surface is roughened. In the background, the yellowish ground moraine cover (■) expands further across softly shaped hills of reddish sandstone. On the horizon the mountain terrain reaches an altitude of ca. 5580 m asl. (- - - -) indicates the surface of the High Glacial inland ice sheet at ca. 6500 m asl. (see also Fig. 2.144). Analog photo 16/10/2014 by M. Kuhle

latter will not yet interconnect with one another for a long time. Their development can only have taken place relatively recently, namely for no longer than since the onset of deglaciation.

Further till covers, which can be directly identified based on their gully texture, are located along the discussed N-S profile section at 31° 59′ 23.15″N 89° 14′ 47.16″E at ca. 4620 m asl. They are accompanied here by dead-ice depressions: at 31° 56′ 59.34″N 89° 16′ 22.81″E at ca. 4750 m asl; at 31° 56′ 43.98″N 89° 17′ 36.56″E at ca. 4780 m asl, as well as at 31° 55′ 45.50″N 89° 17′ 35.69″E at ca. 4760 m asl.

The highest prehistoric lake level of the Siling Co runs at ca. 4600 m asl, i.e., ca. 56 m above the current lake level. On the studied eastern shore, the prehistoric lake levels and their shore terraces are formed by—presumably decameter-thick—till. The low mountain range located SE of the lake, which reaches up to ca. 4970 m asl, is covered by remains of till. Farther SE, ca. 78 km distant from the Siling Co, there is a particularly evenly shaped roche moutonnée. It is 200 m high and lies at 31° 15′ 12.40″N 89° 53′ 18.76″E. Its summit point lies at ca. 5530 m asl.

Fig. 2.198 View from 32° 51′ 8.71″N 88° 52′ 26.27″E, from 4880 m asl, looking toward the WSW across an 8 km wide ground moraine plain (■). The ground moraine is largely clayey, gravelly, and stony. It mainly contains up to fist-sized components that represent many different petrographies (types of stone). The moraine contains only relatively few larger-sized boulders (e.g., above ■). The horizon shows low mountain ranges with a maximum altitude of ca. 5370 m asl. They consist of roches moutonnées (▲ left) as well as higher but equally rounded mountain ridges (▲ middle and right). Their ground moraine covers are transformed by the temporary snow cover; among other things, this has led to the formation of slope gullies by meltwaters. (- - - -) indicates the surface of the Late Glacial inland ice sheet at ca. 6000 m asl. Analog photo 17/10/2014 by M. Kuhle

The mountain area of Baingoin lies 10–25 km farther south. It consists of granite and shows glacigenic forms that prove the continuation of the above-reconstructed inland ice cover for this area as well. Its culmination at 5400 m asl (31° 24′ 7.10″N 89° 54′ 46.46″E) is an abraded, glacial, streamlined hill with a relative altitude of 250–550 m. 7–9 km farther north is another glacigenic landscape that takes the form of depressed basins with small lakes. The basins are coated with till and alternate with hill slopes and crests. Their glacial abrasions have here in some places of the granite rock been roughened. In some places, small rock castles have resulted from this. Such granite tors are a generally well known feature of all areas around the world once covered by prehistoric inland ice sheets. They have been described for Finland and England. They are there referred to as tors and have further developed since the deglaciation period. The causes for their generation are seen in the relief joints within the massive-crystalline rock. Oriented along these rock structures, the postglacial weathering and roughening took place in this ca. 4800 m high plateau region. The existence of tors was also observed in NE Tibet. At ca. 4900–5000 m asl at 38° 4′ 14.04″N 96° 27′ 51.03″E, tors have developed within the late glacial to

Fig. 2.199 View from 32° 33′ 0.18″N 88° 56′ 2.56″E, from 4840 m asl, looking toward the SE across a pebbly-sandy-clayey ground moraine cover (■). Particularly prominent are the structures resulting from glacigenic scouring of the ground moraine: its striations, i.e., the alternation of small moraine ridges and moraine furrows to the left of (■ on the far left) (cf. Figs. 2.141, 2.166, and 2.190). Up to 450 m high low mountain ranges surmount the plateau (foreground); they are glacigenically abraded up to their culmination at ca. 5290 m asl and thus have perfectly rounded shapes (☗). (- - - -) indicates the upper level of the High Glacial inland ice surface. Analog photo 17/10/2014 by M. Kuhle

Fig. 2.200 View from 31° 28′ 25.79″N 89° 59′ 22.76″ E, from 4880 m asl, looking toward the SSW onto a glacially streamlined hill (☗). It consists of granite and is located well over 7 km NNW from the central settlement of Baingoin (also called Banke or Bankog). (■) marks the ground moraine cover of the hill. (♠) show up to several-meter-thick moraine boulders at and on the ground moraine surface. (Δ) mark the striation of the ground moraine surface through glacigenic scouring (cf. Figs. 2.190, 2.199, 2.202, and 2.205). Analog photo 17/10/2014 by M. Kuhle

Holocene fluctuation range of the margin of the Dunde glacier. Their postglacial age has thereby also been proven for Tibet.

The superordinate, terrain-overriding late glacial inland ice glaciation is here in the mountain area of Baingoin also, again, evidenced by the widespread, commonly occurring glacially scoured till sheets. Figure 2.200 shows this scouring on a glacially streamlined hill that reaches an altitude of 4930 m asl. Granite bedrock underlies the till.

The investigation of the surrounding hill and low-mountain landscape of the valley chamber of Baingoin, between 4720 and 4900 m asl, yielded further relevant pieces of evidence. In some places there are well-preserved abrasion forms in the granite as well as up to meter-thick till covers on the slopes and valley bottom. These features prove the existence of a glacier ice sheet that completely filled and covered the relief.

2.9.2 The Ice Age Glaciation Between the Mountain Area West of Baingoin (31° 24′ 7.10″N 89° 54′ 46.46″E) and Pung Co as Well as Nam Co

The primary context of this section is Fig. 2.1: No. 37. East of Baingoin, for a horizontal distance of ca. 33 km, only the petrography of the rock changes, and the formation of the landscape remains glacigenic. This can be proven extensively by ground-polish forms and till sheets in many places. The mountain country here reaches a maximum altitude of 5390 m asl; 15 km south of Baingoin, it reaches up to ca. 5620 m asl. The valley and basin bottoms lie at ca. 4800–4600 m asl. On the one hand, the fluvial overprinting following the deglaciation consists only in the gullification of till covers; on the other, there is evidence of the dislocation of till and the resedimentation of the material as flat fan formations, as well as minor incisions at depths of no more than 1 or 2 m. This degree of incision within the till covers and fan formations has only been generated by the constant force of the main valley streams. The flat fan formations are only found at the mouths of mostly steeper side valleys.

In some places there are big, subangular or angular boulders lying on the till cover. Figure 2.201 shows meter-sized basalt boulders accompanied by angular, smaller components. The boulders are lying with spaces between them and are embedded in a fine-grained matrix. In many places, just like here, the valley slopes consisting of meter-thick till ascend and reach up to rounded mountain slopes of ca. 100–350 m. On their summits the till is interrupted by numerous islands of exposed bedrock.

Even though it is beyond the scope of this book to remark exhaustively on all glacigenic indicators in this region, we will provide a more detailed description of a small group of roches moutonnées located 14–15 km farther SE. As is shown by Fig. 2.202, they bear the features of classical whaleback forms. Their parallel, longitudinal orientations and matching differences of flat thrust slopes and steeper lee slopes indicates an outflow direction of the inland ice from roughly west to east. The low mountain range lying directly to the SW, which reaches up to ca. 5200 m asl, was also transformed in this way. It exhibits till sheets and numerous tongue-shaped basins and dead-ice lakes that are of post-Late Glacial age (post Stage IV; see Fig. 2.5).

These remnants of till cover were found on site, for example, at 31° 9′ 25.31″N 90° 29′ 31.44″E between 4680 and 4770 m asl. They were located here on several gullies with sharp-edged slopes. These gullies incise the till at a depth of several decimeters. The main gully reaches a depth of 1–2 m and thereby proves this also to be the minimum thickness of the till layer.

Till remains are also found farther SW, for example ca. 750 m south of the Bam Co (lake, Mo Hu) at 31° 8′ 0.68″N 90° 33′ 23.98″E between 4600 and 4700 m asl. They reach a thickness of several meters. They spaciously cover a glacigenically abraded hill consisting of sedimentary bedrock. In some places, outcrops of the sedimentary rock are, like small islands, exposed between those till remains. More recently, a few slope gullies have been cut into the till, regressing from the foot of the slope. They thereby expose the till cover in a linear structure. This fluvial transformation (i.e., incision) has only been possible since the deglaciation.

A roche moutonnée, which reaches an altitude of up to 4710 m asl, can be found at 31° 7′ 7.69″ N 90° 34′ 49.13″E, ca. 2.2–3.4 km south of the Bam Co. Seen from the west, it is 60 m high, and is covered with till. This is shown by the characteristic striation pattern, with exaration rills caused by abrasions on subglacial permafrost. They run along the longitudinal axis of the roche moutonnée, which runs from SSW to NNE. If judged by the shape of the roche moutonnée, the Late Glacial direction of the ice flow was roughly from south to north. However, because the ground surface is composed of sedimentary rock and forms a hogback in this location (ridge formed by more erosion-resistant stone material),

Fig. 2.201 View from 31° 17′ 37.30″N 90° 20′ 26.08″ E, from 4650 m asl, looking toward the SE across a valley bottom covered by ground moraine (■ in the foreground). Conspicuous are the more than meter-sized basalt boulders lying embedded in the surface, which prove a local ground moraine. However, lying between these boulders are also smaller, polymictic, more rounded components; they are embedded in the fine-grained material matrix of this ground moraine and were transported over long distances. (□) marks the valley bottom, which has been glaciofluvially transformed since the deglaciation and into which the river has eroded a terrace up to 3 m deep (above □). (■ in the background) mark several-meter-thick ground moraine remnants, which have been dissected by slope gullies since the deglaciation in the course of the Holocene. An up to 5010 m high low mountain range landscape shows abrasions both at its valley flanks (◓ left) as well as higher up on its mountain ridges (◓ right). On the adjacent high plains one finds ground moraines that are coated with ice wedge nets or -polygons (below - - - - on the right). (- - - -) indicates the High Glacial inland ice sheet surface. Analog photo 18/10/2014 by M. Kuhle

the shape of the roche moutonnée very likely reflects the interference between the ice flow and the underlying morphology of the bedrock. One must, moreover, assume mutual impacts between these two forces.

The basic profile of the 26 km long Bam Co displays the characteristics of a glacigenic depression with erosion in a N-S direction. It could have contained a Late Glacial glacier tongue, running from N to S or vice versa, as well as a correspondingly big body of dead ice.

Massive till covers lie on an extensive, relatively flat slope (10°–20°) at 31° 8′ 47.51″N 90° 40′ 54.57″E at an altitude of between 4680 and 4950 m asl. The location lies 7 km east of the Bam Co beyond a 5200 m high mountain chain. These till sheets begin to thin out at ca. 4950 m asl, leaning against slopes that reach up to 5200 m asl. At 5200 m lies the culmination of the respective hill crest, which shows a glacially abraded form. The till was sampled on the valley bottom at ca. 31° 9′ 54.19″N 90° 42′ 11.57″E at

Fig. 2.202 A classically shaped roche moutonnée (lowermost ▲ in the mid-background) 7.5 km west of the Bam Co or Mo Hu (Tso, lake, whose shore is located at 31° 11′ 25.77″N 90° 30′ 52.34″E at 4565 m asl), at 31° 11′ 58.93″N 90° 26′ 20.73″E. It was photographed looking toward the SW. It rises up to ca. 4624 m asl and protrudes ca. 4 m above the ground moraine (■ in foreground). This roche moutonnée is of the "whaleback" type, and belongs to a whole group of five similar roches moutonnées. Behind the aforementioned feature are two considerably larger ones: the larger roche moutonnée lies at 31° 11′ 8.66″N 90° 25′ 50.42″E and reaches an elevation of ca. 4642 m asl. It is ca. 22 m high (▲ left) and is covered by ground moraine (■ background); the other, adjacent roche moutonnée (▲ right) lies at 31° 11′ 17.85″N 90° 25′ 18.16″E and protrudes 10 m, between 4620 and 4630 m asl. All five roches moutonnées are oriented in a W-E direction. The one in the foreground and the one mid-way to the right in the background have flat collision slopes in the west and steep lee slopes in the east, thereby suggesting that the direction of flow of the inland ice was from roughly W to E. (Δ) marks ground moraine striations on the slope due to glacigenic scouring, observed in many places (see Fig. 2.200). (- - - -) indicates the surface level of the High Glacial inland ice sheet. Analog photo 18/10/2014 by M. Kuhle

4700 m asl, where the material has accumulated in debris flow fans previously washed out and re-deposited from higher regions upslope.

Some 4.5 km farther ENE at 31° 9′ 41.09″N 90° 43′ 30.72″E between ca. 4690 and 4820 m asl one finds erratic till material (Fig. 2.203). Because of the gaps ("windows") in the till, the glacigenically polished bedrocks below are visible. Along the valley profile, 0.8–1.6 km upvalley, decameter-thick till deposits are documented for both valley slopes. On the orographic left side, the formation of a 1.6 km long landslide is indicated. On the opposite, orographic right side, fresh, isolated slope gullies have been formed in the loose rock of the till cover; their youth indicates a fairly recent, geomorphological change of regime only since the deglaciation.

In the following, we focus on inland ice indicators a few decakilometers farther north (Sect. 2.4.5.1). They date back to the Late Quaternary to Holocene prehistory and provide further proof of the large scale of the Tibetan inland glaciation. Thirty-three kilometers to the NNE, 6 km west of the Pung Co (Tso, lake), light-colored till mantles dark-colored hilltops of a low mountain range located at 31° 26′ 6.84″N 90° 51′ 2.53″E at 4780 m asl. Here the till cover also only exhibits a very marginal, fresh progress of gully formation (see above).

A further till cover reaching the culmination of a hilltop at 4800 m asl was found at 31° 25′ 25.36″N 90° 52′ 49.47″E. This is a location halfway between the one just mentioned and the Pung Co.

Fig. 2.203 This location lies 11.7 km east of the Bam Co or Mo Hu (Tso, lake). It is viewed here from 31° 9′ 22.05″N 90° 43′ 42.11″E at 4700 m asl, looking toward the NNW. A glacigenic, striated rocky outcrop abrasion ("Schichtkopfstreifenschliff") is found on the dark bedrock (▲). On top of it lies light-colored and thus clearly erratic ground moraine (■). Its cover is interrupted at ca. 31° 9′ 41.09″N 90° 43′ 30.72″E between ca. 4690 and 4820 m asl. (▲) mark debris-flow (mudflow) fans of dislocated ground moraine formed since the deglaciation. (□) marks a fluvial gravel floor with currently ongoing moor formation. (- - - -) indicates the High Glacial inland ice surface. Analog photo 18/10/2014 by M. Kuhle

Seven kilometers farther north, at 31° 29′ 11.57″N 90° 52′ 38.34″E, till covers a mountain slope of dark rock between ca. 4790 and 4680 m asl across several kilometers, i.e., again very extensively.

This excursion to the north shall be concluded in the up to 5270 m high low mountain range relief west of southern end of the Pungo Co. Figure 2.204 documents till remains in the small valleys leading down from a 200 m high group of hills at 31° 23′ 21.33″N 91° 0′ 56.68″E. They display fine-featured, sharp-edged, and consequently fresh erosion incisions (gullies). Due to fluvial (retrogressive) erosion, this light-colored, thin-layered loose rock material is in some places displaced all the way down to the dark bedrock underneath. The hill slopes that flank these valleys show striations of the layer edges ("Schliffkopfstreifenschliff") in a NE-SW direction.

Only a little south but 60 km farther west—among small lakes and tarns typical of tills—several-meter-sized granite boulders were recorded at 31° 20′ 10.69″N 90° 18′ 18.18″E at ca. 4620 m asl. The boulders are distributed across an even terrain, with several 100 m to 1.5 km distant from glacigenically abraded and rounded granite hills. They were observed in WNW-ESE direction over a distance of ca. 1.6 km along the

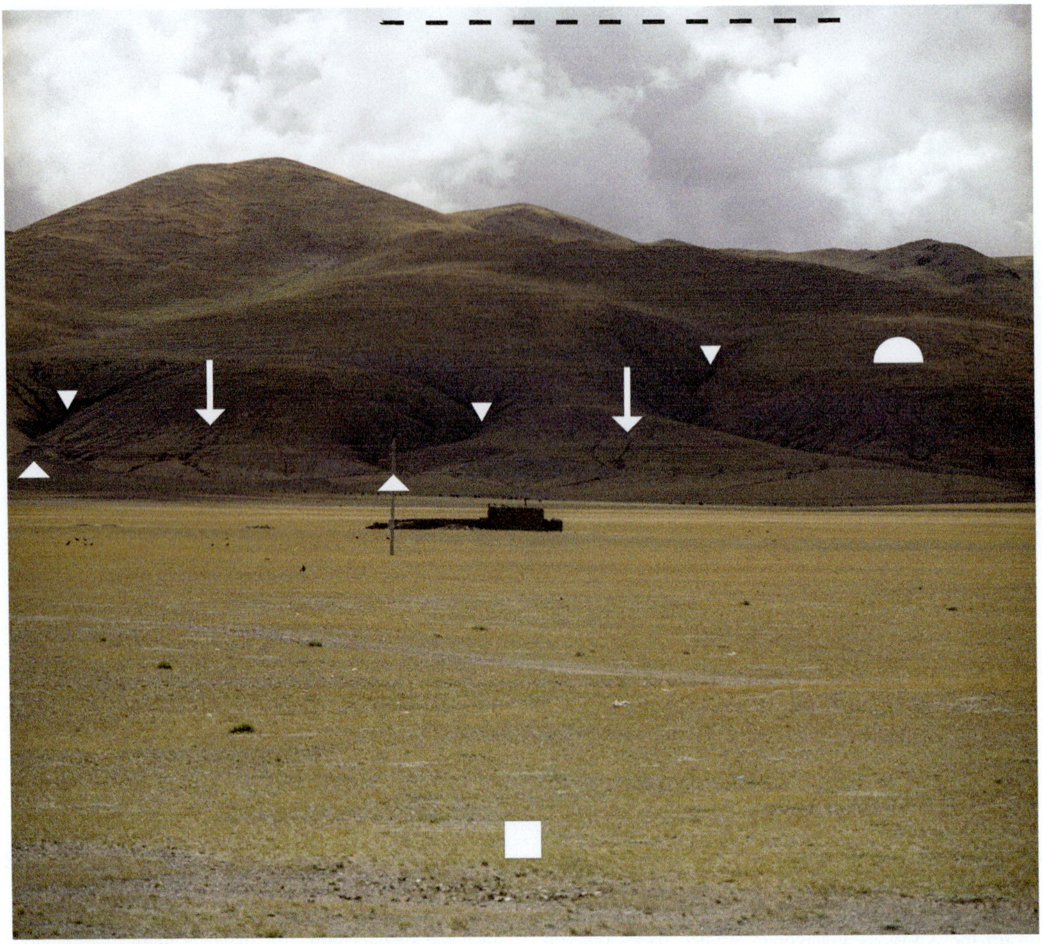

Fig. 2.204 View from 31° 24′ 45.83″N 90° 59′ 24.65″ E, at 4576 m asl, looking toward the SE onto a mountain ridge 6.6 km west of the southern end of the Pung Co. The mountain ridge lies at 31° 23′ 21.33" N 91° 0′ 56.68″E and reaches an altitude of 4860 m asl. (■) marks a nearly horizontal ground moraine plain. Its material consists of polymictic, angular, subangular, as well as rounded gravel components. Its biggest fraction reaches head size. (◓) marks glacigenic flank-polishing with a ground moraine cover. The latter exhibits exaration rills or -furrows (above ↓ left), which resulted from glacigenic scouring (cf. Figs. 2.200, 2.202, and 2.205). The light-colored, thin ground moraine cover on the slope is disrupted by fine, fresh (i.e., young) slope gullies (↓), in some places incised down to the underlying dark bedrock. (▼) mark small valleys from whose depth contours the ground moraine has been removed. Their material has been redeposited at the mouth of the valley in alluvial fans (▲). (- - - -) indicates the High Glacial inland ice surface. Analog photo 24/09/2014 by M. Kuhle

surface of the till. It is most likely that they are local moraine boulders that have been dislocated and mainly horizontally transported from the surrounding granite region. In order to explain their transport through local glaciation from the west, a ΔELA of ca. 360–400 m would be required. The present author does indeed assume that a glacier tongue advanced from the west; however, he further assumes that this expansion originated from the remains of an ice dome left over from the Late Glacial (Stage III–IV, Fig. 2.5) inland ice as part of its melting recession. This melting process was climatically postponed and presumably lasted into the

Holocene. At the time of this advance of the local outlet glacier tongue, the ELA was already running higher than the necessary glacier accumulation zone. Such glacier-tongue advances resulted from the still-existing, remaining glacier-ice domes with relatively large ice thicknesses. They reached down into the surrounding terrain which by then already was glacier-free. Such Postglacial glacier-tongue advances—without a climatic link to a simultaneous ELA depression—have already been described for the de-iced landscapes in southern Tibet (see Sect. 2.4.7.4.6).

About 10 km SE at 31° 15′ 32.50″N 90° 21′ 16.15″E at ca. 4660 m asl, another local till was found, here also with up to several-meter-sized granite boulders. On its surface are also small lakes and tarns as well as swampy areas. Here also we encounter moraine boulders that were transported from neighboring hill groups during the melting of the late Late Glacial inland ice sheet (Stage III–IV, Fig. 2.5). The till cover discussed here forms the uppermost, i.e., youngest layer on top of other, older layers of till dating back to the inland ice Stages 0–II (Fig. 2.5).

3.5–4 km ESE, at 31° 14′ 30.95″N 90° 23′ 17.42″E, between ca. 4630 and 4700 m asl, till extends across the ca. 5 km wide, basin-shaped bottom of a main-valley chamber. It contains polymictic boulders, among them basalt boulders with lengths of several decimeters.

The ca. 4870 m high saddle at 31° 5′ 16.43″N 90° 37′ 26.37″E is a pass once completely overridden by ice. This is indicated by till deposits (“cushions”) along its flanks. Among other things, these can be identified based on the formation of fresh gully slopes. These tills reach an altitude of 4900–4940 m asl.

11–12 km ENE, at ca. 31° 9′ 53.94″N 90° 42′ 11.56″E, a flat fan of loose rock extends between ca. 4700 and 4820 m asl from the bottom of the main valley up and into a trough-shaped valley located farther south. The polymictic coarse-rubble fraction of this loose rock contains up to decimeter-sized, angular boulders. The picture emerges of a local till that was dislocated by melting water. Such a secondary dislocation process happened here during the late Late Glacial deglaciation. A corresponding loose rock fan is found 5–6 km ESE at 31° 8′ 22.92″N 90° 45′ 21.86″E at ca. 4710 m asl. It is partially covered by large, light-colored boulders in different densities. Similar to some of the fans discussed previously, this fan deposit emerges from a flat trough-shaped valley in a low mountain range landscape of ca. 540 m relief. This landscape is composed of completely rounded hill and mountain tops.

At 31° 3′ 20.27″N 90° 50′ 27.89″E, a glacial knob rising to 4810 m asl culminates above the flatly layered till sheet. North from it is a basin-like depression that lies more than 10 m below the base level of the main-valley bottom. This basin contains a silted small lake.

Compatible with such an abraded basin landscape are the (in many places) adjacent rock thresholds. Such an abrasion threshold is found, for example, at 31° 1′ 26.89″N 90° 53′ 7.84″E, in 4830–4900 m asl. This landform shows clear glacigenic abrasions. Further abrasion thresholds as well as glacial knobs culminate nearby: for example, at 31° 1′ 52.57″N 90° 52′ 29.82″E and a little further E at 31° 0′ 8.74″N 90° 54′ 57.78″E, at 4794 and 4914 m asl, respectively. These preserved landforms have developed in the granite bedrock and have been extensively covered with till.

At ca. 30° 59′ 15.47″N 90° 57′ 19.03″E between ca. 4740 and 4800 m asl runs a meter-thick till cover with up to meter-sized boulders transported for at least a kilometer (Fig. 2.205). The large boulders are too far distant from the slopes of the glacially streamlined hills and roches moutonnées to be explained in terms of solifluctional transport. There are no indications of debris flows either. On the one hand, there is not a sufficient catchment area that could facilitate glacial lake outburst floods (GLOFs). On the other, there are also no traces of any debris-flow drainages or relevant accumulation forms (see Fig. 2.205). However, only these features would be meaningful indicators for the

Fig. 2.205 View from 30° 59′ 15.47″N 90° 57′ 19.03″ E, from ca. 4760 m asl, looking toward the SW across a ground moraine cover (■ in the foreground). The moraine has a thickness of at least several meters. This minimal thickness must be extrapolated because the moraine fills the km-wide terrain or rocky ground of the valley chamber with an evenly leveled surface. It contains isolated, independently lying, up to several-meter-large, angular and subangular boulders (◉). These are polymictic boulders some of which are erratic. The angular boulders and those in the decimeter size range with roughened surfaces consist of quartzite (◉). The enclosing hills are glacially streamlined knobs (▲). They are coated with ground moraine (■ in the background). (↑) indicate nearly slope-parallel furrows in the surface of the ground moraine, which were formed by glacigenic scouring (cf. Fig. 2.141) and then in some places further transformed since the deglaciation through the influence of solifluction. (- - - -) indicates the High Glacial inland ice surface. Analog photo 21/09/2014 by M. Kuhle

action of debris flows. Consequently, there is no physical alternative to the glacigenic transport of these large boulders (see Sect. 2.4.5.1).

At 30° 57′ 38.71″N 90° 58′ 9.37″E is a roche moutonnée—like hill that is covered by erratic till material up to its summit at ca. 4870 m asl. The light-colored loose rock is in some places 3–4 m thick and lies on dark sedimentary bedrock. A dug-up exposure at 4803–4807 m asl proves this fact.

Further till covers, nearly uninterrupted by relief, are found at 30° 57′ 24.97″N 90° 58′ 32.21″E between ca. 4750 and 4810 m asl. They extend gently from the valley bottom up to the foot of the slopes of the surrounding hills. On the till surface one observes polymictic, fist- to several-decimeter-sized boulders. They lie isolated from one another and are embedded in a pelitic matrix. The enclosing and glacigenically rounded hills reach an altitude of 5100 m asl. They are likewise covered with till. This moraine is here in many places coarser, containing larger boulders.

At 30° 56′ 58.46″N 90° 59′ 16.92″E we find clear evidence of glacial polishing. These glacigenic bedrock polishes (presumably on granite) are exposed by a gap (window) in the till cover between ca. 4797 and 4830 m asl over a horizontal distance of 260 m. A little farther north, the relevant glacigenic ground polishings are also documented free of moraine cover (30° 57′ 16.48″N 90° 59′ 7.36″E, 4820 m asl).

Fig. 2.206 View from 30° 54′ 15.41″N 91° 1′ 23.72″E, from ca. 4745 m asl, toward the south across the Holocene surf zone (nearshore) of the Nam Co (Tso, lake, also Tengri Nor). The up to 1.16 km wide, eastern shoreline of the lake has been formed from thick ground moraine cover (■). The ground moraine contains polymictic and partly erratic boulders (◉). They reach sizes of more than a meter (see person for scale) and are rounded, although most are subangular and faceted. Among them are granite boulders (they were sampled). In the background is a Holocene cliff (↙) in the quarzite bedrock (also sampled). After deglaciation the lake shore of the Nam Co was formed through limnic undercutting, which in turn was reinforced through the impact of shore ice, i.e., the formation of a frost cliff during winter. In the background, behind the Nam Co, the various eastern foothills of the Nyainquentanglha Shan with their glacigenic, triangular slopes (▲ black) are recognizable. (▲ white) mark glacigenically abraded and rounded hill- and mountain slopes. (- - - -) indicates the surface of the High Glacial inland ice sheet, which entirely covered the landscape of the plateau. Analog photo 21/09/2014 by M. Kuhle

A little farther south (30° 56′ 48.07″N 90° 59′ 22.45″E), a perfect roche moutonnée forms the continuation of the longitudinal profile. It rises up to 4840 m asl and has a N-S longitudinal axis.

From the valley bottom in this location continues—in a SSE direction—a head of a valley or a mountain threshold which shows a gradient from 4750 to ca. 4890 m asl. At the top (30° 56′ 21.95″N 90° 59′ 57.85″E) we reach a pass once overtopped by Late Glacial ice. This is proven by its till covering. However, not only this indistinct pass, which is only 150 m below its surrounding hills, but also these very hills are covered with till (Fig. 2.87). The same can be said for the wider surroundings to the west and east, respectively (see Sect. 2.4.5.1).

The corresponding southern slope of the discussed pass descends down to the Nam Co's lake level of 4729 m asl. In a SE direction this slope runs onto the foreshore of the NE lakeshore of the Nam Co. Along this slope, we find stones and boulders on the till surface with sizes in the decimeter and meter range. The largest boulder charted here consists of limestone, and is compact, rounded, and 1.2 m long. It was found lying in a mild depression at 30° 56′ 2.62″N 91° 0′ 21.78″E at 4790 m asl on a till cover that is here several meters thick. The surface of the boulder is roughened as a result of frost weathering and corrosion.

The different thicknesses of the till covers reaching up to the rounded hill tops can be estimated by examining the cliffs of the Nam Co shore. Here, we find many places in which till covers between 1 and 6 m deep have been exposed and cross-sectioned.

At the foot of the southern slope of the glacially overtopped pass and the discussed surrounding hill landscape lies the foreshore of the Nam Co. It continues for 18 km along the eastern shore of the lake. Of this total distance, 12 km were investigated in more detail. Along the 0.38–1.16 km wide shoreline, till covers were found. They consist of polymictic, at times erratic boulders with sizes in the meter range (Fig. 2.206). They lie isolated from one another, embedded in a fine-grained matrix. Among the erratic material are faceted granite boulders. The bedrock underneath is quartzite. Such occurrences of granite erratics can be found, for example, at 30° 54′ 15.41″N 91° 1′ 23.72″E at 4745 m asl and at 30° 52′ 24.57″N 91° 2′ 52.99″E at 4742 m asl, including the entire 4.26 km long profile section in between. In contrast, there are no traces of debris flows anywhere to be found (Fig. 2.206). The kilometer-wide distance from the mountain slopes and the extremely mild gradient of those loose rock surfaces (see Fig. 2.206) exclude all possibilities but the glacigenic transport of these boulders. The prehistoric water lines (drainage areas) and lakeshore positions are the only local forces that have gradually and mildly transformed the till sheet through eluviation and erosion of pelites lying close to the surface (Fig. 2.206).

In the area to the north and east of the Nam Co, we find indications of ground polishings dating back to the former inland ice. In some places, the bedrock in this area consists of granite as well as sedimentary rock such as limestone and quartzite. Classical landforms indicative of glacigenic ground polish ("Grundschliff") are preserved here. We thus find homogenously formed ground and hillslope abrasions, glacially streamlined hills, roches moutonnées, and glacial knobs imposed on granite at (e.g.) 30° 57′ 16.48″N 90° 59′ 7.36″E and 30° 56′ 48.07″N 90° 59′ 22.45″E, on sedimentary rock such as limestone at (e.g.) 30° 56′ 1.82″N 90° 59′ 51.76″E north of the lake, but also on the extremely hard quarzite at 30° 53′ 25.14″N 91° 2′ 23.36″E east from the Nam Co. Another representative example is shown in Fig. 2.207. This ca. 3.4 km long hill east of the Nam Co is a particularly evenly shaped example of numerous neighboring hills. Its culmination reaches the same altitude level as the likewise adjacent (cf. Figs. 2.85, 2.86, and 2.87) glacigenic abrasion that runs in an ENE-WSW direction on the northern slope of the Nyainqentanglha (Kuhle 1991). We are here presumably dealing with the same, once continuous, tertiary peneplain level. What remains currently unclear, however, is which of the inland ice stages between Stage 0–IV (Fig. 2.5) is indicated by the S-N longitudinal orientation of the glacigenically transformed hill discussed here—the same S-N orientation as that of the glacial ice stream more generally.

2.9.3 Summary of Sect. 2.9

A local mountain glaciation has been reconstructed for the 6120 m high massif (33° 20′ 22.18″N 88° 45′ 10.64″E) up to the Holocene Stage V (Neoglacial in Fig. 2.5). It reached a maximum ΔELA of around 250 m. Previously an inland glaciation existed that mostly covered the terrain during the late Late Glacial (Stage IV) and in any case completely during the Late Glacial Stages III to I as well as the High Glacial Stage 0. This means that local terminal moraines could no longer be formed, or rather, were no longer formed, so that today only till sheets cover the plateau region. In the area under discussion, this glaciation without terminal moraines still persisted at an ΔELA of ca. 700 m or even lower (Fig. 2.5).

Along its foothills, the ca. 6290 m high Zaqungngomar Massif (33° 3′ 27.21″N 88° 32′ 30.05″E) is likewise surrounded by at least meter-thick till covers that contain erratic material. Only after the inland ice had receded during the Holocene, did a new small glacial advance, now manifested on the basis of moraine material, occur at a ΔELA of ca. 240 m.

Fig. 2.207 This hill (▲ in the middle below - - - -) ascends to ca. 5050 m asl, culminating at 30° 49′ 27.15″ N 91° 9′ 25.08″E. It is a glacially streamlined hill that rises ca. 250–270 m above the ground moraine plain (■ white). It is located ca. 10 km east of the Nam Co and was photographed looking toward the east. The almost nearly rounded hill is—very much like its surroundings—covered by ground moraine (■ black). (Δ) marks fine-textured ground moraine furrows and ground moraine rib-like lineaments in longitudinal orientation with the glacially streamlined hills. The hill's longitudinal axis runs from S to N. Its flat collision slope is on the south and its steeper lee slope on the north. This suggests a local outflow direction of the inland ice from south to north. (three ▲ on the very left and one ▲ on the very right) mark further glacial knobs. (- - - -) indicates the terrain-covering, High Glacial inland ice surface. Analog photo 18/10/2014 by M. Kuhle

The inland ice indices of Stages IV to 0, which alternate between glacigenic erosion formations on the hills and glacigenic accumulations in the depressions—where the latter also exhibit characteristics of glacigenic scouring—can be found in uninterrupted sequence southward to the Siling Co and across the mountain region west of Baingoin (31° 24′ 7.10″N 89° 54′ 46.46″E) as well as east up to the Pum Co and Nam Co.

Further neoglacial (i.e., Holocene to Historical) glacial stages are for natural reasons not found in this currently unglaciated region.

2.10 Summary of Sects. 2.8 and 2.9: The Outcome of the Field Evidence Regarding the High to Late Glacial Glaciation Between the Northern Slope of the Culmination of Central Tibet at the Dogai Coring and the Southern Slope of Its Culmination at Pung Co and Nam Co

The primary context of this section is provided by Fig. 2.1: No. 37. Throughout Sects. 2.8 and 2.9 (Fig. 2.1: No 37) the empirical findings of the field investigations have been presented along a total profile length of ca. 610 km: i.e., along a N-S profile of ca. 330 km and a NNW-SSE profile to the south of ca. 225 km, in addition to an excursion ca. 55 km to the north. The profile, which was inspected repeatedly and at least twice in many places, reached a width of 50–60 km. The lowest-lying terrain surfaces are located at 5200 m asl at the culmination of Central Tibet, at 4830 m asl in the north, and at 4530–4730 m asl in the south. The highest adjoining mountain ranges, whose slopes adjoin without an anti-gradient, reach altitudes of 6120 m, 6290 m (Zaqungngomar), 6500 m (Purog Kangri), and 7120 m asl (Nyainquentanglha). Consequently, there exists a mean surface altitude of ca. 5000 m asl.

Glacigenic forms have been found along all sections of the profile. This means that glacigenic abrasion forms, sediments, and accumulations have been shown to exist in all locations. This evidence is based on the following indicators:

1. Erosion forms in the bedrock: rounded mountain tops, rounded hills with polished rock surfaces, glacially streamlined hills with exaration rills and glacigenic abrasions of the layer edges ("Schichtkopfstreifenschliffe"), roches moutonnées (whaleback forms), and alternations between striation basins and striation thresholds.
2. Accumulations of loose rock: big boulders in locations and positions within the relief that exclude their transport via any other method than by glacigenic transport, i.e., by means of glacier ice; erratic boulders lying either directly on bedrock of another type, or more indirectly by being embedded in matrices of fine-grained material; they were transported long distances across terrain without steep, even slopes and occasionally with anti-gradients (upslope); accumulations of polymictic boulder and also smaller loose rock components with both mixed petrography and differently rounded shapes ranging from subangular to faceted to rounded; erratic tills and non-erratic tills, which can be divided into local moraines and those transported over great distances.
3. Combinations of accumulation and erosion forms: striated till covers that are transformed by lineaments such as exaration rills and glacigenic scouring are found both in isolated positions as well as in transitional areas between bedrock outcrops (e.g., bassets transformed by ground polish). They are indicative of subglacial permafrost that has hardened the surface of the till and thereby made it more susceptible to scouring. Therein lies a special characteristic of the Central Tibetan tills: Even today—during the interglacial period—the plateau is still subjected to permafrost, which must necessarily have been in place during the ice age (Kuhle 2014a).

These indicative forms give proof to a superordinate, relief-overriding, large-scale inland glaciation of Central Tibet. The ice thickness probably exceeded 1500 m.

The temporal chronology of the formation of those forms must have been as follows. During Stage 0 (Fig. 2.5; Würm Ice Age, LGP, LGM, MIS 3-2), a predominant ground polish happened due to the maximal ice thickness. During this period mostly glacigenic erosion forms were generated. During the Late Glacial, i.e., during Stages I–IV (Fig. 2.5), until the large-scale deglaciation period, glacigenic accumulations and accumulation forms were deposited on the glacially abraded surfaces. During the Holocene, during the Neoglacial Stages V–'VII (Fig. 2.5), the Late Glacial till covers of the today dry (glacier-free) hill and mountain slopes were denuded by small local glaciers. For this reason, today there are numerous glacially streamlined hilltops that protrude above the Late Glacial till covers.

The characteristic glacigenic basin and threshold landscape of Central Tibet is the genetic prerequisite for the large-scale, interglacial formation of lakes. This landscape is a polyglacial phenomenon. It was formed during numerous Pleistocene alternations of High Glacial ground polishings (i.e., glacigenic deep erosions) and an Early to Late Glacial filling of these depressions with till sheets (see above). This chronology is supported by lake datings that are available for Tibet (Sect. 2.5.). A tectonic genesis (tectogenesis) of the depressions and numerous lake basins is unlikely. Such micro-scale tectonics has at least never been proven, as there is no example of it anywhere else on earth. For Tibet, especially, this hypothesis must be deemed highly unlikely and can be excluded with almost absolute certainty due to the enormous thickness of the Earth's crust in this continental region. Lake formations on earth are mostly linked to prehistoric glacial landscapes or to karst. Karst formation, however, is dependent on local evaporites (soluble rock), which are scarcely present in our study area. And even in those small regions where they are present, their subrosion (dissolution) is prevented by permafrost.

The freeing of the surface of Central Tibet from the prehistoric inland ice cover must have led to a geomorphological change of regime. The large-scale, denudative abrasion through glacigenic ground polishing as well as the distribution of till sheets, which both substituted for and leveled out the relief, ceased completely with the deglaciation of the landscape. The subaerial surface of the mountain slopes was from now on exposed to linear erosion by running water. In our region, however, this process has so far only led to the sporadic incision of small, sharp-edged and isolated slope gullies. Moreover, this happened only in the soft debris covers. This provides evidence of a fairly recent deglaciation of the inland ice cover, i.e., only during the Late Glacial to Early Holocene. At the same time, it proves the prior existence of a fairly late inland glaciation of Central Tibet during the Last Glacial Cycles (LGP, MIS 3-2, Würm Ice Age, Stage 0).

The question of how big the ΔELA (snowline depression) must have been at a minimum in order to glaciate the entire region can be answered as follows. As was discussed in Sect. 2.8.3, it needed a ΔELA of less than 350 m in the most central regions of Central Tibet. If we now expand here from 180 km to a total horizontal distance of ca. 610 km, a larger ΔELA is necessary for the total glaciation of the area. This is so because the average surface altitude of this expanded region lies ca. 300–400 m lower, namely at ca. 5000 m asl. This means that for a total glaciation a ΔELA of ca. 550 m could have been sufficient (400/2 = 200; 350 + 200 = 550). From this we can infer with considerable probability the existence of a last inland ice stage during the Late Glacial Stage IV (Fig. 2.5).

In this assessment we also have to take into consideration the effect of self-reinforcement of glacier formation through an increase in the surface altitude of the ice. This can occur quite abruptly when the ELA (snowline) comes close to that base surface or even reaches it. The last time this must have been the case was during Stage IV. It is even more certain, however, that this closely matches reality with respect to the older Late Glacial Stages III, II, and I. All increases of the ΔELA that were reconstructed for the entirety of High Asia for the aforementioned three stages in comparison with the youngest Stage IV can be studied in Fig. 2.5.

2.11 Cross-Referencing Ice-Margin and Glacier Reconstructions from the Last Ice Age Until the Present with Empirical Findings from Field Research in Other Regions of High Asia not Covered in This Book

The complete reconstruction of the inland glaciation of High Asia has been recorded in local detail in 30 maps (map sheets) with a scale of 1:1,000,000 in Kuhle (2004) and (2011a).

Figure 2.1 shows 37 different study areas. They are located between the Himalayas in the south and the Sajan in the north as well as the Zagros in the west, and the Minya Konka and Xuebao in the east of High Asia. As test areas representative of the glaciation of High Asia, the field study sites Nos. 36, 17, 24, 23, 32, 31, 37, 5, and 10 (Fig. 2.1) were taken into consideration in this book, although in varying detail.

For the two regions Nos. 36 and 37, the most recent findings from the expeditions in 2013 and 2014, respectively, were presented here for the first time.

Likewise, for those other regions, field and laboratory data were included into this book's discussion that had not previously been analyzed and published in detail. The new lab data included, for example, later analyzed rock and loose rock samples. The selection of study areas has here been focused on the most central and most arid (low-precipitation) areas of the entire region. All authors who have commented on the palaeoclimatic situation of High Asia have argued, and still do, that these areas are the most unlikely areas to have been covered by glacier ice. It is for this reason that those areas are very suitable for the present investigation. If it is possible to prove traces of a glaciation, it must necessarily be sufficient proof of a minimal ice cover in High Asia. All other regions not explicitly dealt with in this book are located either farther north where it is colder, or further east and south where it is considerably more humid (higher in precipitation). It follows that the glaciation in all these other regions must have been more extensive—as it is still today—than formerly in the here chosen, most central regions.

In order to make these older empirical findings accessible, which have scarcely or not at all been dealt with in this book but which are closely linked to our current topic "Glaciation of High Asia: from the Last Ice Age to the present," the following body of literature is relevant:

The glaciation of the region	No. 1 (Fig. 2.1)	Kuhle (1974), (1976), (2007b)
	No. 2 (Fig. 2.1)	Kuhle (1980), (1982a), (1983), (2001a), (2011b), (2012b), (2013a, b)
	No. 3	Kuhle (1982b), (1982c), (1987a, c, d), (1988d, e), (2003), (2013a)
	No. 4	Kuhle (1987a), (1989), (1998a), (2001a), (2005a), (2011a), (2013a)
	No. 5	Kuhle (1986a), (1987a), (1988c), (1988d, e), (1989), (1991), (1998a), (1999), (2001c), (2002b)
	No. 6	Kuhle (1988b), (1989), (1994a, b), (1998a), (2005b), (2012b)
	No. 7	Kuhle (1988d), (1988f, g), (1996c), (1997), (1998a), (2001a), (2006a), (2014c)
	No. 8	Kuhle (1994b), Grosswald et al. (1994)
	No. 9	Kuhle (1990a), (1998a), (2013a)
	No. 10	Kuhle (1991), (1998a, d), (2013a), (2014c)
	No. 11	Kuhle (1994b), (2013a), (2014c), Grosswald et al. (1994), Kuhle et al. (1997)
	No. 12	Kuhle (2003), (2013a), Kuhle and Kuhle (2010)
	No. 13	Kuhle (2001a), (2011a), (2013a)
	No. 14	Kuhle (1998a), (2013a)
	No. 15	Kuhle (1997), (1998a, c), (2006a), (2014c)
	No. 17	Kuhle (1998a), (2001a), (2004), (2013a), (2014c)
	No. 18	Kuhle (2001a), (2006b), (2011a), (2013a), (2014c)
	No. 19	Kuhle (1987c), (1987d), (1997), (2003), (2013a)

(continued)

(continued)

	No. 20	Kuhle (1997), (1998a), (2001a), (2005a), (2013a), (2013b), (2014b, c)
	No. 21	Kuhle (2004), (2011a)
	No. 22	Kuhle (1998a), (2001), (2006a), (2011a), (2013a), (2014c)
	No. 23	Kuhle (1999), (1998b), (2004), (2005c), (2013a)
	No. 24	Kuhle (2001a), (2010), (2013a), (2014c)
	No. 25	Kuhle (2001a), (2009), (2013a)
	No. 26	Kuhle (2001a), (2009), (2013a)
	No. 27	Kuhle (2001a), (2004), (2006a), (2009), (2013a), (2014c)
	No. 28	Kuhle (2011a), (2013a), (2014c)
	No. 29	Kuhle (2001a), (2004), (2006a), (2014c)
	No. 30	Kuhle (2007a), (2009), (2011a), (2013a)
	No. 31	Kuhle (2012c), (2013b), (2014b, c)
	No. 32	Kuhle (2007a), (2011b), (2012b), (2013a, b)
	No. 33	Kuhle (2012a), (2013a)
	No. 34	Kuhle (2012a), (2013a)
	No. 35	Kuhle (2012a), (2013a)

The ensuing presentation of the glaciation chronology from the ice age to the present (Chap. 3) is based on and summarizes all the findings from High Asia. It therefore not only reiterates the findings presented in this book but also includes those presented in the references just cited.

References

Akçar N, Ivy-Ochs S, Schlüchter C (2008) Application of in-situ produced terrestrial cosmogenic nuclides to archaeology: a schematic review. Quat Sci J 57:226–238

Ambolt N, Norin E (1982) Reports from the scientific expedition to the North-Western Provinces of China under the Leadership of Dr. Sven Hedin—The Sino-Swedish Expedition. Publication 54 I. Geography 5 Sven Hedin Central Asia Atlas, Memoir on Maps, vol III, Fasc. 1, The Pamirs, K'unlun, Karakoram and Chang T'ang regions, compiled by E. Norin; The Sven Hedin Foundation; Etnografiska Museet, Stockholm, pp 1–61

Avouac JP, Dobremez JF, Bourjot L (1996) Paleoclimatic interpretation of a topographic profile across middle Holocene regressive shorelines of Longmu Co (Western Tibet). Palaeogeogr Palaeoclimatol Palaeoecol 120(1–2):93–104

Balco G (2011) Contributions and unrealized potential contributions of cosmogenic-nuclide exposure dating to glacier chronology, 1990–2010. Quat Sci Rev 30: 3–27

Bertolotti M (2013) Celestial messengers—cosmic rays: the story of a scientific adventure. Springer, Berlin

Blisniuk PM, Sharp WD (2003) Rates of Late Quaternary normal faulting in central Tibet from U-series dating of pedogenic carbonate in displaced fluvial gravel deposits. Earth Planet Sci Lett 215(1–2):169–186. https://doi.org/10.1016/s0012-821x(03)00374-1

Briner JP, Miller GH, Davis PT, Finkel RC (2006) Cosmogenic radionuclides from fjord landscapes support differential erosion by overriding ice sheets. Geol Soc Am Bull 118:406–420

Burbank J, Fort M (1985) Bedrock control on glacial limits: examples from the Ladakh and Zanskar ranges, North-Western Himalaya, India. J Glaciol 31 (108):143–149

Cain JC (1987) The earth as a magnet. In: Akasofu S-I, Kamide Y (eds) The solar wind and the earth, Tokyo, pp 55–69

Campbell WH (2003) Introduction to geomagnetic fields, 2nd edn. Cambridge University Press, New York

Chevalier M-L, Hilley G, Tapponier P, Van Der Woerd J, Liu-Zeng J, Finkel RC, Ryerson J, Li H, Liu X (2011) Constraints on the late Quaternary glaciations in Tibet from cosmogenic exposure ages of moraine surfaces. Quat Sci Rev 30:528–554

Chi J-M, Ren B-H (1980) Glaciers in China. Lanzhou Inst. of Glaciology and Cryopedology, Academia Sinica, Shanghai

Dainelli G (1922–1934) Studi sul Glaciale. Risultati Geologici e Geografici, Relazioni scientifiche della Spedizione Italiana De Fillipi nell' Himalaya, Caracorum e Turchestan Cinese (1913–1914), vol I–III, Ser. II. Zanchelli, Bologna, Italy, pp 1–658 (in Italian)

Ding Y (1987) The study of the thermo-hydrological environment of glacial development in the Karakoram N-side, Lanzhou (unpublished)

Dortch JM, Owen LA, Caffee MW, Brease P (2009) Late Quaternary glaciations and equilibrium line altitude variations of the McKinley River region, central Alaska Range. Boreas 39:233–246

Dreimanis A, Vagners UJ (1971) Bi-modal distribution of rock and mineral fragments in basal tills. In: Goldthwaite RP (ed) Till, A symposium. Ohio State University Press, Columbus, OH, pp 237–250

Drew F (1875) The Jummoo and Kashmir territories: a geographical account. E. Stanford, London

Dunai TJ (2010) Cosmogenic nuclides—principles, concepts and applications in the earth surface sciences. Cambridge University Press, Cambridge

Finsterwalder R (1937) Die Gletscher des Nanga Parbat. Glaziologische Arbeiten der Deutschen Himalaya Expedition 1934 und ihre Ergebnisse. Z Gletscherk 25:57–108

Fu P, Stroeven AP, Harbor JM, Hättestrand C, Heyman J, Caffee MW, Zhou L (2013) Paleoglaciation of Shaluli Shan, southeastern Tibetan Plateau. Quat Sci Rev 64:121–135

Gansser A (1964) Geology of the Himalayas. Wiley, London, pp 1–289

Gasse F, Arnold M, Fontes JC, Fort M, Gibert E, Huc A, Bingyan Li, Yuanfang Li, Qing Liu, Mélières F, Van Campo E, Fubao Wang, Qingsong Zhang (1991) A 13,000-year climate record from western Tibet. Nature 353:742–745

Gasse F, Fontes JC, Van Campo E, Wei K (1996) Holocene environmental changes in Bangong Co basin (Western Tibet). Part 4: discussion and conclusions. Palaeogeogr Palaeoclimatol Palaeoecol 120(1–2):79–92

Godwin-Austen HH (1868) Notes on the Pangong Lake District of Ladakh, from a journal made during a survey in 1863. J Asiatic Soc Bengal 37, part 2 (Calcutta)

Goehring BM, Brook EJ, Linge H, Raisbeck GM, Yiou F (2008) Beryllium-10 exposure ages of erratic boulders in southern Norway and implications for the history of the Fennoscandian Ice Sheet. Quat Sci Rev 27:320–336

Gosse JC, Phillips FM (2001) Terrestrial in situ cosmogenic nuclides: theory and application. Quat Sci Rev 20:1475–1560

Granger DE, Muzikar PF (2001) Dating sediment burial with in situ-produced cosmogenic nuclides: theory, techniques, and limitations. Earth Planet Sci Lett 188:269–281

Groom DE, Mokhov NV, Striganov SI (2001) Muon stopping power and range tables 10 MeV–100 TeV. At Data Nucl Data Tables 76(2)

Grosswald MG, Kuhle M, Fastook JL (1994) Würm Glaciation of Lake Issyk-Kul Area, Tian Shan Mts.: a case study in glacial history of Central Asia. GeoJournal 33 (2–3, Tibet and High Asia, Results of the Sino-German and Russian-German Joint Expeditions (III), Ed: Kuhle M):273–310

Gu Z, Liu J, Yuan B, Liu T, Liu R, Liu Y, Zhang G (1993) The changes in monsoon influence in the Qinghai-Tibetan Plateau during the past 12,000 years. Geochemical evidence from the L. Selin sediments. Chin Sci Bull 38(1):61–64

Gunn JP (1930) Report on the Kumdan Dam and Shyok Flood of 1929. Government of Punjab Publication, Lahore

Hartmann GA, Pacca IG (2009) Time evolution of the South Atlantic Magnetic Anomaly. Anais Acad Bras Ciênc 81(2):243–255 (Annals of the Brazilian Academy of Sciences). ISSN 0001-3765 www.scielo.br/aabc

Heimsath AM (2006) Eroding the land: steady-state and stochastic rates and processes through a cosmogenic lens. In: Siame LL, Bourlès DL, Brown ET (eds) In situ-produced cosmogenic nuclides and quantification of geological processes. Geological Society of America Special paper 415, pp 111–129

Heydemann A, Kuhle M (1988) The petrography of southern Tibet–results of microscopic and X-ray analyses of rock samples from the 1984 expedition area (Transhimalaya to Mt. Everest N Slope). GeoJournal 17(4). In: Kuhle M, Wang W (eds) Tibet and High Asia. Results of the Sino-German Joint Expeditions (I), pp 615–625

Hättestrand C, Stroeven AP (2002) A relict landscape in the centre of Fennoscandian glaciation: geomorphological evidence of minimal Quaternary glacial erosion. Geomorphology 44:127–143

Herzschuh U, Zhang C, Mischke S, Herzschuh R, Mohammadi F, Mingram B, Kürschner H, Riedel F (2005) A late Quaternary lake record from the Qilian Mountains (NW China): evolution of the primary production and the water depth reconstructed from macrofossil, pollen, biomarker, and isotope data. Global Planet Change 46:361–379

Herzschuh U, Winter K, Wünnemann B, Li S (2006) A general cooling trend on the central Tibetan Plateau throughout the Holocene recorded by the Lake Zigetang pollen spectra. Quat Int 154(155):113–121

Herzschuh U, Kramer A, Mischke S, Zhang C (2009) Quantitative climate and vegetation trends since the late glacial on the northeastern Tibetan Plateau deduced from Koucha Lake pollen spectra. Quat Res 71:162–171

Heyman J, Stroeven AP, Caffee MW, Hättestrand C, Harbor J, Li Y, Alexanderson H, Zhou L, Hubbard A (2010) Palaeoglaciology of Bayan Har Shan, NE Tibetan Plateau: the case of a missing LGM expansion. In: Heyman J (ed) Palaeoglaciology of the Northeastern Tibetan Plateau. PhD thesis, Stockholm University

Heyman J, Stroeven AP, Harbor JM, Caffee MW (2011) Too young or too old: evaluating cosmogenic exposure dating based on an analysis of compiled boulder exposure ages. Earth Planet Sci Lett 302:71–80

Höfer HV (1879) Gletscher- und Eiszeitstudien. Sitzungsbericht der Akademie der Wissenschaften. Mathematisch-physikalische Klasse I. 79:331–367

Hövermann J, Süssenberger H (1986) Zur Klimageschichte Hoch- und Ostasiens. Berl Geogr Studien 20:173–186

Huntington E (1905) The mountains of Turkestan. Geogr J 25

Ivy-Ochs S, Kober F (2008) Surface exposure dating with cosmogenic nuclides. Quat Sci J 57:179–209

Kaufmann G, Lambeck K (1997) Implications of Late Pleistocene Glaciation of the Tibetan Plateau for present-day uplift rates and gravity anomalies. Quat Res 48:267–279

Kirchner N, Greve R, Stroeven AP, Heyman J (2011) Paleoglaciological reconstructions for the Tibetan Plateau during the last glacial cycle: evaluating numerical ice sheet simulations driven by GCM-ensembles. Quat Sci Rev 30:248–267 von Klebelsberg R (1948/49): Handbuch der Gletscherkunde und Glazialgeologie, vols. 1 and 2. Springer, Wien, pp 1–1028

Klein J, Gosse J (1996) Terrestrial factors that influence production rates. Radiocarbon 38(1):161–162

Kleman J (1994) Preservation of landforms under ice sheets and ice caps. Geomorphology 9:19–32

Kleman J, Hättestrand C (1999) Frozen-bed Fennoscandian and Laurentide ice sheets during the Last Glacial Maximum. Nature 402:63–66

Kong P, Na C, Brown R, Fabel D, Freeman S, Xiao W, Wang Y (2011) Cosmogenic 10Be and 26Al dating of paleolake shorelines in Tibet. J Asian Earth Sci 41:263–273

Kramer A, Herzschuh U, Mischke S, Zhang C (2010) Late glacial vegetation and climate oscillations on the southeastern Tibetan Plateau inferred from Lake Naleng pollen profile. Quat Res 73:324–335

Kuhle M (1974) Vorläufige Ausführungen morphologischer Feldarbeitsergebnisse aus den SE-Iranischen Hochgebirgen am Beispiel des Kuh-i-Jupar. Z Geomorphol N.F. 18(4):472–483

Kuhle M (1976) Beiträge zur Quartärgeomorphologie SE-Iranischer Hochgebirge. Die quartäre Vergletscherung des Kuh-i-Jupar. Göttinger Geogr Abh 67 I:1–209; II:1–105

Kuhle M (1980) Klimageomorphologische Untersuchungen i. d. Dhaulagiri- u. Annapurna-Gruppe (Zentraler Himalaya). Tagungsber. u. wiss. Abh. 42. Deutsch Geographentages 1979:244–247

Kuhle M (1982a) Der Dhaulagiri- und Annapurna-Himalaya. Z Geomorphol 41, Vol. I, Vol. II, Abb. 1–184, Stuttgart, 1–229

Kuhle M (1982b) Erste Deutsch-Chinesische Gemeinschaftsexped. nach Tibet und in die Massive des Kuen-Lun-Gebirges (1981). Ein Expeditions- u. vorl. Forschungsber. In: Hagedorn H, Giessner K (eds) Tagungsber. und wiss. Abhandlungen des 43. Deutschen Geographentages. Steiner, Wiesbaden, Mannheim, pp 63–82

Kuhle M (1982c) Was spricht für eine pleistozäne Inlandvereisung Hochtibets? Sitzber. Braunschweig. Wiss. Ges., Sonderh., 6, Göttingen, pp 68–77

Kuhle M (1982d) Periglazial- und Glazialformen und -prozesse in NE-Tibet. Sitzungsberichte und Mitteilungen der Braunschweigischen Wissenschaftlichen Gesellschaft, Sonderheft 6 (Die Chinesisch/Deutsche Tibet-Expedition 1981, Braunschweig-Symposium vom 14.-16.04.1982), pp 19–24

Kuhle M (1983) Der Dhaulagiri- und Annapurna-Himalaya. Empirische Grundlage. Z Geomorphol Suppl. Bd. 41:1–383

Kuhle M (1985a) Ein subtropisches Inlandeis als Eiszeitauslöser. Südtibet- u. Mt. Everest-Expedition 1984. Georgia Augusta, pp 35–51

Kuhle M (1985b) Permafrost and periglacial indicators on the Tibetan Plateau from the Himalaya Mountains in the south to the Quilian Shan in the north (28–40°N). Z Geomorphol N.F. 29(Bd. 2):183–192

Kuhle M (1985c) Glaciation research in the Himalayas: a new Ice Age theory. Universitas 27(4):281–294

Kuhle M (1986a) Former glacial stades in the mountain areas surrounding Tibet—in the Himalayas (27°–29° N: Dhaulagiri, Annapurna, Cho Oyu and Gyachung Kang areas) in the south and in the Kuen Lun and Quilian Shan (34°–38°N: Animachin, Kakitu) in the north. In: Joshi SC (ed) Nepal, Himalaya. Geo-ecological perspectives. HR Publishers, Kay Kay Printers, New Delhi, pp 437–473

Kuhle M (1986b) Südtibet- und Mt. Everest-Expedition 1984- oder: Ein subtropisches Inlandeis als Eiszeitauslöser? Universitas, 41, 1(476):64–78

Kuhle M (1986c) Schneegrenzberechnung und typologische Klassifikation von Gletschern anhand spezifischer Reliefparameter. Petermanns Geogr Mitt 130:41–51

Kuhle M (1986d) Die Vergletscherung Tibets und die Entstehung von Eiszeiten. Spektrum Wiss 9:42–54

Kuhle M (1987a) Subtropical mountain- and highland-glaciation as ice age triggers and the waning of the glacial periods in the Pleistocene. GeoJournal 14(4):393–421

Kuhle M (1987b) Absolute Datierungen zur jüngeren Gletschergeschichte im Mt. Everest-Gebiet u. die mathem. Korrektur von Schneegrenzberechnungen. Stuttg., Tagungsber. 45., Dt. Geographentag, Berlin 1985, pp 200–208

Kuhle M (1987c) Glacial, nival and periglacial environments in northeastern Qinghai-Xizang-Plateau. In: Hövermann J, Wang Wenjing J (eds) Reports on the North-Eastern part of the Quinghai-Xizang (Tibet) Plateau by the Sino-German Scientific Expedition 1981. Science Press, Beijing, pp 176–244

Kuhle M (1987d) The problem of a Pleistocene inland glaciation of the north-eastern Qinghai-Xizang-Plateau (Tibet). In: Hövermann J, Wang Wenjing J (eds) Reports on the North-Eastern Part of the Quinghai-Xizang (Tibet) Plateau by the Sino-German Scientific Expedition 1981. Science Press, Beijing, pp 250–315

Kuhle M (1988a) Topography as a fundamental element of glacial systems. A new approach to ELA-calculation and typological classification of paleo- and recent Glaciations. In: Kuhle M, Wang Wenjing J (eds) Tibet and High Asia. Results of the Sino-German Joint Expeditions (I). GeoJournal 17 (4):545–568

Kuhle M (1988b) Die eiszeitliche Vergletscherung W-Tibets zwischen Karakorum und Tarim-Becken und ihr Einfluß auf die globale Energiebilanz. Geogr Z 76(3):135–148

Kuhle M (1988c) Geomorphological findings on the build-up of Pleistocene glaciation in southern Tibet and on the problem of inland ice. Results of the Shisha Pangma and Mt. Everest Expedition 1984. In: Kuhle M, Wang Wenjing J (eds) Tibet and High Asia. Results of the Sino-German Joint Expeditions (I). GeoJournal 17(4):457–511

Kuhle M (1988d) The Pleistocene Glaciation of Tibet and the onset of ice ages- an autocycle hypothesis. In: Kuhle M, Wang Wenjing J (eds) Tibet and High Asia. Results of the Sino-German Joint Expeditions (I). GeoJournal 17(4):581–596

Kuhle M (1988e) Subtropical mountain- and highland-glaciation as Ice Age triggers and the waning of the glacial periods in the Pleistocene. Chin Transl Bull Glaciol Geocryol 5(4):1–17 (in Chinese language)

Kuhle M (1988f) The Pleistocene glaciation of Tibet and the onset of ice ages—an autocyclic hypothesis. In: International Congress "The Neogene of Karakoram and Himalayas." March 1988, Leicester, UK; Abstract, 11

Kuhle M (1988g) Letzteiszeitliche Gletscherausdehnung vom NW-Karakorum bis zum Nanga Parbat (Hunza-Gilgit- und Indusgletschersystem). Tagungsber. Wiss. Abh., 46., Dt. Geographentag München 1987, AK- Sitzung "Neueste physisch-geographische Forschungsergebnisse aus Hochtibet und angrenzenden Gebieten," Steiner, Stuttgart, pp 606–607

Kuhle M (1989) Die Inlandvereisung Tibets als Basis einer in der Globalstrahlungsgeometrie fußenden, reliefspezifischen Eiszeittheorie. Petermanns Geogr Mitt 133(4):265–285

Kuhle M (1990a) New data on the Pleistocene glacial cover of the southern border of Tibet: the glaciation of the Kangchendzönga Massif (8585 m, E-Himalaya). GeoJournal 20(4):415–421

Kuhle M (1990b) The cold deserts of High Asia (Tibet and contiguous mountains). GeoJournal 20(3):319–323

Kuhle M (1991) Observations supporting the Pleistocene inland glaciation of High Asia. In: Kuhle M, Xu D (eds) Tibet and High Asia. Results of the Sino-German Joint Expeditions (II). GeoJournal 25 (2–3):133–231

Kuhle M (1993) Eine Autozyklentheorie zur Entstehung und Abfolge der quartären Kalt- und Warmzeiten auf der Grundlage epirogener und glazialisostatischer Bewegungsinterferenzen im Bereich des tibetischen Hochlandes. Peterm Geogr Mittlg 137(3):133–152

Kuhle M (1994a) Present and Pleistocene glaciation on the north-western margin of Tibet between the Karakorum Main Ridge and the Tarim Basin supporting the evidence of a Pleistocene inland glaciation in Tibet. In: Kuhle M (ed) Tibet and High Asia. Results of the Sino-German and Russian-German Joint Expeditions (III). GeoJournal 33(2–3):133–272 (Kluwer, Dordrecht)

Kuhle M (1994b) New findings on the ice-cover between Issyk-Kul and K2 (Tian Shan, Karakorum) during the Last Glaciation. In: Zheng D, Zhang Q, Pan Y (eds) Proceedings of the international symposium on the Karakorum and Kunlun Mountains (ISKKM), Kashi, China, June 1992. China Meteorological Press, Beijing, pp 185–197

Kuhle M (1995) Glacial isostatic uplift of Tibet as a consequence of a former ice sheet. GeoJournal 37 (4):431–449

Kuhle M (1996a) Die Entstehung von Eiszeiten als Folge der Hebung eines subtropischen Hochlandes über die Schneegrenze – dargestellt am Beispiel Tibets. Der Aufschluss 47(4):145–164

Kuhle M (1996b) Neue Ergebnisse zur Eiszeitforschung Hochasiens in Zusammenschau mit den Untersuchungen der letzten zwanzig Jahre. Berl Geogr Abh., Beiheft Nr.4/96, Berlin

Kuhle M (1996c) Rekonstruktion der maximalen eiszeitlichen Gletscherbedeckung im Nanga Parbat Massiv (35° 05′–40°N/74° 20′–75°E). In: Kick W (ed) Beiträge und Materialien zur Regionalen Geographie, 8, Forschung am Nanga Parbat. Geschichte und Ergebnisse. Geographisches Institut der T.U. Berlin, pp 135–156

Kuhle M (1997) New findings concerning the Ice Age (LGM) glacier cover of the East Pamir, of the Nanga Parbat up to the Central Himalaya and of Tibet, as well as the Age of the Tibetan Inland Ice. In: Kuhle M (ed) Tibet and High Asia (IV). Results of investigations into high mountain geomorphology. Paleo-glaciology and climatology of the pleistocene. GeoJournal 42(2–3):87–257

Kuhle M (1998a) Reconstruction of the 2.4 million km^2 Late Pleistocene ice sheet on the Tibetan Plateau and its impact on the global climate. Quat Int 45/46:71–108 (additional Figures in: 47/48, 173–182)

Kuhle M (1998b) New findings on the inland glaciation of Tibet from South and Central West Tibet with evidences for its importance as an Ice Age trigger. In: Tandon OP (ed) Himalayan geology. The role of the Tibetan Plateau in forcing global climatic changes, vol 19(2), pp 3–22

Kuhle M (1998c) The Ice Age glaciation of East-Pamir (36° 40′ -39o10′N/74° 40′ -76° 10′E). In: Stellrecht I (ed) Karakorum – Hindukush – Himalaya: dynamics of change. Culture Area Karakorum, Scientific Studies, 4/1, Köln, Rüdiger Köppe Verlag, pp 29–47

Kuhle M (1998d) Neue Ergebnisse zur Eiszeitforschung Hochasiens in Zusammenschau mit den Untersuchungen der letzten zwanzig Jahre. Petermanns Geogr Mitt 142(3–4):219–226

Kuhle M (1999) Reconstruction of an approximately complete Quaternary Tibetan inland glaciation between the Mt. Everest- and Cho Oyu massifs and the Aksai Chin. A new glaciogeomorphological SE-NW diagonal profile through Tibet and its consequences for the glacial isostasy and Ice Age cycle. In: Kuhle M (ed) Tibet and High Asia (V). GeoJournal 47 (1–2):3–276

Kuhle M (2001a) The maximum Ice Age (LGM) glaciation of the Central- and South Karakorum: an investigation of the heights of its glacier levels and ice thickness as well as lowest prehistoric ice margin positions in the Hindukush, Himalaya and in East-Tibet on the Minya Konka-massif. In: Kuhle M (ed) Tibet and High Asia (VI): glaciogeomorphology and prehistoric glaciation in the Karakorum and Himalaya. GeoJournal 54(1–4) and 55(1):109–396

Kuhle M (2001b) The Tibetan ice sheet; its impact on the palaeomonsoon and relation to the earth's oribital variations. Polarforschung 71(1–2):1–13

Kuhle M (2001c) Reconstruction of outlet glacier tongues of the Ice Age South-Tibetan ice cover between Cho Oyu and Shisha Pangma as a further proof of the Tibetan inland ice sheet. Polarforschung 71(3):79–95

Kuhle M (2002a) A relief-specific model of the Ice Age on the basis of uplift-controlled glacier areas in Tibet and the corresponding albedo increase as well as their positive climatological feedback by means of the global radiation geometry. Climate Res 20:1–7

Kuhle M (2002b) Outlet glaciers of the Pleistocene (LGM) south Tibetian ice sheet between Cho Oyu and Shisha Pangma as potenial sources of former mega-floods. In: Martini P, Baker VR, Garzón G (eds) Flood and megaflood processes and deposits: recent and ancient examples. Special Publication of the International Association of Sedimentologists (IAS), vol 32. Blackwell Scientific, Oxford, pp 291–302

Kuhle M (2003) New geomorphological indicators of a former Tibetan ice sheet in the central and northeastern part of the high plateau. Z Geomorphol N.F. 130 (Suppl):75–97

Kuhle M (2004) The High Glacial (Last Ice Age and LGM) glacier cover in High- and Central Asia. Accompanying text to the mapwork in hand with detailed references to the literature of the underlying empirical investigations. In: Ehlers J, Gibbard PL (eds) Extent and chronology of glaciations, vol 3 (Latin America, Asia, Africa, Australia, Antarctica). Elsevier B.V., Amsterdam, pp 175–199

Kuhle M (2005a) The maximum Ice Age (Würmian, Last Ice Age, LGM) glaciation of the Himalaya—a glaciogeomorphological investigation of glacier trim-lines, ice thicknesses and lowest former ice margin positions in the Mt. Everest-Makalu-Cho Oyu massifs (Khumbu- and Khumbakarna Himal) including informations on late-glacial-, neoglacial-, and historical glacier stages, their snow-line depressions and ages. In: Kuhle M (ed) Tibet and High Asia (VII): glaciogeomorphology and former glaciation in the Himalaya and Karakorum. GeoJournal 62(3–4):193–650

Kuhle M (2005b) The maximum Ice Age glaciation between Karakoram Main Ridge (K2) and the Tarim Basin and its influence on the global energy balance. J Mt Sci 2(1):5–22

Kuhle M (2005c) Glacial geomorphology and ice ages in Tibet and surrounding mountains. Island Arc 14 (4):346–367

Kuhle M (2006a) The past Hunza Glacier in connection with a Pleistocene Karakorum ice stream network during the Last Ice Age (Würm). In: Kreutzmann H, Saijid A (eds) Karakoram in transition. Oxford University Press, Karachi, pp 24–48

Kuhle M (2006b) The reconstruction of Ice Age glaciation of the Himalaya and High Asia by Quaternary geological and glaciogeomorphological methods. In: Saklani P (ed) Himalaya, geological aspects, vol 4, pp 181–214

Kuhle M (2007a) The past ice stream network in the Himalayas and the Tibetan ice sheet during the Last Glacial Period and its glacial-isostatic, eustatic and climatic consequences. Tectonophysics 445(1–2):116–144

Kuhle M (2007b) The Pleistocene glaciation (LGP and pre-LGP, pre-LGM) of SE-Iranian Mountains exemplified by the Kuh-i-Jupar, Kuh-i-Lalezar and Kuh-i-Hezar Massifs in the Zagros. Polarforschung 77(2–3):71–88 (Erratum/Clarification concerning Figure 15, Vol. 78, (1–2), 2008, 83)

Kuhle M (2009) The former glaciation of High-(Tibet) and Central Asia and its global climatic impact—an Ice Age theory with a remark on potential warmer climatic cycles in the future. In: Krugger, Max I., Stern, Harry P (eds) New Permafrost and Glacier Research. Nova Science Publishers Inc., Hauppauge, NY, pp 163–235

Kuhle M (2010) New indicators of a former Tibetan ice sheet and an ice stream network in the surrounding mountain systems: new field observations and datings on the SE-, S- and W-margin of Tibet from the expeditions 2004–2009. In: Leech ML, Klemperer SL, Mooney WD (eds) Online proceedings for the 24th Himalaya-Karakoram-Tibet workshop. U.S. Geological Survey, Open-File Report 2010-1099

Kuhle M (2011a) The High Glacial (Last Ice Age and Last Glacial Maximum) Ice Cover of High and Central Asia, with a critical review of some recent OSL and TCN Dates. In: Ehlers J, Gibbard PL, Hughes PD (eds) Quaternary glaciation—extent and chronology, a closer look. Elsevier B.V, Amsterdam, pp 943–965 (glacier maps downloadable at: http://booksite.elsevier.com/9780444534477/)

Kuhle M (2011b) The High Glacial (LGP, LGM, MIS 3-2) southern outlet glaciers of the Tibetan inland ice through Mustang into the Thak Khola as further evidence of the Tibetan ice. J Nepal Geol Soc (JNGS) 43 (special issue):175–200

Kuhle M (2011c) Ice Age development theory. In: Singh VP, Singh P, Haritashya UK (eds) Encyclopedia of snow, ice and glaciers. Springer, Dordrecht, pp 576–581

Kuhle M (2012a) The former Tibetan ice sheet. In: Müller J, Koch L (eds) Ice sheets: dynamics, formation and environmental concerns. Nova Science Publishers Inc., Hauppauge, NY, pp 173–203

Kuhle M (2012b) The Early and Late Glacial high mountain glaciation surrounding Tibet as topographic-climatic cause of high-energetic glacial lake outburst floods (GLOFs) and their sedimentological consequences in the lower mountain forelands. In: Veress B, Szigethy J (eds) Horizons in earth science research, vol 7. Nova Science Publishers Inc. Hauppauge, NY, pp 197–227

Kuhle M (2012c) High-glacial (LGP, LGM, MIS 3-2) ice cover in the middle Marsyandi Nadi and the Damodar-Himal down to the junction of the Nar Khola and the Marsyandi Khola (N of Annapurna Himalaya). In: Hartmann M, Weipert J (eds) Biodiversity and Natural Heritage of the Himalaya, vol IV (Biodiversität und Naturausstattung im Himalaya), Erfurt, pp 9–46

Kuhle M (2013a) The Uplift of High Asia above the snowline and its glaciation as an Albedo-Dependent cause of the Quaternary Ice Ages. Nova Science Publishers Inc., Hauppauge, NY, pp 1–240

Kuhle M (2013b) The Glacial (LGP, MIS 3-2) Mustang-Thak Khola outlet glacier of the Tibetan ice sheet and the Marsyandi Khola-ice stream network between Damodar- and Manaslu Himalaya (with Nar-, Dudh-, Dana-, Myardi- and Nadi Khola tributary glaciers). Geography International, vol 3 (Tibet and High Asia (VIII) Glaciogeomorphology and Former Glaciation in the Himalaya and Karakoram) Dedicated to Wolf Tietze. Shaker Verlag, Aachen, pp 1–308

Kuhle M (2014a) Zone of optimum development of patterned grounds as a characteristic of semi-arid areas, an indicator of permafrost-lines and a reference to sub-glacial permafrost below the past arid cold-based Tibetan ice sheet. In: Pokrovsky OS (ed) Permafrost: distribution, composition and impacts on infrastructure and ecosystems. Nova Science Publishers Inc., Hauppauge, NY, pp 107–152

Kuhle M (2014b) The Glacial (MIS 3-2) outlet glacier of the Marsyandi Nadi-icestream-network with its Ngadi Khola Tributary Glacier (Manaslu- and Lamjung Himalaya). The reconstructed lowering of the Marsyandi Nadi ice stream tongue down into the Southern Himalaya foreland. J Mt Sci 11(1):236–287

Kuhle M (2014c) Climate or topography—topography of mountains as basis of modern glacier variations, exemplified by 6500 up to 8000 m high summits in High Asia. Evidences of Holocene and Historical as well as current glacier stages. In: Goncalves AJB, Vieira, AAB (eds) Mountains: geology, topography and environmental concerns. Nova Science Publishers Inc., Hauppauge, NY, pp 65–147

Kuhle M, Jacobsen JP (1988) On the geoecology of southern Tibet – measurements of climate parameters including surface- and soil-temperatures in debris, rock, snow, firn and ice during the South Tibet- and Mt. Everest Expedition in 1984. In: Kuhle M, Wang W (eds) Tibet and High-Asia. Results of the Sino-German Joint Expeditions (I). GeoJournal 17 (4):597–615

Kuhle M, Kuhle S (1997) Der quartäre Klimawandel – System oder geschichtliches Ereignis? Überlegungen zur geographischen Methode am Beispiel von Eiszeittheorien. Erdkunde 51:114–130

Kuhle M, Kuhle S (2010) Review on dating methods: numerical dating in the Quaternary of High Asia. J Mt Sci 7:105–122

Kuhle M, Herterich K, Calov R (1989) On the Ice Age glaciation of the Tibetan highlands and its transformation into a 3-D model. GeoJournal 19(2):201–206

Kuhle M, Grosswald MG, Dyurgerov MB (1997) Quaternary glaciation in north-western High-Asia. Data of Glaciological Studies, 81. In: Proceedings of the international symposium "Seasonal and Longterm Fluctuations of Nival and Glacial Processes in Mountains," Tashkent symposium (12-19.09.1993), pp 70–79

Lal D (1991) Cosmic ray labeling of erosion surfaces: in situ nuclide production rates and erosion models. Earth Planet Sci Lett 104:424–439

Lehmkuhl F, Owen LA (2005) Late Quaternary glaciation of Tibet and the bordering mountains: a review. Boreas 34:87–100

Lehmkuhl F, Klinge M, Lang A (2002) Late Quaternary glacier advances, lake level fluctuations and aeolian sedimentation in Southern Tibet. Z Geomorphol N.F. Suppl. Bd. 126:183–218

Li B, Zhu L (2001) "Greatest lake period" and its palaeo-environment on the Tibetan Plateau. J Geog Sci 11(1):34–42

Li B, Li J, Cui Z (eds) (1991) Quaternary glacial distribution map of Qinghai-Xixang (Tibet) Plateau 1:3,000,000. Scientific advisor: Shi Yafeng, Quaternary Glacier and Environment Research Center, Lanzhou University

Li G, Webb GM, Le Roux JA, Zank GP, Wiedenbeck ME (2008) Time-dependent solar modulation of GCRs using a Monte Carlo simulation. In: Pogorelov NV, Audit E, Zank GP (eds) Numerical modelling of space plasma flows/ASTRONUM-2007. ASP conference series, vol 385, pp 31–37

Linge H, Brook EJ, Nesje A, Raisbeck GM, Yiou F, Clark H (2006) In situ 10 Be exposure ages from southeastern Norway: implications for the geometry of the Weichselian Scandinavian ice sheet. Quat Sci Rev 25:1097–1109

Longstaff TG (1910) Glacier exploration in the eastern Karakoram. Geogr J 35 (London)

Lühr H (2013) Das Erdmagnetfeld wird schwächer – Wie gefährlich ist das für uns? In: Olbrich A (ed) Bayerisches Fernsehen: Faszination Wissen. 15.07.2013

Lynch P (2011) and (2012) Magnetic pole reversal happens all the (geologic) time. http://www.nasa.gov/topics/earth/features/2012-poleReversal.html. Last accessed on 02.08.2013

Madsen DB, Haizhou M, Rhode D, Brantingham PJ, Forman SL (2008) Age constraints on the late Quaternary evolution of Qinghai Lake, Tibetan Plateau. Quat Res 69:316–325

Masarik J, Beer J (2009) An updated simulation of particle fluxes and cosmogenic nuclide production in the earth's atmosphere. J Geophys Res 114:D 11103

Masarik J, Reedy RC (1995) Terrestrial cosmogenic-nuclide production systematics calculated from numerical simulations. Earth Planet Sci Lett 136:381–395

Masarik J, Frank M, Schäfer JM, Wieler R (2001) Correction of in situ cosmogenic nuclide production rates for geomagnetic field intensitiy variations during the past 800,000 years. Geochim Cosmochim Acta 65 (17):2995–3003

Mason K, Gunn JP, Todd HJ (1930) The Shyok flood in 1929. Himalayan J 2:35–47

Merchel S, Benedetti L, Bourlès DL, Braucher RA, Dewald T, Faestsermann R, Finkel RC, Korschinek G, Masarik J, Poutivtsey M, Rochette P, Rugel G, Zell KO (2010) A multi-radionuclide approach for in situ produced terrestrial cosmogenic nuclides: 10Be, 26Al, 41Ca from carbonate rocks. Nucl Instrum Methods Phys Res B 268:1179–1184

Meierding TC (1982) Late Pleistocene glacial equilibrium-line altitudes in the Colorado Front Range: a comparison of methods. Quat Res 18:289–310

Morinaga H, Itota C, Isezaki N, Goto H, Yaskawa K, Kusakabe M, Liu J, Gu Z, Yuan B, Cong S (1993) Oxygen-18 and carbon-13 records for the last 14,000 years from lacustrine carbonates of Siling-Co (lake) in the Qinghai-Tibetan Plateau. Geophys Res Lett 20 (24):2909–2912

Mügler I, Gleixner G, Günther F, Mäusbacher R, Daut G, Schütt B, Berking J, Schwalb A, Schwark L, Xu B, Yao T, Zhu T, Yi C (2010) A multi-proxy approach to reconstruct hydrological changes and Holocene climate development of Nam Co, Central Tibet. J Paleolimnol 43:625–648

Nesje A, Dahl SO (2000) Glaciers and environmental change. Arnold, London, pp 1–203

Neve A (1913) Thirty years in Kashmir. London

Nishiizumi K, Kohl CP, Arnold JR, Dorn R, Klein J, Fink D, Middleton R, Lal D (1993) Role of in situ cosmogenic nuclides ^{10}Be and ^{26}Al in the study of diverse geomorphic processes. Earth Process Landforms 18:407–425

Norin E (1932) Quaternary climatic changes within the Tarim Basin. Geogr Rev 22:591–598

Opitz S, Wünnemann B, Aichner B, Dietze E, Hartmann K, Herzschuh U, Ijmker J, Lehmkuhl F, Mischke S, Plotzki A, Stauch G, Diekmann B (2012) Late Glacial and Holocene development of Lake Donggi Cona, north-eastern Tibetan Plateau, inferred from sedimentological analysis. Palaeogeogr Palaeoclimatol Palaeoecol 337–338:159–176

Owen LA (2009) Latest Pleistocene and Holocene glacier fluctuations in the Himalaya and Tibet. Quat Sci Rev 28:2150–2164

Owen LA, Finkel RC, Barnard PL, Haizhou M, Asahi K, Caffee MW, Derbyshire E (2005) Climatic and topographic controls on the style and timing of Late Quaternary glaciations throughout Tibet and the Himalaya defined by ^{10}Be cosmogenic radionuclide exposure dating. Quat Sci Rev 24:1391–1411

Owen LA, Caffee MW, Finkel RC, Seong Yeong Bae (2008) Quaternary glaciation of the Himalayan-Tibetan orogen. J Quat Sci 23:513–531

Owen LA, Robinson R, Benn DI, Finkel RC, Davis NK, Yi C, Putkonen J, Li D, Murray AS (2009) Quaternary glaciation of Mount Everest. Quat Sci Rev 28:1412–1433

Owen LA, Yi C, Finkel RC, Davis NK (2010) Quaternary glaciation of Gurla Mandhata (Naimon'anyi). Quat Sci Rev 29:1817–1830

Owen LA, Chen J, Hedrick KA, Caffee MW, Robinson AC, Schoenbohm LM, Yuan Z, Li W, Imrecke DB, Liu J (2012) Quaternary glaciation of the Tashkurgan Valley, Southeast Pamir. Quat Sci Rev 47:56–72

Pillewitzer W (1958) Neue Erkenntnisse über die Blockbewegung der Gletscher. IAHS AISH Publication 46, 429–436

Purdue University (2008) Cosmic-ray produced nuclide systematics on earth project. http://www.physics.purdue.edu/primelab/CronusProject/cronus, date 22-07-2013

Putkonen J, Swanson T (2003) Accuracy of cosmogenic ages for moraines. Quatern Res 59(2):255–261

Rhodes TE, Gasse F, Riufen L, Fontes J-C, Keqin W, Bertrand P, Gibert E, Mélières F, Tucholka P, Zhixiang W, Zhi-Yuan C (1996) A Late Pleistocene-Holocene lacustrine record from Lake Manas, Zunggar (northern Xinjiang, western China). Palaeogeogr Palaeoclimatol Palaeoecol 120:105–121

Schäfer JM, Tschudi S, Zhao Z, Wu X, Ivy-Ochs S, Wieler R, Baur H, Kubik PW, Schlüchter C (2002) The limited influence of glaciations in Tibet on global climate over the past 170 000 yr. Earth Planet Sci Lett 194:287–297

Schaefer JM, Oberholzer P, Zhao Z, Ivy-Ochs S, Wieler R, Baur H, Kubik PW, Schlüchter C (2008) Cosmogenic beryllium-10 and neon-21 dating of late Pleistocene glaciations in Nyalam, monsoonal Himalayas. Quat Sci Rev 27:295–311

Schroeder-Lanz H (1986) Beobachtungen zur rezenten und eiszeitlichen Vergletscherung am Kongurshan/Pamir, Himmelssee/Tienshan und Koko Nor-Nanshan Qaidam. In: Kuhle M (ed) Göttinger Geogr. Abh., 81, Internat. Sympos. über Tibet und Hochasien, Oct. 1985, Göttingen, pp 39–58

Schröder N (2007) The discrepancy between the method of cosmogenic nuclide exposure dating on moraines and morphodynamics, weathering, glacier dynamics, erosion and global climate. Abs Quat Int 167–168:369

Seidler C (2011) http://www.spiegel.de/wissenschaft/weltall/riesige-sonneneruption-kosmischer-streifschuss-fuer-mutter-erde-a-767415.html

Seidler C (2014) http://www.spiegel.de/wissenschaft/weltall/swarm-esa-satelliten-messen-magnetfeld-der-erde-a-976116.html

Seguin J (2013) Anwendbarkeit und Grenzen der Voraussetzungen kosmogener Nukliddatierungen – Analyse anhand von Literaturauswertung. Unveröffentlichte Bachelor of Arts-Arbeit; Department GHG of the Geographical Institute, University of Göttingen, 1–59

Seong YB, Owen LA, Yi C, Finkel RC (2009) Quaternary glaciations of Muztag Ata and Kongur Shan: evidence for glacier response to rapid climate changes throughout the Late Glacial and Holocene in westernmost Tibet. Geol Soc Am Bull 121(3–4):348–365

Shi Y, Zheng B, Li S (1992) Last glaciation and maximum glaciation in the Qinghai-Xizang (Tibet) Plateau: a controversy to M. Kuhle's ice sheet hypothesis. Z Geomorph N. F., Suppl. Bd. 84, Berlin, Stuttgart, 19–35

Stone JO (2000) Air pressure and cosmogenic isotope production. J Geophys Res 105:23753–23759

Stroeven AP, Fabel D, Hättestrand C, Harbor J (2002) A relict landscape in the centre of Fennoscandian glaciation: cosmogenic radionuclide evidence of tors preserved through multiple glacial cycles. Geomorphology 44:145–154

Thomson T (1852) Western Himalaya and Tibet. London

Visser PC (1935a) Durch Asiens Hochgebirge. Huber & Co. AG., Frauenfeld and Leipzig, pp 1–256

Visser PC (1935b) Gletscherbeobachtungen im Karakorum. Z Gletscherk 22:20–45

Visser PC (1938): Glaziologie. Wissenschaftliche Ergebnisse der niederländischen Expedition in den Karakorum und die angrenzenden Gebiete in den Jahren 1922, 1925 und 1929/30 und 1935. Leiden, pp 1–216

Vita-Finzi C (2013) Solar history—an introduction. Springer, Dordrecht

von Engelhardt W (1973) Die Bildung von Sedimenten und Sedimentgesteinen. In: von Engelhardt W, Füchtbauer H, Müller G (eds) Sediment-Petrologie, III, pp 1–378

Wang H (2001) Effects of glacial isostatic adjustment since the late Pleistocene on the uplift of the Tibetan Plateau. Geophys J Int 144:448–458

Wiche K (1958) Die österreichische Karakorum-Expedition 1958. Mitt Geographischen Ges Wien 100: 280–294

Wischnewski J, Mischke S, Wang Yongbo, Herzschuh U (2011) Reconstructing climate variability on the northeastern Tibetan Plateau since the last Late glacial—a multi-proxy, dual-site approach comparing terrestrial and aquatic signals. Quat Sci Rev 30:82–97

von Wissmann H (1959) Die heutige Vergletscherung und Schneegrenze in Hochasien mit Hinweis auf die Vergletscherung der letzten Eiszeit. Akademie der Wissenschaften und der Literatur Mainz, Mathetematisch-naturwissenschaftliche Klasse 14:1103–1407

Wu G, Xiao X (1991) Yadong to Golmud transect Quinghai-Tibet Plateau, China. Publication of the International Lithosphere Program, No. 189, Global Geoscience Transect, 3, 32

Yamanaka H (1982) Radiocarbon ages of upper quaternary deposits in Central Nepal. Sci Rep Tohoku Univ., 7th ser. (Geogr.) 32(1)

Yanhong W, Lücke A, Zhangdong J, Sumin W, Schleser GH, Battarbee RW, Weilan X (2006) Holocene climate development on the central Tibetan Plateau: a sedimentary record from Cuoe Lake. Palaeogeogr Palaeoclimatol Palaeoecol 234:328–340

Yi C, Li X, Qu J (2002) Quaternary glaciation of Puruogangri—the largest modern ice field in Tibet. Quat Int 97–98:111–121

Zhang H, Wünnemann B, Ma Y, Peng J, Pachur H-J, Li J, Qi Y, Chen G, Fang H, Feng Z (2002) Lake level and climate changes between 42,000 and 18,000 ^{14}C yr B. P. in the Tengger Desert, northwestern China. Quat Res 58(1):62–72

Zhu L, Wu Y, Wang J, Lin X, Ju J, Xie M, Li M, Mäusbacher R, Schwalb A, Daut G (2008) Environmental changes since 8.4 ka reflected in the lacustrine core sediments from Nam Co, central Tibetan Plateau, China. The Holocene 18(5):831–839

3 On the Chronology of the Prehistoric Glacial Ice Cover

The first ice age indicated in Fig. 2.5 dates back to 120–140 ka. It is the second to last ice age, the pre-last High Glacial maximum or the Riß Ice Age (Stage-I). Traces of this older ice age exist in High Asia, for instance in the Dhaulagiri-Annapurna Himalayas. These traces consist of petrified, i.e., fossilized moraine deposits. They have turned loose rock into hard rock, i.e., tillite. The present author has described two such tillite locations north of the Himalayan main range and three such locations south of it (Fig. 2.1: No. 2) (Kuhle 1982, Abb. 184; 1983). Further locations are found in the Zagros Mountains (Fig. 2.1: No. 1) (Kuhle 1976) and at the northern margin of Tibet (Fig. 2.1: No. 3) (Kuhle 1987a, b). Here they indicate ΔELAs (snowline depressions) of ca. 1400 m. These traces of the Riß Ice Age are extremely scarce, however, if compared with the dramatic trace density of the last High Glacial maximum, or the Würm Ice Age (Fig. 2.5, Stage 0). That those older glacier traces are hardly preserved anywhere and therefore negligible is not really strange. For one reason this is because of the immense temporal distance. Since the Würm Ice Age, an Interglacial, another Glacial and a second Interglacial, as well as a Holocene Postglacial have taken place. The geomorphological transformation of these various phases has led to the almost complete destruction of the Riß glacier traces in almost every location. This situation is due to the mountain relief; the vertical distance and steepness have accelerated the entire morphodynamic process.

The time span addressed in this book lies between the Würm Ice Age (i.e., the last ice age) and today. Its geomorphological, Quaternary-geological, and glaciological remains are still well to very well preserved.

Apart from the brevity of this time span—whose individual phases have become shorter and shorter up to the present—this is due to the fact that the snowline (ELA = Equilibrium Line Altitude) has risen, i.e., moved upwards. The glaciers have been reduced and therefore have not overridden and destroyed relics of older stages. Consequently, the entire glacigenic landform and feature ensemble, including the current glacier extensions, have been completely and chronologically preserved. Along with the temporal distance, the degree of the fluvial and periglacial transformation has also decreased. This means that from one stage to another, it has only produced limited, stepwise transformations. As a result, these single, continuously yielding, glacial-geomorphological steps allow one to diagnose a stage-layered and stage-mediated transformation of the glacier forefields spanning the entire period from the last ice age until today. This diagnosis is possible due to the stepwise recession of the ELA, a process interspersed with episodes of glacier advances. The chronology is thus very well preserved because during the stepwise uplifting of the ELA, no older stages

M. Kuhle, *The Glaciation of High Asia*,
https://doi.org/10.1007/978-3-319-77566-1_3

were overridden. It is this linkage between the ELA depressions (ΔELA) and the continuously less far-reaching glacier advances that has yielded such a clear glacier chronology. From a global perspective, this chronology is mostly synchronous. The number of glacier advances since the LGM (Stage 0) worldwide is nearly identical (2005). Only the Antarctic poses an exception. This region behaves anti-cyclically because it was more arid during the ice age than it is today. At the same time, the region is even today, during the Interglacial, cold enough to reach its maximal glaciation.

As explained above, the linkage between the glacier surface reduction and the reduction of the ΔELA, whose empirical values can be derived from the gradually decreasing glacier advances, describes the glacier history since the ice age in a chronological order. Figure 2.5 illustrates this chronology of the glaciation of High Asia from the ice age to the present. The chronology of the renewed glacier advances since the maximal glacier expansion was recorded on the basis of numerous glacier forefields in different mountain ranges and individual mountains (representing all cardinal directions) in all the 36 study areas depicted in Fig. 2.1. Radiocarbon datings allow one to correlate the chronology of the snowline depressions with a certain age.

Up to the current glacier margins, 15 stages can be differentiated (Stage 0 to `VII and `VII to XIV; Fig. 2.5). At least Stages XIII and XIV in High Asia are climatically (i.e., by way of ΔELA) not differentiable any more. Instead, they are mainly topographically determined (Kuhle 2014). The other 13 stages have, since the ice age until today, been linked to ΔELAs of ca. 1300 m to ca. 10–20 m. These are necessarily average ΔELAs derived on the basis of many individual values. Examples of this procedure of the present author are given in Kuhle (1976, 1982, 1986, 1987b, 1988b, 2005). Topography as a fundamental element of glacial systems and the basis for the approach to ELA calculation or the theory of snowline calculations has been verified on 223 current glaciers. These are current glaciers in mountain ranges around the world as well as in the Arctic. To a large extent, High Asian glaciers were taken into consideration here (Kuhle 1988a, Table 1).

3.1 Summary

Evidence for the pre-last High Glacial maximum or Riß Ice Age (Stage-I in Fig. 2.5) is transmitted through lithified moraines in High Asia. Such tillites have only been found very sporadically in the Himalaya, at the northern fringe of Tibet and west of the plateau, in the Zagros Massif (Fig. 2.1: Nos. 1–3). They indicate snowline depressions (ΔELAs) of ca. 1400 m that date back 140,000–120,000 years ago. The reason why the tillites are only sporadically preserved is that in many places destruction occurred over a long time span and due to the geomorphodynamics imposed through the steep relief. In contrast, an immense number of geomorphological remains of the last Ice Age (LGP, LGM), the Würm Glacial (Stage 0 in Fig. 2.5), are preserved well to very well, so that the chronological order of non-reshaped ice fringe areas fully allows the reconstruction of the stepwise ELA uplift to today's glacier margins. The glacial regression, interrupted by ever smaller snowline depressions (ΔELA), proceeded mostly synchronously across the globe. This is proven, among other things, by current global glacial melting. This evidence is particularly reliable because current global glacial melting has been caused by only a relatively slight increase in climate warming. Also, the number of glacial advances between the LGM and the present is almost identical worldwide for equal-sized glaciers. The glacial stages represented in Fig. 2.5, which were reconstructed in almost uninterrupted spatial distribution across High Asia (Fig. 2.1), are correlated with ^{14}C data that allow reasonable estimations of the minimum ages of each of the 15 stages. On average, the maximal snowline depression (ΔELA) in High Asia during the LGM lay at 1300 m (Stage 0).

References

Kuhle M (1976) Beiträge zur Quartärgeomorphologie SE-Iranischer Hochgebirge. Die quartäre Vergletscherung des Kuh-i-Jupar. Göttinger Geographische Abhandlungen, 67, vol I, pp 1–209; vol II, pp 1–105

Kuhle M (1982) Der Dhaulagiri- und Annapurna-Himalaya. Zeitschrift für Geomorphologie, Suppl. 41, vol I, vol II, Abb. 1–184:1–229

Kuhle M (1983) Der Dhaulagiri- und Annapurna-Himalaya. Empirische Grundlage. Zeitschrift für Geomorphologie 41:1–383

Kuhle M (1986) Former glacial stades in the mountain areas surrounding Tibet—in the Himalayas (27°–29° N: Dhaulagiri, Annapurna, Cho Oyu and Gyachung Kang areas) in the south and in the Kuen Lun and Quilian Shan (34°–38°N: Animachin, Kakitu) in the north. In: Joshi SC (ed) Nepal, Himalaya. Geo-ecological perspectives. HR Publishers, Kay Kay Printers, New Delhi, pp 437–473

Kuhle M (1987a) Glacial, nival and periglacial environments in northeastern Qinghai-Xizang-Plateau. In: Hövermann J, Wang Wenjing J (eds) Reports on the North-Eastern part of the Quinghai-Xizang (Tibet) Plateau by the Sino-German scientific expedition 1981. Science Press, Beijing, pp 176–244

Kuhle M (1987b) The problem of a Pleistocene inland glaciation of the north-eastern Qinghai-Xizang-Plateau (Tibet). In: Hövermann J, Wang Wenjing J (eds) Reports on the North-Eastern part of the Quinghai-Xizang (Tibet) Plateau by the Sino-German scientific expedition 1981. Science Press, Beijing, pp 250–315

Kuhle M (1988a) Topography as a fundamental element of glacial systems. A new approach to ELA-calculation and typological classification of paleo- and recent Glaciations. In: Kuhle M, Wang Wenjing J (eds) Tibet and High Asia. Results of the Sino-German joint expeditions (I). GeoJournal 17 (4):545–568

Kuhle M (1988b) Geomorphological findings on the build-up of Pleistocene glaciation in southern Tibet and on the problem of inland ice. Results of the Shisha Pangma and Mt. everest expedition 1984. In: Kuhle M, Wang Wenjing J (eds) Tibet and High Asia. Results of the Sino-German joint expeditions (I). GeoJournal, 17(4):457–511

Kuhle M (2005) The maximum Ice Age (Würmian, Last Ice Age, LGM) glaciation of the Himalaya—a glaciogeomorphological investigation of glacier trim-lines, ice thicknesses and lowest former ice margin positions in the Mt. Everest-Makalu-Cho Oyu massifs (Khumbu- and Khumbakarna Himal) including informations on late-glacial-, neoglacial-, and historical glacier stages, their snow-line depressions and ages. In: Kuhle M (ed) Tibet and High Asia (VII): Glaciogeomorphology and former glaciation in the Himalaya and Karakorum. GeoJournal 62(3–4):193–650

Kuhle M (2014) Climate or topography—Topography of mountains as basis of modern glacier variations, exemplified by 6500 up to 8000 m high summits in High Asia—evidences of holocene and historical as well as current glacier stages. In: Goncalves AJB, Vieira AAB (eds) Mountains: geology, topography and environmental concerns. Nova Science Publishers Inc., Hauppauge, NY, pp 65–147

4 The Overall Picture of the Maximal Ice Expansion in High Asia During the Last Ice Age

The primary context for this section is provided by Figs. 2.1 and 4.1. In this book, the most arid (i.e., low-precipitation) areas as well as the most centrally located areas of Central Tibet were considered as representative examples. The extent of their maximal glaciation can be seen by way of a map ensemble. These are, from west to east, the following maps with a scale of 1:1 mio. (million): TJ12, TJ11, TJ22, TJ21, TH23, TJ32, TJ31, TH33, UJ12, UJ11, and UH13. All these maps are accessible via the internet (2011, glacier maps downloadable at: http://booksite.elsevier.com/9780444534477/).

Also accessible via the same link are maps (1:1 mio.) depicting glacier expansions that were reconstructed for other regions in High Asia, but published separately (Sect. 2.11). Seen from north to south and from west to east, they carry the following sheet-denominations: TK22, TK11, SJ33, SJ32, SJ31, RJ31, RH33, TK21, TJ23, TK31, UJ22, UJ21, UH23, UJ32, UJ31 and UH33 (ibid.).

On the empirical basis of numerous, extensive field investigations, which altogether span a 6.5-year period and have been carried out by the author since 1973, the overall glaciation of High Asia during the Würm Ice Age can be summarized and depicted as in Fig. 4.1. It shows the three glaciation centers of Tibet: I1, I2, and I3. A fourth glaciation center is located WNW in the upland region of the Pamir and Tian Shan (Fig. 4.1a). The glaciation centers of Tibet were elongated ice domes separated from one another by ice-free depressions. In the north, this was the up to 200 km wide depression of the ca. 800 km long Tsaidam Basin, filled with a ca. 200 m deep lake (see above). In the south, a ca. 550 km long, narrow section of the medial Tsangpo Valley separated ice centers I2 and I3. I2 had a length of ca. 3000 km. In the north it connected to the outlet glaciers of the Kuenlun, in the west to the Karakoram ice stream network, and in the east joined together with the ice stream network of the Animach'ing (A'nyèmaqén) and the high mountains of Szechuan with the Minya Konka (Fig. 4.1b). I1 had contact with the ice stream net of the Quilian Shan in the north. I3 extended its outlet glaciers southward all the way through the Himalayas. This ice flow led across transfluence passes ("Transfluenzpässe") and followed lateral antecedent transverse valleys (Fig. 2.122). Simultaneously, the ice stream net of the 2500 km long Himalayan bow was connected with I3. The extensive glacier covers of the Pamir and Tian Shan plateau were connected to one another via the narrow mountain range glaciations of the King Ata Tagh, Koktutau, Dankowa, and Terski Alatau. To the south, a glacier contact existed

The characters and symbols in the figures in this chapter can also be viewed in the online version of the figures at http://extras.springer.com/.

M. Kuhle, *The Glaciation of High Asia*,
https://doi.org/10.1007/978-3-319-77566-1_4

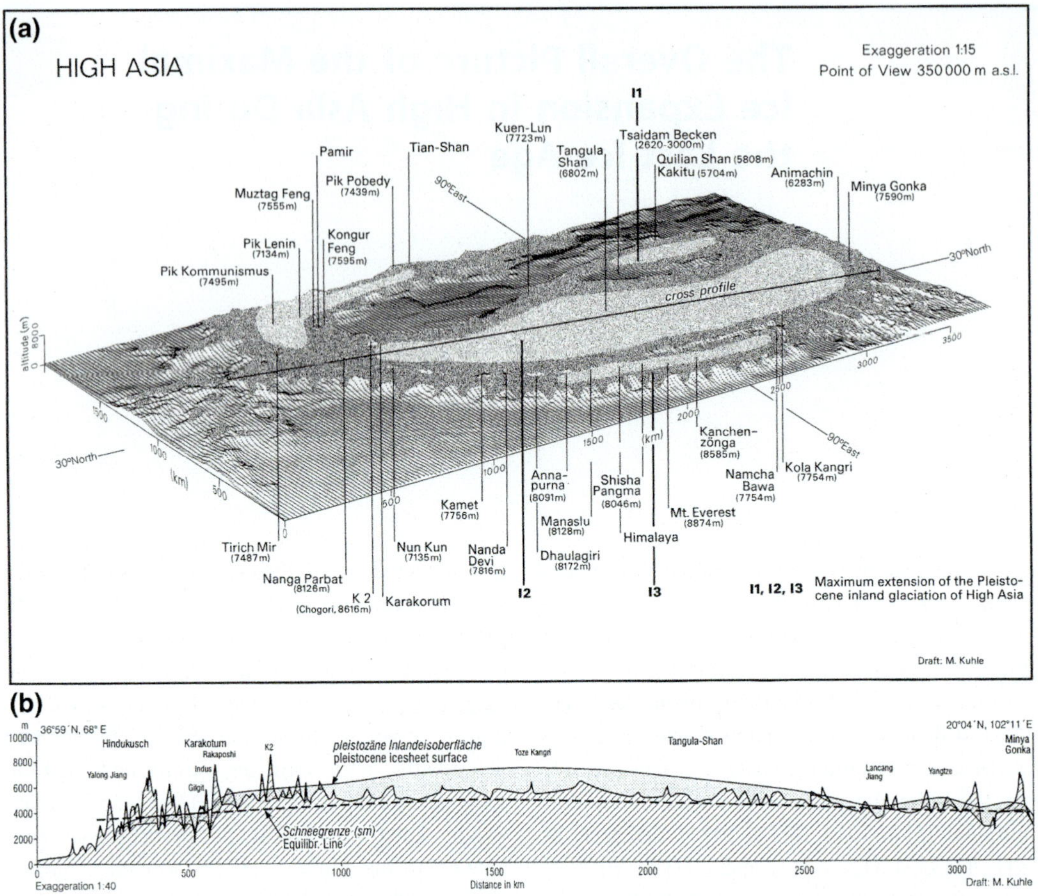

Fig. 4.1 **a** Block diagram showing the reconstructed 2.4 million km^2 ice sheet, or ice stream network, covering the Tibetan Plateau, with the three centers I1, I2, and I3. Only peaks higher than 6000 m asl rise above the ice surface. **b** Cross-profile of the central ice sheet, extending from the Hindu Kush in the west to the Minya Gonka (Konka) in the east (Kuhle 1980, 1982, 1987, 1988, 1990, 1991, 1994, 1997, 1998, 1999, 2001, 2004, 2005, 2011, 2013)

between the ice cover on the Pamir plateau and the ice stream network of the Karakoram. To the Tibetan ice sheet in the southeast there existed a two-way bridging contact: on the one hand, by way of the glaciation of the Kuenlun, but also toward the SSE via the Karakoram ice stream network (Figs. 2.7, 2.16, 2.18 and 2.19).

4.1 The Height of the Prehistoric Inland Ice Surface in Tibet

The largest surface altitude of the Tibetan ice sheet was reconstructed at ca. 7000 m asl (Fig. 4.1b). This value was reconstructed in reference to the Ulugh Muztagh in the northern part of Central Tibet (Sect. 2.4.6.1). For the center of Central Tibet, at the Purog Kangri, surface levels of between 6700–6800 m and 7000 m asl are assumed (Figs. 2.156, 2.178 and 2.180). For both regions, ice thicknesses of up to more than 2000 m are estimated. These inland ice surface levels and thicknesses are for good reasons attributed to the High Glacial (Stage 0 in Fig. 2.5). At the 6500 m high Zangsar Purog-II-Kangri Massif, a minimum surface level of 6420 m asl is evidenced. Joining this area to the south are 7 glacigenically abraded, perfectly rounded mountain peaks with altitudes of between 6260 and 6050 m asl (see also Fig. 2.102). These peak and abrasion forms provide evidence of inland ice levels belonging to the Late Glacial Stages I to II and perhaps even

Stage III (cf. Sects. 2.4.6.1 and 2.4.7.3). This means that dating from the High Glacial until the late Late Glacial, these round peaks were overridden by the inland ice with decreasing thicknesses (Sect. 2.4.7.4.6).

4.2 The Lack of Cirques as Evidence for a Last Inland Glaciation of Tibet During the Last Ice Age (LGP, LGM, Würm Glacial)

Cirques with steep backslopes (slopes below which lies a considerably more flat cirque floor) are formed in Tibet in those mountains where current (today's) glaciers exist (Figs. 2.76, 2.77, 2.80 and 2.195). On slightly lower mountains, which are no longer glaciated precisely because of their lower altitude, cirques are completely absent. These are, among others, the above-mentioned (in Sect. 4.1), completely rounded mountain ridges with altitudes of up to ca. 6260 m asl (see Figs. 2.13–2.15, 2.92, 2.106, 2.108, 2.111, 2.114, 2.128, 2.130, 2.180, 2.185–2.188, 2.191, 2.192, 2.198 and 2.199). With ΔELAs (snowline depressions) of only a few hundred meters, glaciers must still have existed on these mountains and thereby cirques must, theoretically speaking, have been formed. Since these are missing, however, cirque formation must have been prevented by the still-existing inland ice sheet that completely covered these mountains.

Accordingly, it must be established here that cirque formation in Tibet only took place during the Interglacial. This is the case because cirques can only be formed on those mountains that exceed the highest, Interglacial ELA (snowline). During the Glacial (of the Ice Age Era), however, an inland ice sheet had formed that covered the entire mountain terrain. This complete cover prohibited the formation of cirques. The mountains became rounded through abrasion. If the melting of the covering ice sheet outlasted the Late Glacial, cirques also could not have formed, as the mountains that might have been subject to Late Glacial cirque formation were still covered by the remaining inland ice sheet. This total glaciation persisted into the Postglacial, i.e., it persisted into the Holocene. During the Holocene the ELA then moved above the altitude of the respective mountains (Sects. 2.4.7.4.5 "Second example," 2.4.7.4.6, and 2.9.2). The mountains were therefore too low for the local formation of small, Holocene glaciers on the respective slopes and thus for the formation of interglacial cirques.

If no inland ice had existed during the last ice age, the Würm Ice Age (LGP), in Tibet, those very mountains with elevations of between 5600 and 6260 m asl would have cirques today. However, because the 5600–6260 m high mountains are rounded, and do not contain cirques, by necessity an inland ice sheet must have existed in Tibet.

It is for this same reason that the mountains of the Scandes (Northern Europe) in Scandinavia do not have Late Glacial cirques. It is here that the North European inland ice sheet lay during the Last Ice Age (LGP, Würm, MIS 3-2). Its melting also lasted into the Postglacial. Accordingly, the Scandinavian inland ice glaciation also persisted into the Holocene (e.g., Schwarzbach 1974).

4.3 Summary

Through 6.5 years of field reseach carried out by the author in both the driest regions of High Asia and Tibet—which are primarily dealt with in this book due to their relevance to the question regarding the former glaciation—as well as in the humid regions also studied since 1973, it has been established that an extensive, all-encompassing inland ice sheet existed during the last Ice Age. Figure 4.1 shows its longitudinal profile and basic outline, as well as the three glaciation centers of Tibet (I1, I2, and I3), including a fourth center in the region of the Pamir and Tian Shan plateau. The three glaciation centers of Tibet were ice cupolas that were separated from one another from north to south by the lake-filled Tsaidam Basin and the middle section of the Tsangpo Valley. Outlet glaciers of the Tibetan inland ice sheet flowed through the mountain areas surrounding the Tibetan Plateau, including the Himalaya, Karakoram, Kuenlun, Quilinan Shan, and the mountains to the north and south of the Minya Konka—and partly also down into the lowlands. The local autochthonous glaciation of these mountain massifs generated ice-stream networks that consisted of decakilometer-long, vastly dendritic valley glacier arms. The surface of the flat-cupola-shaped ice domes of the central inland ice sheet reached—at least during the LGM (Stage 0)—approx. 7000 m asl, which amounts to a glacier thickness of up to and above 2000 m.

The characteristic formations of small hanging glaciers, the cirques, are only formed in Tibet on those mountains that today rise above the snowline level (ELA). On slightly lower mountains they are missing even though during the late Late Glacial (Stage IV) and Holocene (Stage V), their peaks must have risen above the lowered ELA. Consequently, cirque formation must have been prevented here by an inland ice sheet that completely covered those mountains. As a result, these mountains are perfectly rounded in many places. By the time the peaks were eventually ice-free, the ELA had already been elevated above the level of their peaks, so that no hanging glaciers or cirques could be formed. The absence of interglacial cirque formation on these 5600–6260 m high mountains therefore proves a climatically strongly delayed deglaciation. This was necessarily due to the massive inland ice bodies.

References

Kuhle M (1980) Klimageomorphologische Untersuchungen i. d. Dhaulagiri- u. Annapurna-Gruppe (Zentraler Himalaya). Tagungsber. u. wiss. Abh. 42. Deutschen Geographentages 1979:244–247

Kuhle M (1982) Der Dhaulagiri- und Annapurna-Himalaya. Zeitschrift für Geomorphologie Suppl. 41, vol I, vol II, Abb. 1–184:1–229

Kuhle M (1987) Subtropical mountain- and highland-glaciation as ice age triggers and the waning of the glacial periods in the Pleistocene. GeoJournal 14 (4):393–421

Kuhle M (1988) Geomorphological findings on the build-up of Pleistocene glaciation in southern Tibet and on the problem of inland ice. Results of the Shisha Pangma and Mt. Everest expedition 1984. In: Kuhle M, Wang Wenjing J (eds) Tibet and High Asia. Results of the Sino-German joint expeditions (I). GeoJournal 17(4):457–511

Kuhle M (1990) New data on the Pleistocene glacial cover of the southern border of Tibet: the glaciation of the Kangchendzönga Massif (8585 m, E-Himalaya). GeoJournal 20(4):415–421

Kuhle M (1991) Observations supporting the Pleistocene inland glaciation of High Asia. In: Kuhle M, Xu Daoming (eds) Tibet and High Asia. Results of the Sino-German joint expeditions (II). GeoJournal 25(2–3):133–231

Kuhle M (1994) Present and Pleistocene glaciation on the north-western margin of Tibet between the Karakorum Main Ridge and the Tarim Basin supporting the evidence of a Pleistocene inland glaciation in Tibet. In: Kuhle, M. (Ed.). Tibet and High Asia. Results of the Sino-German and Russian-German joint expeditions (III). GeoJournal 33(2–3):133–272. Kluwer, Dordrecht

Kuhle M (1997) New findings concerning the Ice Age (LGM) glacier cover of the East Pamir, of the Nanga Parbat up to the Central Himalaya and of Tibet, as well as the Age of the Tibetan Inland Ice. In: Kuhle M (ed) Tibet and High Asia (IV). Results of investigations into high mountain geomorphology. Paleo-Glaciology and Climatology of the Pleistocene. GeoJournal 42(2–3):87–257

Kuhle M (1998) Reconstruction of the 2.4 million km^2 Late Pleistocene ice sheet on the Tibetan Plateau and its impact on the global climate. Quatern Int 45/46:71–108 (additional Figures in: 47/48:173–182)

Kuhle M (1999) Reconstruction of an approximately complete Quaternary Tibetan inland glaciation between the Mt. Everest- and Cho Oyu massifs and the Aksai Chin. A new glaciogeomorphological SE-NW diagonal profile through Tibet and its consequences for the glacial isostasy and Ice Age cycle. In: Kuhle M (ed) Tibet and High Asia (V). GeoJournal 47 (1–2):3–276

Kuhle M (2001) The maximum Ice Age (LGM) glaciation of the Central- and South Karakorum: an investigation of the heights of its glacier levels and ice thickness as well as lowest prehistoric ice margin positions in the Hindukush, Himalaya and in East-Tibet on the Minya Konka-massif. In: Kuhle M (ed) Tibet and High Asia (VI): Glaciogeomorphology and Prehistoric Glaciation in the Karakorum and Himalaya. GeoJournal 54(1–4) and 55(1):109–396

Kuhle M (2004) The high glacial (Last Ice Age and LGM) glacier cover in High- and Central Asia. Accompanying text to the mapwork in hand with detailed references to the literature of the underlying empirical investigations. In: Ehlers J, Gibbard PL (eds) Extent and chronology of glaciations, vol 3 (Latin America, Asia, Africa, Australia, Antarctica). Elsevier B.V., Amsterdam, pp 175–199

Kuhle M (2005) The maximum Ice Age (Würmian, Last Ice Age, LGM) glaciation of the Himalaya—a glaciogeomorphological investigation of glacier trim-lines, ice thicknesses and lowest former ice margin positions in the Mt. Everest-Makalu-Cho Oyu massifs (Khumbu- and Khumbakarna Himal) including informations on late-glacial-, neoglacial-, and historical glacier stages, their snow-line depressions and ages. In: Kuhle M (ed) Tibet and High Asia (VII): Glaciogeomorphology and former glaciation in the Himalaya and Karakorum. GeoJournal 62(3–4):193–650

Kuhle M (2011) The High Glacial (Last Ice Age and Last Glacial Maximum) Ice Cover of High and Central Asia, with a critical review of some recent OSL and TCN Dates. In: Ehlers J, Gibbard PL, Hughes PD (eds) Quaternary Glaciation—extent and chronology, a closer look. Elsevier B.V, Amsterdam, pp 943–965 (glacier maps downloadable at: http://booksite.elsevier.com/9780444534477/)

Kuhle M (2013) The uplift of High Asia above the snowline and its glaciation as an albedo-dependent cause of the quaternary Ice Ages. Nova Science Publishers Inc., Hauppauge, NY, pp 1–240

Schwarzbach M (1974): Das Klima der Vorzeit. 3 Auflage, Stuttgart, pp 1–380

5 The Paleoclimatic Significance of the Ice Age Glaciation of High Asia

The Tibetan inland ice cover had a minimum extent of 2.4 million km^2. The glacier covers of the entirety of High Asia were, as shown by Fig. 5.1, even slightly larger. Unlike all the other prehistoric inland glaciations on earth, the Tibetan glaciation occurred at subtropical latitudes. It could develop there only because the Tibetan plateau rose high enough so that it reached the ELA. The nourishment area of the inland ice consequently comprised the entire surface of the plateau. On it developed ice thicknesses of up to more than 2000 m. It was thus a subtropical inland ice sheet with a surface lying at 5000–7000 m asl (Chap. 4.1). The subtropical latitudinal position is responsible for a ca. four times greater global radiation intensity than in all the other former inland ice regions (Bernhardt and Phillips 1958). The present author summarized his radiation budget measurements in High Asia at between 3800 and 6650 m asl as follows: "Approximately 25,000 representative pieces of data on global radiation, return radiation and albedo were obtained. When weather conditions were such that radiation was not impeded by cloud cover, the values of incoming radiation were between 1000 and 1300 W/m^2, which is approximately the solar constant at the upper limit of the atmosphere in relation to the corresponding position of the sun at that time. Theoretical incoming radiation on September 21st, as a mean value, is about 1180 W/m^2 at the latitude of the area under investigation (30°N). With an atmosphere approximately transparent to radiation, incoming radiation at Tibetan altitudes produces an energy input at least four times higher compared to between 60°N to 70°N latitude of the Pleistocene North European inland ice centre" (Kuhle 1998: 100 (1989: 277 in German language)).

When the higher transparency of the atmosphere at high altitudes (Lauscher 1956) of between 5000 and 7000 m asl is added to the effect of its latitudinal position, the global radiation on the Tibetan inland ice surface reached energy levels that were approximately five times larger (Kuhle 1987a: 408, Fig. 22; Kuhle and Jacobson 1988) than those that impinged on lowland ice covers in high-latitude locations. The ca. 2.4 million km^2 large inland ice surface reflected ca. 90% of this shortwave global radiation back into outer space.

The reason the albedo was so extremely high was that nearly the entire inland ice surface belonged to the nourishment area (see above). It was therefore entirely covered by snow (Kuhle 1987a; Kuhle 1989: 278, 279, Abb. 17–19). Whereas the lowland glaciations that had flat and thus very wide ablation areas, and which only

The characters and symbols in the figures in this chapter can also be viewed in the online version of the figures at http://extras.springer.com/.

M. Kuhle, *The Glaciation of High Asia*,
https://doi.org/10.1007/978-3-319-77566-1_5

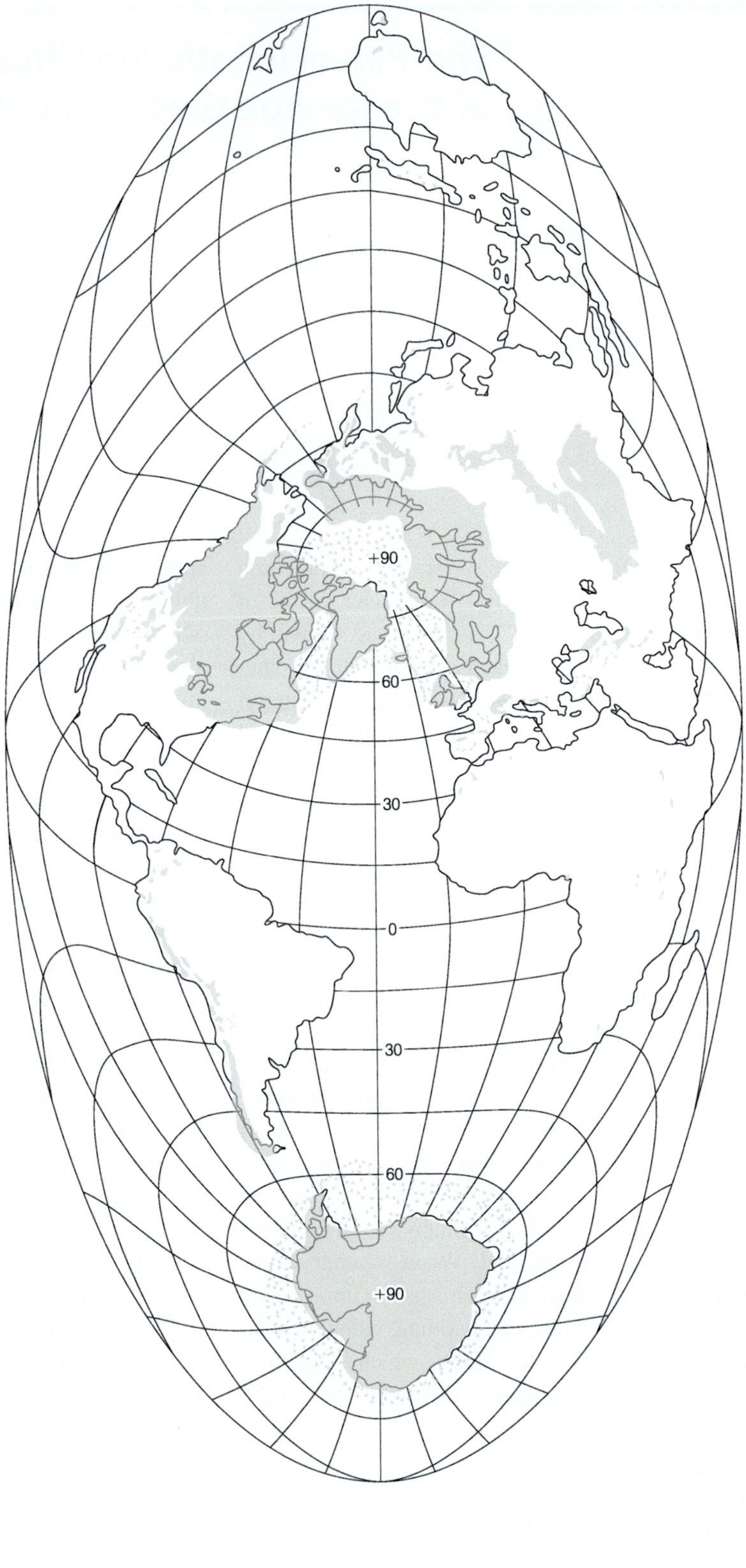

Fig. 5.1 The maximal extent of glaciation during the LGM, based on an equal-area projection. Continental ice is indicated by a dense pattern, marine ice by a sparse pattern. When compared to the lowland ice sheets in the polar regions, the 2.4 million km^2 Tibetan ice sheet is remarkable for its unique, extremely insolated, subtropical location, with an average elevation of 6000 m asl. Based on Broecker and Denton (1990), modified after Kuhle (1980, 1982, 1987a, 1988a, 1990, 1991, 1994, 1997, 1998a, 1999, 2001, 2004, 2005, 2011, 2013)

rose above the ELA at the center, the ablation areas of the Tibetan glaciation were associated with very narrow outlet-glacier tongues. These reached steeply down to ca. 2000 m asl at the northern margin of the plateau and down to below 1000 m asl at the southern margin of Tibet. The peripheral ablation areas consisted of dirty-grey ice with a correspondingly low albedo of ca. 20% (Kuhle 1987a: 409, Fig. 23).

Herein lies another particularly distinctive climate-based characteristic of a subtropical—and therefore plateau-located—inland ice cover. Its peripheral ablation areas are very narrow because the outlet glaciers flow steeply down into very warm regions and are therefore extremely short (Fig. 4.1).

Figure 5.1 makes clear that, in terms of size, this was only a relatively small inland ice sheet that was only ca. 1.25 times larger than the current Greenlandic inland ice. Because of its fivefold-greater energy reflection, however, it represented, in terms of its effect on climate, an inland ice cover almost the size of the Antarctic inland ice (Kuhle 1987a, 1988b, 1989, 1998, 2002, 2009, 2013).

In order to clarify the enormous palaeoclimatic impact of Tibet, we may consider that today the subtropical plateau displays its energy impact in the *opposite* direction, namely by being the most efficient heating surface of the earth's atmosphere. This is due to the low albedo of the today unglaciated, dark surface areas of gravel and rock (Kuhle 1985, 1987a). Accordingly, the present author stated in 1987: "It is the author's opinion that, increasingly raised by the 'sub-tropical latitude factor' and the '… massive elevation of the huge Tibetan highland, which increases the temperature of the free atmosphere in summer more than in any other place on earth' (as Flohn wrote in 1959: 323), the enormous increase in glacier areas becomes [became] the trigger of further drops in temperature" (Kuhle 1987a: 415).

This most efficient heating surface of the earth's atmosphere was thus turned into an equally efficient cooling surface during the ice ages. The crucial step for this was the uplift of the Tibetan Plateau above the ELA. This took place for the first time at the beginning of the Pleistocene (Quaternary) Ice Age Era and then again during each subsequent ice age until the last one, the Würm Ice Age (LGP; Stage 0 in Fig. 2.5).

Based on the present author's proof of the Tibetan inland ice cover, the empirical foundation for a terrain- or relief-specific model of the ice age globally lies in the uplift-generated glacier areas in Tibet and the corresponding albedo increase, as well as their positive climatological feedback by means of the global radiation geometry. This leads to an ice age theory with the following headline: Uplift of High Asia above the snowline and its glaciation: The albedo-dependent cause of the Quaternary ice ages (Kuhle 1987a, 1988b).

Bielefeld (1997) has calculated—with the help of worldwide data from the *Atlas of the Surface Heat Balance of the Continents* (Henning 1989) and the *Heat Balance of the Arctic* (Budyko 1971)—that the Tibetan ice, with its size of 2.4 million km^2, accounted for 32% of global cooling during the entire Last Glacial: "In other words, the existence of a Tibetan ice sheet would further reduce the global loss in radiation by 32%" (Bielefeld 1997: 334). "Even when the compensatory effect of an infra-red emission change caused by a drop of 7° in the earth's surface temperature, a global reduction in radiation of 7%, without the Tibetan ice sheet, and 10%, taking the Tibetan ice into account, must have taken place." (ibid.: 335).

Bielefeld has thus confirmed Kuhle's energetic explanation of an ice age origin through these calculations.

5.1 Modeling of the Maximal Ice Expansion in High Asia for the Estimation of Precipitation Conditions

Modeling must be based on empirical data in order to be meaningful and valid. In the current study areas, the present author has reconstructed the former Equilibrium Line Altitudes (ELAs) on the basis of former glacier extensions and ice margins, which in turn were reconstructed on the basis of moraines. The reconstruction of the former ELA made it possible to determine the ΔELA (snowline depression) in opposition to the current position of the ELA for Tibet. The largest, average ΔELAs reconstructed in this manner, which were reconstructed for numerous test areas in Tibet and along its margins (Fig. 2.1) and which naturally also applied to the inland glaciation of Tibet, were projected onto the high-altitude data of the plateau. In so doing, a precipitation amount of 100 mm per year was estimated for the Tibetan region during the LGM. This is only ca. one-fourth (25%) of the current precipitation.

The model relates to I2 and I3 in Fig. 4.1, i.e., to Central and southern Tibet. The area north of I2 was not empirically included. The modeling of the ice sheet was developed by Klaus Herterich and Reinhard Calov from the Max-Planck-Institute for Meteorologie in Hamburg (Kuhle et al. 1989).

The first test run of the model was conducted using an average Tibetan prehistoric ELA of 4250 m asl relative to the present altitude of the plateau. Figure 5.2 shows the result after 10,000 years.

An inland ice cover emerges that covers the entirety of Tibet (Kuhle 1989), with local ice thicknesses amounting to 1.0–1.8 km. This, however, is less than the maximal ice thickness actually observed. Therefore, the amount of precipitation of no more than 100 mm/yr is too low, or else the time span of 10,000 years too short. It must be assumed that the thickness of the continental, arid, and therefore very cold plateau ice, with average annual temperatures of below −20 °C, must have become considerably larger still. Only then could the cold-brittle ice outflow have outbalanced the accumulation of the inland ice on the plateau. Only when the peripheral glacier outflow of the Tibetan ice equaled the nourishment by snow in the center could a balanced mass balance have existed. Only then would have ended the increase in ice thickness up to the center of the inland ice.

Figure 5.3 considers the functions of terrain, ice thicknesses, and flow properties after the first 10,000 years of inland ice. It indicates a correspondingly complex structure of ice runoff conditions.

This figure also shows the surface height of the cupola-shaped inland ice. With its three largest runoff centers, the inland ice cupola rises up to more than 6500 (6.5 km) m asl. The most far-spanning, largest cupola surface encompasses an expanse at 6500 m asl between west of the Zangsar-Purog II Massif (Sects. 2.4.6, 2.4.6.1, 2.4.7.1, and 2.4.7.3) and east of the Gladongding Massif (Kuhle 1991). In its center lies the Purog Kangri (Purog I) and its broader surroundings (Sects. 2.8 and 2.8.3). After only 10,000 years, this central inland ice cupola here already had attained an Early Würm Glacial ice build-up with a width of 270–300 km. This largest cupola surface must, for the previously mentioned reasons, have been still higher, broader, as well as longer during the LGM (Stage 0 in Fig. 2.5). The number of Tibetan ice domes must naturally have been reduced with an increase of the inland ice cover's thickness. At the time of the maximum stage (Stage 0), the largest and highest central dome had "swallowed" some of the neighboring ice domes, i.e., had integrated them into its own cupola geometry.

Figure 5.4 proves this scenario in reverse chronological fashion. With otherwise unchanged parameters, the average ELA is at 4500 m asl, i.e., 250 m higher than in the test run of Figs. 5.2 and 5.3. This corresponds to the elevation of the ELA from the High Glacial (Stage 0) to the early Late Glacial of Stages I to II (which still existed ca. 17 ka to 15 ka ago; Fig. 2.5). The above-mentioned highest ice dome of Central Tibet is here dissolved into three small ice domes that still rise above 6500 m asl. Such a stage, however, not only existed during the last Late Glacial, but also during the Early Glacial. It must analogously also have happened before during the previous Quaternary ice ages: in each case once during the growth of the ice and once again during the meltdown of the ice sheet (Kuhle 1995).

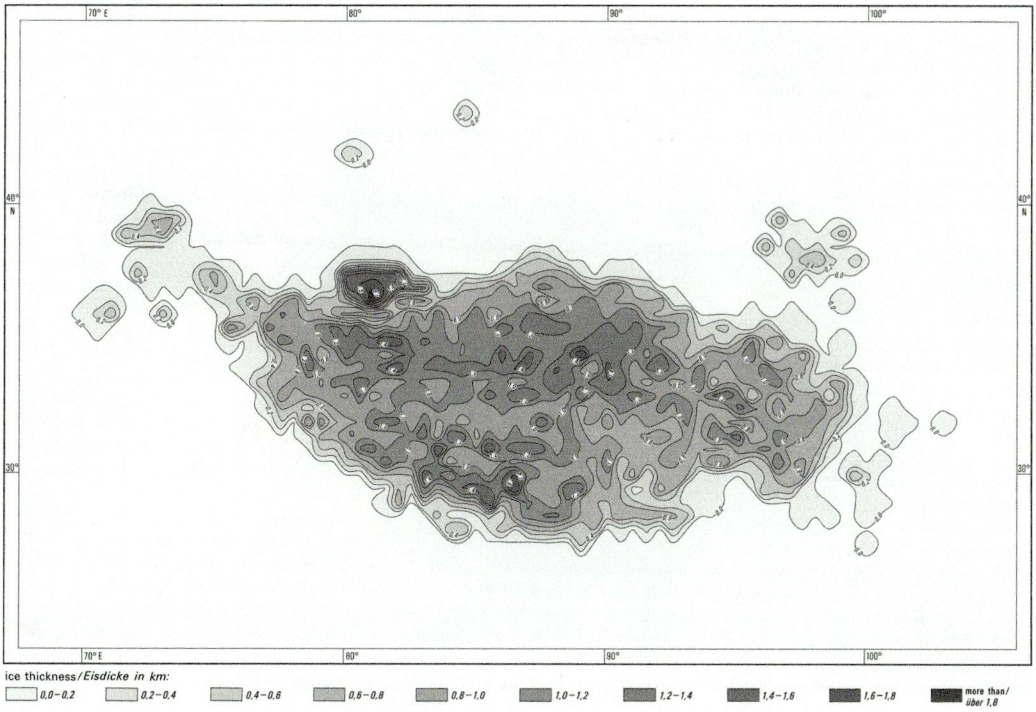

Fig. 5.2 The glacial ice thickness in Tibet after 10,000 years, assuming an average precipitation of 100 mm/yr and an average ELA at 4250 m asl. The model, prepared by Herterich and Calov, is based on the empirical data obtained by the present author in the respective research areas (see Fig. 5.2). These include the lowest High Glacial (LGM = Stage 0) glacier margins, reconstructed with the help of end moraines, and the ascertained ELA depression (ΔELA) as well as the estimation of temperature and precipitation (b max) in Tibet and its surrounding mountain ranges. The model is based in detail on: (1) the graph of the annual snow balance b (b max = 100 mm/y) as a function of the height above the ELA; (2) the contour lines (m asl) of the glacial snow surface (ELA = 4250 m asl) in Tibet; this fits the observed heights of the glacial equilibrium line (according to Kuhle et al. 1989: 204–205); and (3) t = 10^3 yr. Cartography: E. Höfer and A. Flemnitz. After Kuhle 1997: 120, Fig. 46

Even with the elevated ELA (Fig. 5.4) and persistent low precipitation of 100 mm/yr, Tibet still glaciates completely after 10,000 years. Furthermore, inland ice thicknesses of more than 1000 m are reached during this time span. This result of the modeling is significant, because, for one, it shows that the Tibetan ice could emerge during the Early Würm within only 10,000 years. And, moreover, because it confirms the present author's claim that it could very well have functioned as the forerunner and pacemaker for all the individual Pleistocene ice ages (Kuhle 1987a, 1988b, 1989, 1998).

It is conspicuous that not only in the first (Figs. 5.2 and 5.3) but even in the second modeling the Tsangpo Valley was completely filled with glacier ice (Fig. 5.4). Figure 4.1 (between I2 and I3), in contrast, leaves the eastern section of the Tsangpo Valley ice-free, namely in the section between north of the Shisha Pangma and the Namcha Barwa (Kuhle 1988a: 508, 1988b: 582). For precisely this valley section, the present author had not yet possessed sufficient field data that could definitively prove a glacier filling (Kuhle 1991). Herein then lies the discrepancy between empirical evidence and modeling.

This section must be concluded with a remark relating to the northern margin of the Tibetan Plateau. Today this region certainly only records

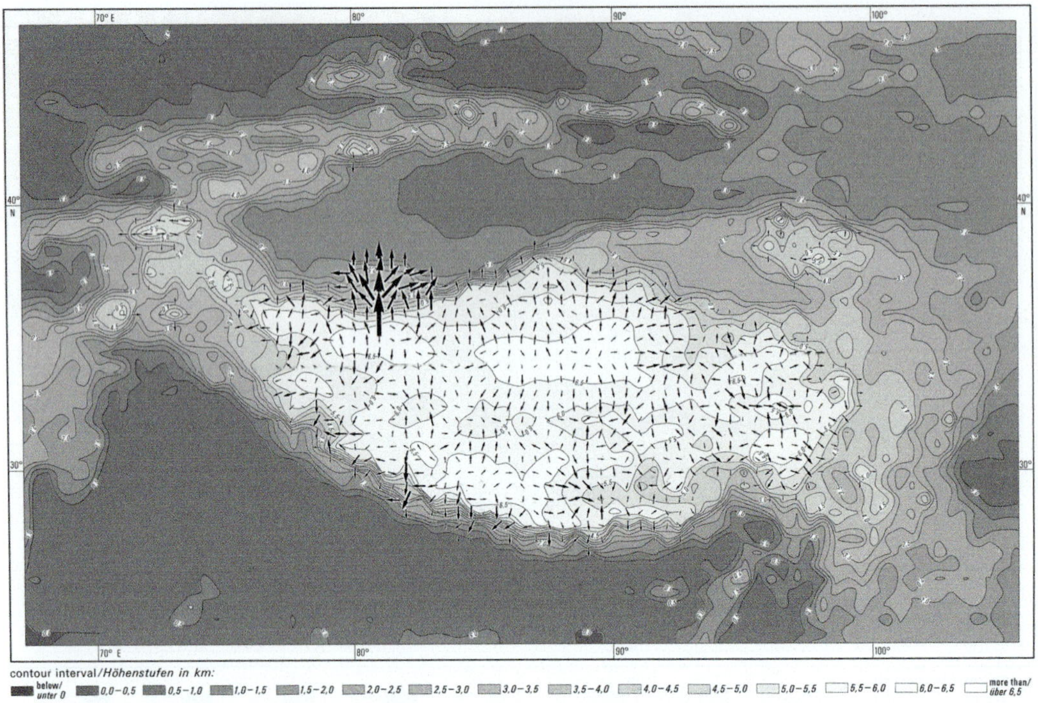

Fig. 5.3 Surface altitudes and ice flow (the arrows and their different thicknesses indicate the flow vectors) of the High Glacial Tibetan ice sheet at an average ELA of 4250 m asl (Stage 0; see Figs. 2.5 and 5.2), modeled by Herterich and Calov according to ELA reconstructions in High Asia of the present author (see Fig. 5.2). Precipitation (b max) = 100 mm/y; integration time of the ice accumulation = 10,000 years. Cartography: E. Höfer and A. Flemnitz. After Kuhle 1997: 122, Fig. 47

low precipitation of even below 200 mm/yr. During the Ice Age, however, it was much more humid here than the 100 mm/yr specified in the models. In the Tarim Basin, the Gobi, as well as in the Tsaidam Basin (Qaidam Pindi), lakes existed (Sect. 2.5.). Lake Tsaidam, for example, was at least 410 km long and 110 km wide, and was calculated to be 35–47 ka old based on radiocarbon dating (Kuhle 1987b: 307; 1998: Fig. 15). In other words, we can reconstruct hygric conditions that are not unfavorable for glaciation. The locally very differentiated distribution of precipitation has already been discussed elsewhere by the author (Kuhle 1997: 254).

The likewise favorable, very low temperatures can be made clear by way of a comparison. While on the North European mainland in the region of the High Glacial inland ice there is practically no interglacial permafrost today, such permafrost still exists in Tibet across vast expanses.

5.2 Summary

The Tibetan inland ice cover, located in subtropical latitudes, had a total area of 2.4 million km^2. Moreover, there existed further glacial areas in High Asia. The uplift of Tibet above the snowline (ELA) led to a 5000 to 7000 m high glacial nourishment area equaling the size of the entire plateau. Its enormous paleoclimatic impact is based on the large energy efficacy of its surface. Due to its position in subtropical latitudes and its high altitude, incoming global radiant energy is approximately five times higher than in all the other regions with glacial inland ice in the Nordic lowlands. While due to their low albedo, the dark rock and

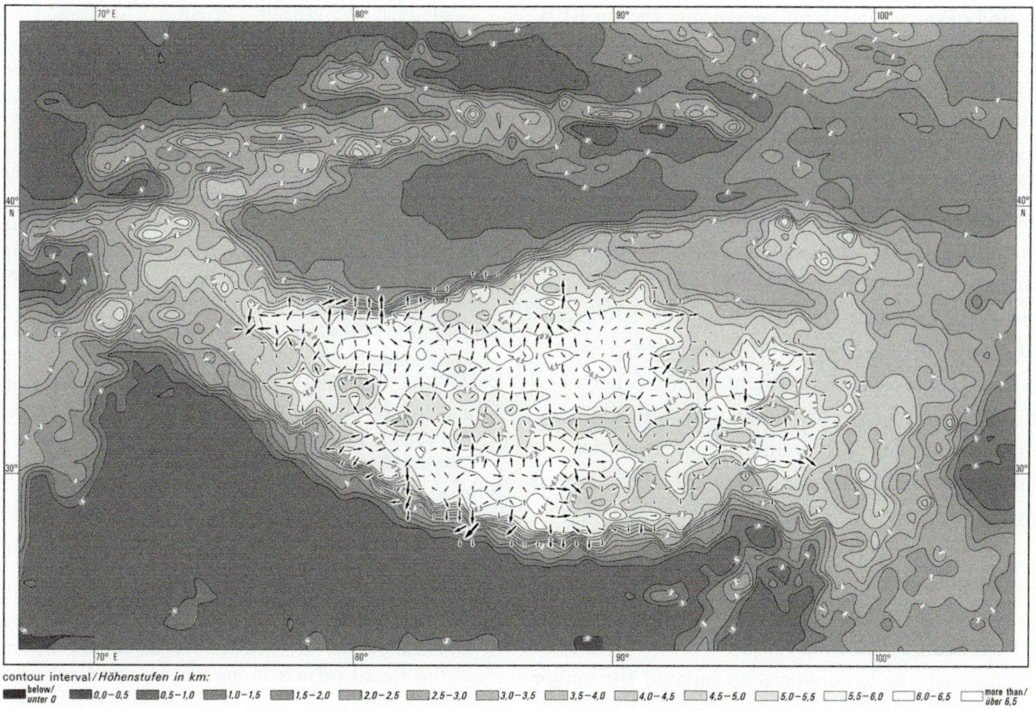

Fig. 5.4 Surface altitudes and ice flow (the arrows and their different thicknesses indicate the flow vectors) of the Tibetan ice sheet during its Early and Late Glacial (Stage I to II, see Fig. 2.5) extension at an average ELA at 4500 m asl, modeled by Herterich and Calov according to ELA reconstructions in High Asia by the present author (see Fig. 5.2). Precipitation (b max) = 100 mm/y; integration time of the ice accumulation = 10,000 years. Cartography: E. Höfer and A. Flemnitz. After Kuhle 1997: 124, Fig. 48

rubble surfaces of today's ice-free plateau make Tibet the most efficient heating surface of the earth's atmosphere, its glacial ice cover equally made Tibet the most efficient cooling surface on the planet, thus reflecting enormous irradiation energy (up to 90% of the total received) directly back into space in the form of shortwave radiation. The high albedo of the snow-covered and firn-ice surfaces thus prohibited today's transformation into thermal (heat) radiation. Due to its subtropical position and high altitude, the Tibetan ice sheet and adjacent glacial areas thus had a cooling effect equivalent to a Nordic lowland inland ice cover of ca. 12–13 million km^2—which equals the size of Antarctica. The 2.4 million km^2 sized Tibetan ice sheet thereby accounted for 32% of the entire Last Glacial cooling (i.e., that produced by all of the earth's Last Glacial ice covers in aggregate). This necessarily leads to the ultimate conclusion of an ice-age-era triggering effect through the uplift of High Asia above the snowline, which in turn justifies a relief-specific and in terms of global radiation geometry well-founded theory of ice age genesis.

The reconstruction of the former glacial extensions as well as Equilibrium Line Altitudes (ELAs), which were based entirely on empricial field data, has been used for the modeling of precipitation estimations. This calculation was also carried out together with Herterich and Calov by means of a 3-D model (Kuhle et al. 1989). The largest snowline depressions (ΔELAs) during the inland glaciation of Tibet, which had been assessed in Tibet and along its fringe areas in numerous study areas (Fig. 2.1), were projected onto the altitude data of the plateau region. During this exercise, an annual precipitation of 100 mm was assumed, which is only about 25% of the current precipitation in Tibet. The first testing of the model, with an average former ELA for Tibet of 4250 m asl, yielded a Tibetan inland ice cover that after 10,000 years displayed locally differentiated ice thicknesses of 1.0–1.8 km. The second testing of the model, with an average former ELA for Tibet of 4500 m asl, still yielded a completely covering Tibetan inland ice sheet that after 10,000 years developed ice thicknesses still above 1.0 km. The ELA of 4250 m asl corresponds to Stage 0 (MIS 3-2, Würm Glacial, LGP) and the ELA of 4500 m asl corresponds to a Late Glacial snowline ascent during Stages I to II (ca. 17–15 Ka ago). The testings still yield ice thicknesses well below the actual maximal ice thicknesses and surface levels of the Tibetan inland ice cover for two reasons:

1. because the large lakes north of Tibet suggest a more humid climate, with precipitation exceeding 100 mm/y;
2. because the ice cover accumulated for longer than the 10,000 years specified in the model.

Still, in both test trials, the surface level of the Tibetan inland ice already reached altitudes of more than 6500 m asl. In reality, it must have been considerably higher than that.

At any rate, the results of the modeling show clearly that the Tibetan ice cover could develop during the Early Würm within only 10,000 years, and that consequently it could have played the role assigned to it by the present author as the probable forerunner and pacemaker of all individual Pleistocene ice ages, not only energentically but also in terms of timing.

References

Bernhardt F, Philipps H (1958) Die räumliche und zeitliche Verteilung der Einstrahlung, der Ausstrahlung und der Strahlungsbilanz im Meeresniveau. Die Einstrahlung. Abhandlung des Meteorologisch-Hydrologischen Dienstes 15:1–227

Bielefeld B (1997): Investigation into albedo–controlled energy loss during the last glaciation. In: Kuhle M (ed) Tibet and High Asia IV, Results of investigations into high mountain geomorphology, Paleo-Glaciology and climatology of the Pleistocene (Ice Age Research) GeoJournal 42(2–3):329–336. Kluwer Academic Publishers, Dordrecht, Netherlands

Broecker WS, Denton GH (1990) What drives glacial cycles? Sci Am 1:48–56

Budyko MI (1971) Heat Balance of the Arctic. Bonner Meteorologische Abhandlungen, Bonn

Flohn H (1959) Bemerkungen zur Klimatologie von Hochasien. Aktuelle Schneegrenze und Sommerklima. Akademie der Wissenschaften und der Literatur Mainz, Abhandlungen der Mathematisch-Naturwissenschaftlichen Klasse 14 (Jahrgang 1959), pp 309–331

Henning D (1989) Atlas of the surface heat balance of the Continents. Gebruder Borntraeger, Berlin

Kuhle M (1980) Klimageomorphologische Untersuchungen i. d. Dhaulagiri- u. Annapurna-Gruppe (Zentraler Himalaya). Tagungsber. u. wiss. Abh. 42. Deutschen Geographentages 1979:244–247

Kuhle M (1982): Der Dhaulagiri- und Annapurna-Himalaya. Zeitschrift für Geomorphologie, Suppl. 41, vol I, vol II, Abb. 1–184:1–229

Kuhle M (1985) Glaciation research in the Himalayas: a new Ice Age theory. Universitas 27(4):281–294

Kuhle M (1987a) Subtropical mountain- and highland-glaciation as ice age triggers and the waning of the glacial periods in the Pleistocene. GeoJournal 14(4):393–421

Kuhle M (1987b) The problem of a Pleistocene inland glaciation of the north-eastern Qinghai-Xizang-Plateau (Tibet). In: Hövermann J, Wang Wenjing J (eds) Reports on the North-Eastern Part of the Quinghai-Xizang (Tibet) Plateau by the Sino-German Scientific Expedition 1981. Science Press, Beijing, pp 250–315

Kuhle M (1988a) Geomorphological findings on the build-up of Pleistocene glaciation in southern Tibet and on the problem of inland ice. Results of the Shisha Pangma and Mt. Everest Expedition 1984. In: Kuhle M, Wang Wenjing J (eds) Tibet and High Asia. Results of the Sino-German joint expeditions (I). GeoJournal 17(4):457–511

Kuhle M (1988b) The Pleistocene Glaciation of Tibet and the onset of ice ages- an autocycle hypothesis. In: Kuhle M, Wang Wenjing J (eds) Tibet and High Asia. Results of the Sino-German joint expeditions (I). GeoJournal 17(4):581–596

Kuhle M (1989) Die Inlandvereisung Tibets als Basis einer in der Globalstrahlungsgeometrie fußenden, reliefspezifischen Eiszeittheorie. Petermanns Geogr Mitt 133(4):265–285

Kuhle M (1990) New data on the Pleistocene glacial cover of the southern border of Tibet: the glaciation of the Kangchendzönga Massif (8585 m, E-Himalaya). GeoJournal 20(4):415–421

Kuhle M (1991) Observations supporting the Pleistocene inland glaciation of High Asia. In: Kuhle M, Xu Daoming (eds) Tibet and High Asia. Results of the Sino-German joint expeditions (II). GeoJournal 25 (2–3):133–231

Kuhle M (1994) Present and Pleistocene glaciation on the north-western margin of Tibet between the Karakorum Main Ridge and the Tarim Basin supporting the evidence of a Pleistocene inland glaciation in Tibet. In: Kuhle M (ed) Tibet and High Asia. Results of the Sino-German and Russian-German joint expeditions (III). GeoJournal 33(2–3):133–272. Kluwer, Dordrecht

Kuhle M (1995) Glacial isostatic uplift of Tibet as a consequence of a former ice sheet. GeoJournal 37 (4):431–449

Kuhle M (1997) New findings concerning the Ice Age (LGM) glacier cover of the East Pamir, of the Nanga Parbat up to the Central Himalaya and of Tibet, as well as the Age of the Tibetan Inland Ice. In: Kuhle M (ed) Tibet and High Asia (IV). Results of Investigations into High Mountain Geomorphology. Paleo-Glaciology and Climatology of the Pleistocene. GeoJournal 42(2–3):87–257

Kuhle M (1998) Reconstruction of the 2.4 million km^2 Late Pleistocene ice sheet on the Tibetan Plateau and its impact on the global climate. Quatern Int 45/46:71–108 (additional Figures in: 47/48:173–182)

Kuhle M (1999) Reconstruction of an approximately complete Quaternary Tibetan inland glaciation between the Mt. Everest- and Cho Oyu massifs and the Aksai Chin. A new glaciogeomorphological SE-NW diagonal profile through Tibet and its consequences for the glacial isostasy and Ice Age cycle. In: Kuhle M (ed) Tibet and High Asia (V). GeoJournal 47 (1–2):3–276

Kuhle M (2001) The maximum Ice Age (LGM) glaciation of the Central- and South Karakorum: an investigation of the heights of its glacier levels and ice thickness as well as lowest prehistoric ice margin positions in the Hindukush, Himalaya and in East-Tibet on the Minya Konka-massif. In: Kuhle M (ed) Tibet and High Asia (VI): Glaciogeomorphology and Prehistoric Glaciation in the Karakorum and Himalaya. GeoJournal 54(1–4) and 55(1):109–396

Kuhle M (2002) A relief-specific model of the Ice Age on the basis of uplift-controlled glacier areas in Tibet and the corresponding albedo increase as well as their positive climatological feedback by means of the global radiation geometry. Clim Res 20:1–7

Kuhle M (2004) The High Glacial (Last Ice Age and LGM) glacier cover in High- and Central Asia. Accompanying text to the mapwork in hand with detailed references to the literature of the underlying empirical investigations. In: Ehlers J, Gibbard PL (eds) Extent and chronology of glaciations, vol 3 (Latin America, Asia, Africa, Australia, Antarctica). Elsevier B.V., Amsterdam, pp 175–199

Kuhle M (2005) The maximum Ice Age (Würmian, Last Ice Age, LGM) glaciation of the Himalaya—a glaciogeomorphological investigation of glacier trim-lines, ice thicknesses and lowest former ice margin positions in the Mt. Everest-Makalu-Cho Oyu massifs (Khumbu- and Khumbakarna Himal) including informations on late-glacial-, neoglacial-, and historical glacier stages, their snow-line depressions and ages. In: Kuhle M (ed) Tibet and High Asia (VII): glaciogeomorphology and former glaciation in the Himalaya and Karakorum. GeoJournal 62(3–4):193–650

Kuhle M (2009) The former glaciation of High-(Tibet) and Central Asia and its global climatic impact—an Ice Age theory with a remark on potential warmer climatic cycles in the future. In: Krugger Max I, Stern Harry P (eds) New permafrost and glacier research. Nova Science Publishers Inc., Hauppauge, NY, pp 163–235

Kuhle M (2011) The High Glacial (Last Ice Age and Last Glacial Maximum) Ice Cover of High and Central Asia, with a critical review of some recent OSL and TCN Dates. In: Ehlers J, Gibbard PL, Hughes PD (eds) Quaternary glaciation—extent and chronology, A Closer Look. Elsevier B.V, Amsterdam, pp 943–965 (glacier maps downloadable at: http://booksite.elsevier.com/9780444534477/)

Kuhle M (2013) The uplift of High Asia above the snowline and its glaciation as an albedo-dependent cause of the quaternary Ice Ages. Nova Science Publishers Inc., Hauppauge, NY, pp 1–240

Kuhle M, Jacobsen JP (1988) On the geoecology of southern Tibet—measurements of climate parameters including surface- and soil-temperatures in debris, rock, snow, firn and ice during the South Tibet- and Mt. Everest expedition in 1984. In: Kuhle M, Wang Wenjing (eds) Tibet and High-Asia. Results of the Sino-German joint expeditions (I). GeoJournal 17 (4):597–615

Kuhle M, Herterich K, Calov R (1989) On the Ice Age glaciation of the Tibetan highlands and its transformation into a 3-D model. GeoJournal 19(2):201–206

Lauscher F (1956): Strahlungs- und Wärmehaushalt. Berichte des Deutschen Wetterdienstes 4, Nr. 22, pp 21–29, Bad Kissingen

6 The Nourishment Areas of the High Glacial, Late Glacial, and Present Glaciation of High Asia and the Resulting Ablation Areas—A Summary

Figures 4.1 and 6.1 summarize the book, focusing on the climatically most important test areas of High Asia. Figure 4.1 depicts the empirically reconstructed maximum glaciation that consists of both glacier nourishment and ablation areas.

Figure 6.1 concentrates on the accumulation areas. To begin with, it assumes an uplift of Tibet and/or a lowering of the ELA (snowline) by 1200 meters in altitude in relation to the current (today's) surface altitude (Fig. 6.1: I). Whether the lowering of the ELA extends beyond (below) the terrain or whether we are dealing with an elevation of the landscape above the ELA is not relevant for the extent of the resulting glaciation. 1200 m was selected because it is the same value that prevailed in all other mountain regions on earth during the Last Ice Age (Würm Ice Age).

For the mountain ranges of High Asia, the present author has in fact reconstructed an average value of 1300 m based on the empirical evidence (cf. Stage 0 in Fig. 2.5). Nevertheless, to be on the safe side, an ΔELA of 1200 m was chosen for the calculation in Fig. 6.1: I.

In Fig. 6.1: II, the vertical shift of the ELA below the present plateau level amounts to 600 m. This intermediary stage comes closest to the late Late Glacial Stage IV (Fig. 2.5).

Figure 6.1: III depicts the current relation between the terrain altitude and the ELA.

The black areas are terrain features that are located above the ELA. Their reduction from stages I to III accordingly shows the recession of the glacier feeding areas from the Last Ice Age until the present. *In this way one only takes the rock surface of Tibet into consideration, not the feedback surface elevation caused by the glacier build-up. As a result, the calculation tends to underestimate the real extent of glaciation.*

Nevertheless, at a lowering of the ELA by 1200 m in altitude in comparison to today's altitude in relation to the terrain, still more than ca. 80% of the Tibetan area comes to lie above the snowline (Fig. 6.1: I). Because the glacier covering includes both the glacier accumulation area above the ELA and glacier ablation areas below the ELA, the entirety of Tibet was covered with ice as a result of this lowering of the snowline.

At the late Late Glacial lowering by 600 m in comparison to today's ELA, approximately one-fourth to one-third (ca. 29%) of Tibet's surface area ends up above the snowline (Fig. 6.1: II). If complemented by the glacier ablation surface area, a glacier surface area of ca. 50% results. If, however, one further takes into consideration the self-reinforcing feedback of the glaciation through

The characters and symbols in the figures in this chapter can also be viewed in the online version of the figures at http://extras.springer.com/.

M. Kuhle, *The Glaciation of High Asia*,
https://doi.org/10.1007/978-3-319-77566-1_6

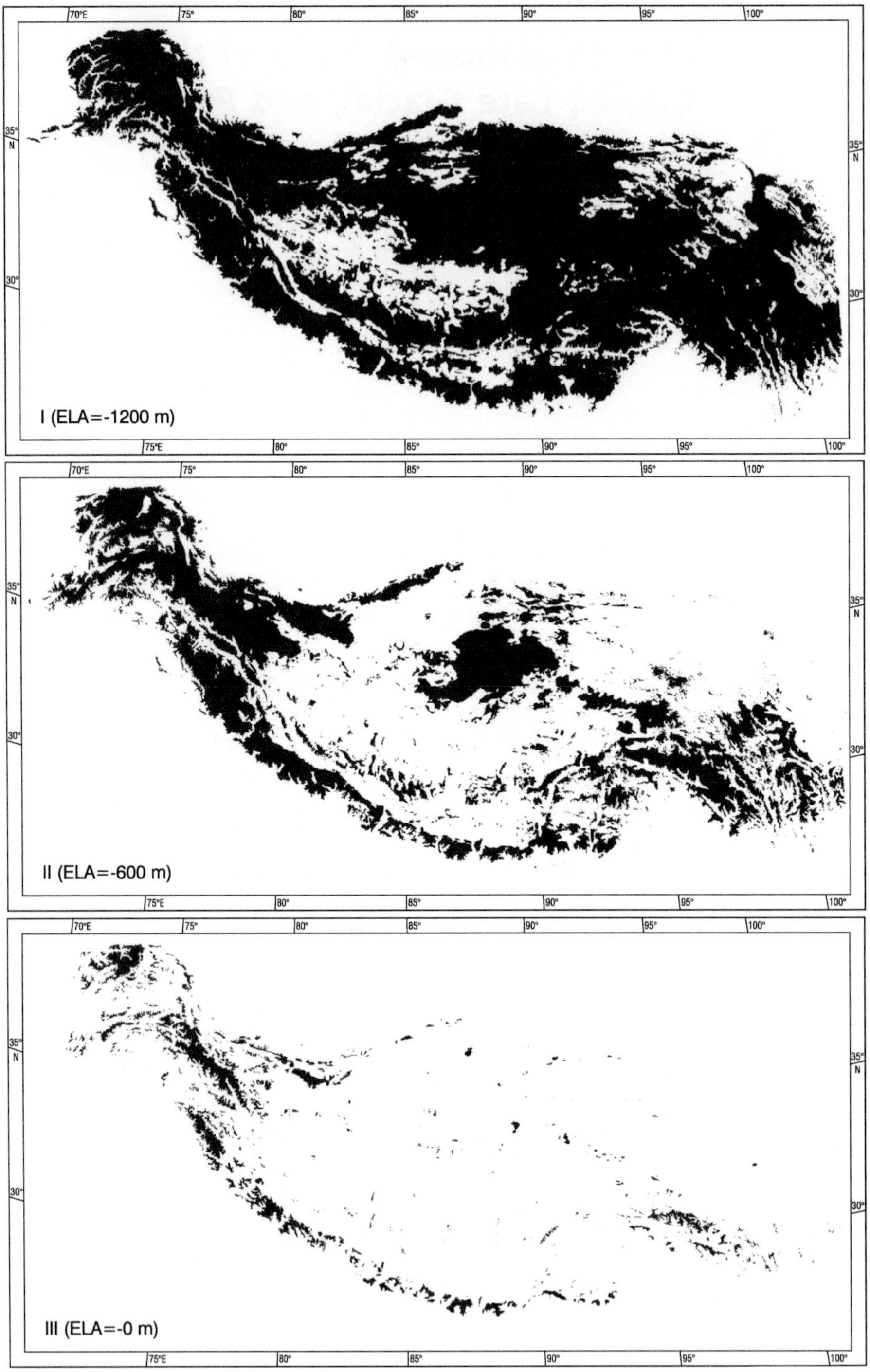
70°E
75°
80°
85°
90°
95°
100°
35° N
30°
I (ELA=-1200 m)
75°E
II (ELA=-600 m)
III (ELA=-0 m)

◀ **Fig. 6.1** (I) The last ice age (Stage 0, Würm, LGP: see Fig. 2.5; MIS 3-2) extent of the glacier feeding areas in and around Central Tibet, south of the Tsaidam Basin (without Tian Shan and Quilian Shan), at an ELA depression (ΔELA) of up to −1200 m in relation to the relief (topography). (II) The Early Glacial (see Kuhle 2012) and Late Glacial Stage (Stage I to IV, see Fig. 2.5) extent of the glacier feeding areas in and around Central Tibet linked with the relief (topography) at an ELA depression of up to −600 m. (III) The present extent (Stage XIII, 1980–2000; see Fig. 2.5) of the glacier feeding areas in and around Central Tibet linked with the relief (topography) at an ELA depression of up to −0 m. These maps have been created in reference to numerous, separate test areas, which taken together completely cover Tibet. By adding the glacier ablation areas during state I (= High Glacial = LGM = Stage 0, see Fig. 2.5), the emerging picture is that the surface of Tibet was completely covered by a unified ice sheet (see Fig. 4.1). At least half of the Tibetan surface—probably much more than that—was covered with glacier ice during state II (Early Glacial and Late Glacial = Stage I–IV, see Fig. 2.5). Project, concept, and draft: M. Kuhle; plot and calculation program: H. Gieseler; area calculations: R. Staschel; cartography: E. Höfer and A. Flemnitz. After Kuhle 1997: 119, Fig. 45

its surface build-up, one is confronted even here with an almost complete inland glaciation of Tibet.

The self-reinforcement results from the fact that the glacier nourishment area is steadily increased and heightened by accumulations of glacier ice, firn-ice, firn, and snow. The effect of the build-up above the ELA in turn reinforces a more effective glacier nourishment through the lower temperatures of the higher feeding area surface. The effect of the build-up of the ablation areas up to the ELA and partially above it is both the cooling of the ablation area surface as well as the enlargement of the nourishment areas. This process is exponentially increased on the even-leveled Tibetan Plateau, because with the increased lowering of the ELA larger lowland areas of the Tibetan Plateau become also covered by the glaciation. They are thus made accessible through the described self-reinforcement (Kuhle 1987, 1988). It is for this reason that an ELA lowering of 500 m in comparison to today with the same relation to the terrain surface of Tibet was probably already sufficient to trigger the build-up of an extensive Tibetan ice sheet.

Figure 6.1: II shows that the Purog Kangri Massif and its wider surroundings in the center of Central Tibet are a potential, continuous, glacier nourishment area. This is supported by Kuhle's Quaternary-geological and -geomorphological investigations (see Sects. 2.8 and 2.8.3).

Figure 6.1: III illustrates the feeding areas of today's glaciers. Here the snowline is the least deeply lowered into the relief. The ELA only extends below the higher mountain summits. The resulting feeding areas belong to the comparatively small mountain glaciers. The largest among them are valley glaciers, which in the Himalayas reach lengths of 15–30 km. The present author could empirically investigate, for example, the following glaciers: Gangotri (29.8 km), Ngozumpa (20.85 km), Langtang Ri-South (ca. 19.1 km), Yalung (ca. 19 km), Kangshung (18.7 km), Kangchenjunga-North (ca. 18.2 km), Khumbu (18 km), Tashain (16.27 km), Barun (16.11 km), Himlung-South (ca. 16 km), West Rongbuk (15.91 km), Central Rongbuk (15.67 km), Nangpa (15.3 km), Diamir (15.16 km), and Rhakiot (14.83 km). In the Karakoram the largest valley glaciers reach lengths of 40–73 km. The following represents a few of the longest examples that were likewise investigated by the present author: Baltoro (58.80 or 58.71 km), Batura (57.24 km), Biafo (53.25 km), Hispar (49.2 km), Chogolungma (ca. 47 km), and Skamri (40.80 km). The third-largest valley glacier of High Asia is, after the Siachen Glacier (73 km), the 61.84 km long South Inylcheck Glacier (Kuhle 2014). It is not located in the Karakoram but in the eastern Teskey Alatau in the Tian Shan (Fig. 4.1: Pik Pobedy, Tian Shan).

The Historical to Holocene (Postglacial) fluctuations of the here-mentioned valley glaciers of the Karakoram and Tian Shan, which relate to the development between Fig. 6.1: II and III, were also investigated and stratigraphically classified by Meiners (1997).

The largest current glacier ice cap—a sort of plateau glacier—in the center of Tibet is the Purog Glacier. It extends 25.7 km in a SW-NE direction (Fig. 2.176) and 29.86 km in a NW-SE direction. Its compact layout has a surface area of ca. 420 km^2 (Sect. 2.7). Even its feeding area is still well recognizable in Fig. 6.1: III.

6.1 The Transformation of Glacier Forms, Glacier Budgets, and Glacier Types from the Last Ice Age to the Present

First and foremost, these were transformations of the inland ice (Fig. 4.1) into valley and hanging glaciers (Fig. 6.1: III) (cf. empirical findings in Kuhle 1987; 1988: Figs. 4, 5, and 9). With the rise of the ELA and its increasing elevation from the High Asian terrain, the glacier nourishment area went through the phases I to III depicted in Fig. 6.1. During the transition from I to II, conditions changed from those shown in Fig. 2.211 to those in Fig. 2.212. In the course of the development from II to III, the Tibetan inland ice disintegrates during the Postglacial. The most spacious deglaciation of Tibet thus took place during the Early to Middle Holocene. The Tibetan mountain groups today only display small ice caps (Fig. 2.176) as well as valley and hanging glaciers. The outlet glaciers and ice stream networks of the mountain ranges that surround Tibet have been replaced by locally (autochtonously) nourished valley and hanging glaciers (Figs. 2.21, 2.29, 2.76, 2.77, 2.102, 2.174, 2.175, 2.177 and 2.178). High Asia has thus passed through a 'glacier typology dependent on the vertical distance' (Kuhle 1987); namely: from the inland ice, the largest glacier type, all the way to the valley glacier (Figs. 2.78, 2.79 and 2.117) and in the most extreme cases even to the mountain-face glacier (mountain flank ice), the smallest type (Fig. 2.102).

While during the Ice Age the primary snow precipitation falling on the flat inland ice surface nourished the ice masses by more than 90%, it is today the secondary nourishment that mainly feeds the extant glaciation—which nowadays only persists as remnants in the mountain ranges. This is nourishment through avalanches (see empirical field data in Kuhle 1986a, 1987, 1988: Figs. 4 and 5; 2001: 110, 118, 130, 131, 136, 179, 180, 185/186, 229/230, 248, 285; Kuhle 2006a) (see Figs. 2.79 and 2.80). While today denudation via avalanches provides the most intensive geomorphodynamics (Fig. 2.80), its equivalent was the ground polish of the very massive glacier covers during the Ice Age. It was this very same glacier ice mass that naturally reduced the height of the rock faces towering above and thereby the width of the altitudinal belt within which avalanches could track (empirical findings in Kuhle 1987).

While the temperatures on the surface of the Tibetan inland ice probably hardly ever exceeded −10 °C, the current glacier surfaces do, during summer, reach the altitudinal zone between −10 °C and 0 °C. While during the Ice Age the higher mountains that rose above the inland ice and its outlet glaciers (Fig. 4.1b) were cold-dry rock peaks, their walls today are full of decameter-thick ice balconies and flank ice (empirical field data: ibid.; Kuhle 1986b, c, 1988). While today only the very highest peaks of the Himalayas and Karakoram reach into the pergelid altitudinal zone to above 6800–7400 m asl, this pergelid altitudinal zone was lowered nearly in parallel with the snowline by ca. 1300 m during the Ice Age. Its lower limit lay at ca. 5500–6100 m asl. Today the pergelid altitudinal zone is, for example, reached at Mount Everest at above 7400 m asl (see empirical measurement data in Kuhle 1986b, c; Kuhle and Jacobson 1988). During the winter months, the monsoonal summer snow is blown off the top like drifting dry sand, such that an ice-free, rocky summit is left exposed (Kuhle 1986c: Photos 1–3). During the Ice Age this circumstance also applied to all the peaks in Tibet and its peripheral mountain ranges that rose above the 5500–6100 m mark without being overridden by the inland ice and its outlet glaciers. These peaks included: Kailas (Figs. 2.118 and 2.119), Gurla Mandatha, Zangqung Kangri, Shakangsham (Shakham) Kangri (Fig. 2.102), Loinbo Kangri (Fig. 2.122), 4 summits of Nyainqentanglha (Fig. 2.79), Qungmogangze (Qiongmo Kangri), Nenang, Norin Kang (Langkazi), Namcha Barwa, and Jialabali (Ghyala Peri). Today, by contrast, the cold and therefore dry snow is—at a temperature down to −46°C—only blown off the slopes of the very highest peaks (cf. empirical measurement data in ibid.: 159; Kuhle 1986b). This "drifting snow" is then deposited on the lee side of the surrounding valley bottoms. Through this drifting process it is partly lost to the nourishment of the glaciers. During the Ice Age, by contrast, this physically dry snow was blown onto random locations on the surface of the Tibetan ice sheet—in the very same way that this occurs still today in the Antarctic. Today ice-free rock faces, ridges, and peaks tower above the Antarctic inland ice (among many others Schatz Ridge, Flamingo, and Mt. Vinson in the Ellsworth Mountains; Mont Belgica)—as they once did during the Ice Age above the continental-arid inland ice of Tibet.

References

Kuhle M (1986a) Schneegrenzberechnung und typologische Klassifikation von Gletschern anhand spezifischer Reliefparameter. Petermanns Geogr Mitt 130:41–51

Kuhle M (1986b) The upper limit of glaciation in the Himalayas. GeoJournal 13(4):331–346

Kuhle M (1986c) Die Obergrenze der Gletscherhöhenstufe—Oberflächentemperaturen und Vergletscherung der Himalayaflanken von 5000–8800 m. Z Gletscherk Glazialgeol 22(2):149–162

Kuhle M (1987) A numerical classification of glaciers by means of specific relief parameter. J Glaciol Geocryol 9 (3, Sept. 1987, Chinese Society of Glaciology and Cryopedology, Academia Sinica):207–214

Kuhle M (1988) Topography as a fundamental element of glacial systems. A new approach to ELA-calculation and typological classification of paleo- and recent glaciations. In: Kuhle M, Wang Wenjing J (eds) Tibet and High Asia. Results of the Sino-German Joint Expeditions (I). GeoJournal 17(4):545–568

Kuhle M (2001) The maximum Ice Age (LGM) glaciation of the Central- and South Karakorum: an investigation of the heights of its glacier levels and ice thickness as well as lowest prehistoric ice margin positions in the Hindukush, Himalaya and in East-Tibet on the Minya Konka-massif. In: Kuhle M (ed) Tibet and High Asia (VI): Glaciogeomorphology and Prehistoric Glaciation in the Karakorum and Himalaya. GeoJournal 54(1–4) and 55(1):109–396

Kuhle M (2006) The past Hunza Glacier in connection with a Pleistocene Karakorum ice stream network during the last Ice Age (Würm). In: Kreutzmann H, Saijid A (eds) Karakoram in transition. Oxford University Press, Karachi, pp 24–48

Kuhle M (2012) The Early and Late Glacial high mountain glaciation surrounding Tibet as topographic-climatic cause of high-energetic glacial lake outburst floods (GLOFs) and their sedimentological consequences in the lower mountain forelands. In: Veress B, Szigethy J (eds) Horizons in earth science research, vol 7. Nova Science Publishers, Inc., Hauppauge, NY, pp 197–227

Kuhle M (2014) Climate or topography—topography of mountains as basis of modern glacier variations, exemplified by 6500 up to 8000 m high summits in High Asia—evidences of Holocene and historical as well as current glacier stages. In: Goncalves AJB, Vieira AAB (eds) Mountains: geology, topography and environmental concerns. Nova Science Publishers Inc., Hauppauge, NY, pp 65–147

Kuhle M, Jacobsen JP (1988) On the geoecology of southern Tibet—measurements of climate parameters including surface- and soil-temperatures in debris, rock, snow, firn and ice during the South Tibet- and Mt. Everest expedition in 1984. In: Kuhle M, Wang Wenjing J (eds) Tibet and High-Asia. Results of the Sino-German joint expeditions (I). GeoJournal 17 (4):597–615

Meiners S (1997) Historical to postglacial glaciation and their differentiation from the Late Glacial period on examples of the Tian Shan and the N.W. Karakorum. In: Kuhle M (ed) Tibet and High Asia, Results of investigations into high mountain geomorphology, Paleo-glaciology and climatology of the Pleistocene (Ice Age Research IV). GeoJournal 42(2–3):259–302

Postscript by Sabine Kuhle

The publication of this book was prolonged by the sudden death of my late husband, on 25 April 2015. The manuscript had been left complete in my hands; however, at the time of submission to the publisher, the English translation did not yet meet current standards. I am thus extremely grateful to Andrew Bond, who took it upon himself to provide us with a meticulous proof-reading of the original manuscript. Through him, readability and linguistic precision have much improved. All remaining errors are our own.

M. Kuhle, *The Glaciation of High Asia*,
https://doi.org/10.1007/978-3-319-77566-1

Index

M. Kuhle, *The Glaciation of High Asia*,
https://doi.org/10.1007/978-3-319-77566-1

Printed by Printforce, the Netherlands